he mathematization of all sciences, the fading of traditional scientific boundaries, e impact of computer technology, the growing importance of computer modelling nd the necessity of scientific planning all create the need both in education and esearch for books that are introductory to and abreast of these developments. The im of this series is to provide such textbooks in applied mathematics for the student cientist. Books should be well illustrated and have clear exposition and sound pedaogy. Large number of examples and exercises at varying levels are recommended. 'AM publishes textbooks suitable for advanced undergraduate and beginning gradate courses, and complements the Applied Mathematical Sciences (AMS) series, vhich focuses on advanced textbooks and research-level monographs.

Texts in Applied Mathematics

Volume 87

Sylvie Méléard

Random Models in Biology, Ecology and Evolution

Springer

Sylvie Méléard
Centre de Mathématiques Appliquées
École Polytechnique
Palaiseau, France

ISSN 0939-2475 ISSN 2196-9949 (electronic)
Texts in Applied Mathematics
ISBN 978-3-662-73482-7 ISBN 978-3-662-73483-4 (eBook)
https://doi.org/10.1007/978-3-662-73483-4

Mathematics Subject Classification: 60J10, 92D15, 92D25, 60J70, 60J85, 92D40, 60H10

Translation from the French language edition: "Modèles aléatoires en Ecologie et Evolution" by Sylvie Méléard, © Springer-Verlag Berlin Heidelberg 2016. Published by Springer-Verlag Berlin Heidelberg. All Rights Reserved.
This book is a translation of the original French edition "Modèles aléatoires en Ecologie et Evolution" by Sylvie Méléard, published by Springer-Verlag GmbH, DE in 2016. The translation was done with the help of an artificial intelligence machine translation tool. A subsequent human revision was done primarily in terms of content, so that the book will read stylistically differently from a conventional translation. Springer Nature works continuously to further the development of tools for the production of books and on the related technologies to support the authors.

This Springer imprint is published by the registered company Springer-Verlag GmbH, DE, part of Springer Nature.
The registered company address is: Heidelberger Platz 3, 14197 Berlin, Germany

In memory of my mother, a passionate biologist,

To my wonderful granddaughters Colombe and Meije

Preface

This book is intended for master's students in applied mathematics or theoretical biologists who wish to expand their knowledge of probabilistic modeling tools. It originated from a course given to third-year students at the École polytechnique (France), but over the years it has grown to far exceed the scope of the course. Its aim is to provide the reader with rigorous tools for modeling biological phenomena subject to random fluctuations. It focuses on stochastic models built from individual behaviors. I would like to warmly thank those who have contributed, through their enthusiasm, their passion for this scientific field, and many enlightening discussions, to the genesis of this work: Vincent Bansaye, Sylvain Billiard, Nicolas Champagnat, Camille Coron, Régis Ferrière, Christophe Giraud, Pierre-Henri Gouyon, Amaury Lambert, Viet Chi Tran, and Amandine Véber.

École polytechnique, France

Sylvie Méléard
sylvie.meleard@polytechnique.edu

Contents

Chapter 1
Introduction

> *... in after years I have deeply regretted that I did not proceed far enough at least to understand something of the great leading principles of mathematics, for men thus endowed seem to have an extra sense.*
>
> Charles Darwin 1809–1882

In his famous work "On the Origin of Species by Means of Natural Selection, or the Preservation of Favoured Races in the Struggle for Life", published in 1859, Darwin revolutionized biology with his theory of evolution and natural selection. Here is the essence of it, extracted from his book:

> *As many more individuals of each species are born than can possibly survive; and as, consequently, there is a frequently recurring struggle for existence, it follows that any being, if it vary however slightly in any manner profitable to itself, under the complex and sometimes varying conditions of life, will have a better chance of surviving, and thus be naturally selected. From the strong principle of inheritance, any selected variety will tend to propagate its new and modified form.*

In this book, we do not claim to survey all the models for such complex phenomena but our goal is to provide the basic mathematical tools for modeling various biological mechanisms.

S. Méléard, *Random Models in Biology, Ecology and Evolution*, Texts in Applied Mathematics 87, https://doi.org/10.1007/978-3-662-73483-4_1

1.1 Introduction to the Course

> *People who have mathematical computation and statistical skills, and establish a collaboration and work on real biological problems, have the chance of doing some very, very significant things for human welfare.*
>
> Jaroslav Stark, Imperial College
> Science, 2004

Population biology ranges from microscopic studies, such as the search for sequences on a strand of DNA, the study of molecular exchanges in a cell, the evolution of cancerous tumors, and the invasion of parasites into a cell, to macroscopic problems concerning the behaviors of large groups of individuals and their interactions (extinction of populations, balance of ecosystems, invasion of a population by another, meta-populations), or population genetics problems (search for common ancestors to samples of individuals, phylogenies). At all levels, randomness occurs and stochastic models can be used to describe biological phenomena. Even though the population seems to present a number of deterministic characteristics, it is composed of individuals whose behavior is highly variable. For instance, each individual moves in a different direction, each bacterium has its own mechanism of cell division, each DNA replication can generate a mutation... This individual variability is a fundamental idea of evolutionary biology. Indeed, even though in many situations the size of the population may be large enough for the biological system to be described only by a deterministic model, in most cases the biological phenomenon, at least in a phase of its development, is determined by a small number of individuals or is strongly influenced by environmental, spatial or temporal fluctuations. For example, random fluctuations can play a decisive role at the initial stage of an epidemic or in the initial development of a cancerous tumor. The approach we want to promote, from a probabilistic point of view, consists in modeling first the (stochastic) individual behaviors and then deducing macroscopic behaviors at the population level.

The aim of this book is to define and study a wide range of probabilistic tools that can provide a better understanding of certain biological dynamics over time, such as the dynamics of a population size or the dynamics of an allele proportion. We are going to introduce various stochastic processes, i.e. random functions indexed by time, which all satisfy a fundamental hypothesis known as the Markov property: the random behavior of the dynamics in the future will depend on its past only through the information given by its current state. This assumption is widely accepted by biologists, even though it is an approximation of reality. In the models, time can be discrete or continuous. Time can be discrete ($n \in \mathbb{N}$) when the model concerns, for example, a generational or periodic phenomenon (seasonal, annual) or if the observation times are discrete. The basic probabilistic tools will then be discrete-time Markov chains. Three classical models shall illustrate this point: random walks to describe spatial movements of individuals, Bienaymé–Galton–Watson processes that model the dynamics of population sizes, and the Wright–Fisher process which describes the genealogy of alleles. But time can also be physical time ($t \in \mathbb{R}_+$). We shall consider two distinct types of processes. In the first case, the processes indexed by continuous time will have continuous trajectories (for example, to model the motion of individuals), the prototype of which is Brownian motion. In the second case, the processes shall have piecewise constant trajectories (with jumps), the prototype of which is the Poisson process. They can model a population size dynamics taking into account all individual births and deaths leading to the branching processes or birth and death processes in continuous time, which we shall study in detail. We shall develop stochastic calculus tools and the concept of stochastic differential equations driven by a Brownian motion or by a Poisson point measure.

When the population size is very large, it becomes difficult to describe the microscopic behavior of the population and to take into account each individual birth or death. We shall have to change the scale of the population size and the time scale, leading to approximations that are more easily mathematically tractable from a theoretical and computational point of view. For some scalings, we shall obtain deterministic approximations, which historically were the first models to describe population dynamics, such as the logistic equation or the Lotka–Volterra systems. For other scalings, we shall obtain random approximations defined as solutions to certain stochastic differential equations, such as the Feller equations in the case of population dynamics or the Wright–Fisher equations in the case of population genetics. Computations on these diffusion processes lead us to deduce some information on the behavior of populations. We shall also define, starting from the Wright–Fisher model, a process for modeling genealogies, namely the Kingman coalescent, which has become fundamental in population genetics in recent years.

The last part of this book presents some recent applications to biological questions of the mathematical tools developed.

1.2 The Importance of Modeling

A vital next step will be to promote the training of scientists with expertise in both mathematics and biology.

Bruce Alberts, Editorial
Science, 2003

Mathematical models can help us to better understand the dynamics of biological phenomena. Biological systems are extremely complex: multi-scale problems, multiple interactions, no conservation laws... Therefore, it is necessary to greatly simplify the biological systems to obtain mathematically tractable models, focusing on the biological phenomenon we are trying to understand. A difficulty of this approach is therefore to achieve a good trade-off between the biological realism of the model and the feasibility of computations, and this work can only be achieved through effective collaboration between biologists and mathematicians. The mathematical models allow us to answer quantitative questions such as the extinction probability, the time to extinction, the invasion probability, the sizes at equilibrium, the resilience of an ecosystem... and to predict how it will evolve in the future. And it is important to address the validation of the model. In the case when experimental data are available, validation of the models can be studied by a statistical approach or by calibration of the model parameters. We will not address these issues in this book.

The interest of a model also lies in its "universality". Different biological problems (for example, with different size scales: genes - cells - bacteria - individuals - colonies) can have similar random behaviors (in terms of reproduction, death, migration) and can be studied with similar mathematical tools. The same mathematical object can be used for different biological purposes. Thus, a simple random walk can model the dynamics of a population size on $\mathbb{N}$ or a displacement on $\mathbb{Z}$. In the first case, if this random walk reaches 0 (extinction of the population), it will stay there (if there is no immigration). In the second case, the random walk can reach 0 an infinite number of times.

The models developed in this book are "elementary" in the sense that the processes studied have values in finite-dimensional spaces. Taking into account the structure of a population, according to a certain number of variables such as age, one or more phenotypes, genotype, spatial position, etc., can require infinite-dimensional models and much heavier mathematical tools, which we will not address here. We refer, for example, to [8] for infinite-dimensional population models.

1.3 Mathematical Modeling in Ecology and Evolution

Let us comment on the biological questions we have in mind, and how they are related to the chapters of this book.

1.3.1 Ecology and Evolution

Ecology is the study of interactions between living beings (animals, plants, microorganisms,...) and with the environment that surrounds them and of which they are themselves a part, such as their habitat. Ecology particularly studies the complex hierarchy of ecosystems, the biological mechanisms associated with the extinction of species, the dynamics of biodiversity, the adaptation of populations. To its development contribute the more specialized disciplines of population dynamics and genetics. Population Dynamics has long been studied from a mathematical point of view. The first quantitative studies of populations are due to Malthus (1798), then to Verhulst (1838). From their work emerged the idea that population growth is limited by resources and by its environment. This idea took on its full force in the theory developed by Darwin (1859). He defended the fundamental concept of natural selection and modification of living species over time. The Theory of Evolution is based on the simple idea that individuals best adapted to their environment will develop their offspring (through heredity) better than the others and their genotypic and phenotypic characteristics will be widespread in the population. Adaptation results from individual variability that appears in reproduction mechanisms (mutation, recombination) or transfer, and by interactions between individuals and their environment. Interactions can take multiple forms. We can mention the competition of individuals of a species for the sharing of resources, the links in a trophic network (food chain) such as prey-predator or host-parasite interactions, and the struggle of females for males.

1.3.2 Dispersion and Colonization

The spatial structuring of a population is the result of individual dispersion behaviors, such as the progression of a herd, the extension of an epidemic, the dispersion of pollen grains, eggs, seeds, etc. These motions can be due to the search for food, the need to find a partner, fleeing from a predator or the need to colonize areas favorable to survival or reproduction. In all cases, the spatial structure impacts the population dynamics and its probability of extinction. It is therefore linked to the demographic evolution of the population and to its genetic distribution. Over the past twenty years, many works have been developed in these directions. They require the introduction of sophisticated mathematical objects, but the basic spatial models are developed in this book. We will consider stochastic processes modeling how individuals move or disperse their offspring. They can be developed at all biological scales, from viruses to mammals.

Chapters 2 and 4 study Markovian stochastic processes modeling individual motions, in discrete or continuous time. The individual can move in a discrete space such as $\mathbb{Z}$, $\mathbb{Z}^2$ or $\mathbb{Z}^3$, or in a continuous space such as $\mathbb{R}$, $\mathbb{R}^2$ or $\mathbb{R}^3$, or in a subdomain of these spaces. The choice of the dimension is related to the nature of the biological problem, dimension 1 for a motion along a road, a river, a blade of grass or a strand of DNA, dimension 2 for a motion on the ground, a wall, a seabed, dimension 3 otherwise. Whether due to natural boundaries (sea, mountain range, cell membrane), or limitation of the resource area or traps, populations often move in closed domains and their behavior at the boundary of the domain is important. It can be absorbing if individuals die or get stuck at its contact or reflective if individuals turn back. The probabilities of reaching barriers will be studied in detail.

1.3.3 Population Dynamics

Population dynamics studies the quantitative development of populations with asexual or sexual reproduction. Points of interest are the mechanisms of self-regulation in a population, the extinction of a population or the existence of possible stationary or quasi-stationary states, the interactions between different species, such as the links of predation between prey and predators, or more generally the structure of trophic networks or models of cooperation and/or competition.

The models of unstructured populations in Chapters 3 and 5 have the advantage of simplicity, but are of course very naive in order to account for biological diversity. Nevertheless, they highlight various behaviors, depending on demographic parameters, and give us computational tools: computation of the probability of persistence of a population or its extinction and the mean time of extinction. They will also allow us to evaluate the size of a population in the long term (in a persistence case) and possibly study its composition between types, when individuals can be of different types.

We will highlight two major classes of models: those that rely on general reproduction laws but do not assume interaction between individuals, called branching processes, and those for which the reproduction law is simpler but take into account competition between individuals, called birth and death processes. The tools for studying these objects are different.

1.3.4 Population Genetics

To understand the evolution of species, it is necessary to understand the internal mechanisms of genealogies within a species. Population genetics took off from the work of Mendel (1822–1884), who was at the origin of the rules of genetic transmission (commonly called Mendel's laws), defining the way in which genes are transmitted from generation to generation. It was initially developed mathematically by biologists Fisher, Haldane and Wright between 1920 and 1940. Population genetics focus on the study of the fluctuation of alleles (the different versions of a gene) over time in populations. These fluctuations can be due to the influence of genetic drift (the variability due to the randomness of individual birth and death), mutations, migrations and natural selection.

In Chapter 6, we will present the classical models describing the dynamics of the genetic information of individuals over time, assuming that the size of the observed population remains constant. This point of view is the historical point of view of Wright and Fisher and is still relevant for population geneticists. This amounts to assuming that population dynamics are in a stationary state. We will also consider the inverse perspective: reconstructing the genealogical lineages of a group of observed individuals, tracing back in time the history of these individuals. By changing the time scale, we will highlight a limit object of a new nature: a stochastic process that allows the reconstruction of the successive common ancestors of a sample set of individuals, called the Kingman coalescent.

Once these elementary building blocks are laid, it becomes possible to approach the modeling of individual dynamics by taking into account the genetic information they contain and the interactions they develop. These questions are at the heart of fundamental research that is developing today between applied mathematics and biodiversity modeling. This is why, rather than giving overly simplistic biological examples, we prefer to develop in detail, in Chapter 7, results obtained in recent research articles, motivated by current biological questions.

Chapter 2
Spatial Populations and Discrete Time

The mathematics is not there till we put it there.

Sir Arthur Eddington (1882–1944),
The Philosophy of Physical Science

In this chapter, we use stochastic processes $(X_n)_{n\in\mathbb{N}}$, indexed by discrete time $n \in \mathbb{N}$, to model random dynamics. We introduce two classes of models that take into account different dependency structures and that generalize sequences of independent random variables: Markov chains and discrete time martingales. These two classes converge in the study of random walks, which constitute a central theme of this chapter. An important motivation will be the modeling of spatial dynamics, but we shall also see models of population size dynamics.

2.1 Random Walks and Markov Chains

We recall the main properties of random walks and discrete time Markov chains, focusing on issues specific to population models. For more details, we refer to [23], [65], [38] and [32].

The simplest random model used to describe the random movement of an individual is that of a random walk on the lattice $\mathbb{Z}^d$. Time is discrete. An individual can go from a point $x = (x_1, \ldots, x_d)$ to another point of the lattice with a certain probability. The successive movements of the individual are independent of each other. More specifically, we denote by X_n the individual's position at time n, and by $Z_n = X_n - X_{n-1}$ their n-th displacement.

S. Méléard, *Random Models in Biology, Ecology and Evolution*,
Texts in Applied Mathematics 87, https://doi.org/10.1007/978-3-662-73483-4_2

Definition 2.1.1 The sequence of random variables $(X_n)_{n\in\mathbb{N}}$ is called a *random walk* on the lattice $\mathbb{Z}^d$ if

$$X_n = X_0 + \sum_{k=1}^{n} Z_k,$$

and the successive displacement random variables $Z_k \in \mathbb{Z}^d$ are independent and identically distributed. If each displacement can only be made towards one of its immediate neighbors, the random walk is said to be *simple*. If, in addition, the displacements towards each of the immediate neighbors are made with the same probability $\frac{1}{2d}$, the walk is said to be *symmetric*.

Figure 2.1 shows simulations of simple symmetric random walks in dimensions 1 and 2. In the first case, the walk can move forward or backward by an amplitude of 1 with probability $\frac{1}{2}$. In the second case, the individual can jump towards one of its immediate neighbors with a uniform probability of $\frac{1}{4}$.

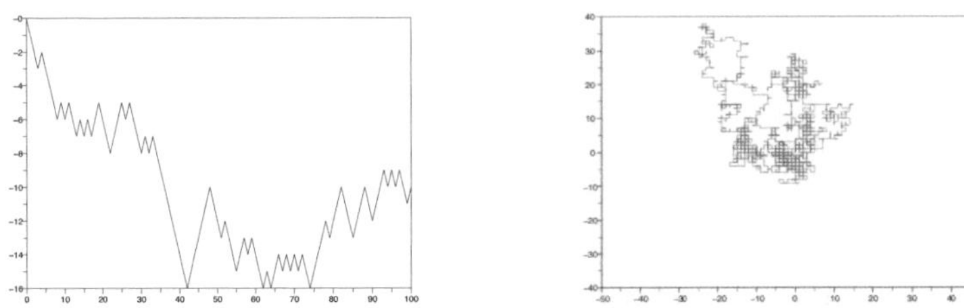

Fig. 2.1 Random walks in dimension 1 and in dimension 2.

A random walk satisfies the following two fundamental properties:

- For all n and k in $\mathbb{N}$, the random variable $X_{n+k} - X_n$ is independent of $X_n, X_{n-1}, \ldots, X_0$.
- For all n and k, $X_{n+k} - X_n$ has the same distribution as $X_k - X_0$.

The increments of the random walk are therefore independent and stationary.

Definition 2.1.2 A process $(X_n)_{n\in\mathbb{N}}$ is a *process with stationary independent increments* (abbreviated as *PSII*) if for all integers $0 \le n_1 < n_2 < \cdots < n_k$, the random variables $(X_{n_1}, X_{n_2} - X_{n_1}, \ldots, X_{n_k} - X_{n_{k-1}})$ are independent and if $X_{n_2} - X_{n_1}$ has the same distribution as $X_{n_2-n_1} - X_0$.

Remark 2.1.3 In this book, we shall essentially restrict ourselves to random walks in dimension one. In the rest of this chapter, the state space considered will be $\mathbb{Z}$.

The random walks introduced above are a simple particular case of stochastic processes satisfying the Markov property, called *Markov chains*. The Markov property describes a characteristic of many random phenomena, for which the future random evolution depends on the past only through the state of the process at the present time. In the rest of the course, we shall see a large number of processes satisfying the Markov property.

Definition 2.1.4 A sequence of random variables $(X_n)_{n\in\mathbb{N}}$ with values in $\mathbb{Z}$ satisfies the *Markov property* if for all $n \in \mathbb{N}$, $j, i_0, \ldots, i_n \in \mathbb{Z}$, such that

$$\mathbb{P}(X_n = i_n, \ldots, X_0 = i_0) > 0,$$

we have

$$\mathbb{P}(X_{n+1} = j | X_n = i_n, \ldots, X_0 = i_0) = \mathbb{P}(X_{n+1} = j | X_n = i_n).$$

The conditional law of X_{n+1} knowing $X_0, \ldots, X_n$ is equal to its conditional law knowing X_n. If this law does not depend on n, we say that the Markov chain is *homogeneous* (in time).

The law of a homogeneous Markov chain is characterized by its transition probability measures

$$P_{i,j} = \mathbb{P}(X_1 = j | X_0 = i), \quad i, j \in \mathbb{Z},$$

and by its initial condition since it can be easily proved that

$$\mathbb{P}(X_0 = i_0, \ldots, X_n = i_n) = \mathbb{P}(X_0 = i_0)\, P_{i_0,i_1} \cdots P_{i_{n-1},i_n}.$$

Later, we will denote by $\mathbb{P}_i$ the law of the Markov chain starting from state i: for any event A,

$$\mathbb{P}_i(A) = \mathbb{P}(A | X_0 = i).$$

Definition 2.1.5 The transition probability measures $(P_{i,j})_{i,j\in\mathbb{Z}}$ define a matrix, called a *transition matrix*, with positive coefficients such that the sum of each row is equal to 1: for all $i, j \in \mathbb{Z}$,

$$P_{i,j} \geq 0; \qquad \sum_{j\in\mathbb{Z}} P_{i,j} = 1.$$

Let us introduce the sequence of σ-fields $\mathcal{F}_n = \sigma(X_0, \ldots, X_n)$ generated by the chain $(X_n)_n$. The σ-field $\mathcal{F}_n$ is generated by the sets of the form

$$\{(X_0, X_1, \ldots, X_n) = (i_0, \ldots, i_n)\}$$

for all integers $i_0, \ldots, i_n$. It describes the information given by the process up to time n. The σ-fields $\mathcal{F}_n$ are increasing (with respect to inclusion) as the information grows over time. The sequence $(\mathcal{F}_n)_n$ is called the *filtration generated by the Markov chain* $(X_n)_n$.

Remark 2.1.6 The sequence $(X_n)_n$ satisfies the Markov property if and only if for any bounded function f on $\mathbb{Z}$, and for any natural number n, we have

$$\mathbb{E}(f(X_{n+1}) | \mathcal{F}_n) = \mathbb{E}(f(X_{n+1}) | X_n).$$

Indeed, it is enough to verify that for all $i_0,\dots,i_n$,

$$\begin{aligned}&\mathbb{E}\big(f(X_{n+1})|(X_0,\dots,X_n)=(i_0,\dots,i_n)\big)\\&=\sum_{j\in\mathbb{Z}} f(j)\,\mathbb{P}\big((X_{n+1}=j|(X_0,\dots,X_n)=(i_0,\dots,i_n)\big)\\&=\sum_{j\in\mathbb{Z}} f(j)\,\mathbb{P}(X_{n+1}=j|X_n=i_n)=\mathbb{E}(f(X_{n+1})|X_n=i_n).\end{aligned}$$

Examples 1) *Simple random walk on* $\mathbb{Z}$. The displacements Z_n are independent and equidistributed with a law on $\{-1,1\}$ assigning probabilities $1-p$ and $p\in(0,1)$ to -1 and 1, respectively. The simple random walk is a homogeneous Markov chain with transition matrix $(P_{i,j})$ satisfying

$$\begin{aligned}P_{i,i+1}&=p;\\P_{i,i-1}&=1-p;\\P_{i,j}&=0 \text{ if } j\neq i+1,\ i-1.\end{aligned}$$

If $p=\frac{1}{2}$, the random walk is symmetric.

This model can, for example, describe the vertical displacement of a plankton particle in the Atlantic Ocean, where the average depth is about 3300 m. If we consider its position at discrete times, we can assume that it moves vertically like a one-dimensional random walk. The depth is sufficiently large compared to the size of the cell for its displacement to be considered as possible over all $\mathbb{Z}$.

2) *Resource dynamics model.* At each moment $n\geq 0$, a quantity R of resources is created and a random quantity Z_n is consumed by a certain group of individuals. The random variables $(Z_n)_n$ are assumed to be independent and identically distributed with integer values, and $p_k=\mathbb{P}(Z_1=k)$ for all integers k. If the level of resources at time n is denoted by X_n, it satisfies the following dynamics:

$$X_n=(X_{n-1}+R-Z_n)^+ \quad \text{for all } n\geq 1.$$

The sequence $(X_n)_n$ forms a homogeneous Markov chain with values in $\mathbb{N}$, with transition matrix (P_{ij}) satisfying for all $i,j\in\mathbb{N}$,

$$\begin{aligned}P_{i,0}&=\mathbb{P}(Z_n\geq i+R)=\sum_{k\geq i+R} p_k;\\P_{i,j}&=\mathbb{P}(i+R-Z_n=j)=p_{R+i-j}\ \text{ if } j\leq R+i-1;\\P_{i,j}&=0\ \text{ otherwise.}\end{aligned}$$

3) *Branching process.* Branching processes were introduced to model the successive generations of a population. They will be studied in Chapter 3. Let X_n be the size of a population at time n and Y_i^n the integer-valued random variable giving the number of offspring of each individual $i\leq X_n$ at time n. The random variables $(Y_i^n, i,n\in\mathbb{N})$ are independent and identically distributed. The dynamics of the

population size is described for $n \geq 0$ by

$$X_{n+1} = \sum_{i=1}^{X_n} Y_i^n, \tag{2.1.1}$$

where the sum is null on $\{X_n = 0\}$. The sequence $(X_n)_n$ forms a homogeneous Markov chain with integer values and with transition matrix $(P_{i,j})$ satisfying

$$P_{0,0} = 1;$$
$$P_{i,j} = \mathbb{P}\left(\sum_{k=1}^{i} Y_k^0 = j\right) \text{ for all } i, j \in \mathbb{N},\ i \geq 1.$$

The inductive property (2.1.1) makes it difficult to obtain an explicit equation for the transition matrix terms. We shall see in Chapter 3 that a good tool to exploit this equation is the generating function of X_n.

4) *Birth and death process.* Birth and death processes are also Markov processes modeling population sizes. As before, X_n denotes the size of a population at time n. At each time n, a single individual can be born or dies with probabilities depending on X_n. Conditionally on $\{X_n = i\}$, the birth and death probabilities are b_i and d_i, respectively. When $b_0 = d_0 = 0$ and $d_i > 0$ for $i \geq 1$, the process takes its values in $\mathbb{N}$ unless a constraint is imposed on the population size, for example, that it cannot exceed the threshold N. In this case, $b_N = 0$. This constraint could be a way of modeling a competition among individuals for resource sharing: a fixed amount of resources will not allow the population to grow beyond a certain threshold. The transition probabilities of the birth and death process are equal to

$$\begin{aligned} P_{i,i+1} &= b_i; \\ P_{i,i-1} &= d_i; \\ P_{i,i} &= 1 - b_i - d_i; \\ P_{i,j} &= 0 \text{ if } i \geq 1,\ j \neq i+1,\ i-1,\ i. \end{aligned}$$

Furthermore, $P_{0,0} = 1$, $P_{0,j} = 0$ for $j \neq 0$ and in the case of population size with threshold N, $P_{N,N+1} = b_N = 0$.

Let us return to the general situation where we consider any homogeneous Markov chain and for $n \in \mathbb{N}^*$, let $(P_{i,j}^{(n)})$ be the matrix defined by

$$P_{i,j}^{(n)} = \mathbb{P}(X_n = j \mid X_0 = i).$$

Proposition 2.1.7 *The matrix $P^{(n)}$ is equal to the product of matrices P^n.*

Proof We show this property by induction on n. It is obviously true for $n = 1$. Assuming now that $P^{(n-1)} = P^{n-1}$, one has

$$
\begin{aligned}
P_{i,j}^{(n)} &= \mathbb{P}(X_n = j \mid X_0 = i) = \sum_{k\in\mathbb{Z}} \mathbb{P}(X_n = j\,, X_{n-1} = k \mid X_0 = i) \\
&= \sum_{k\in\mathbb{Z}} \mathbb{P}(X_n = j \mid X_{n-1} = k\,, X_0 = i)\,\mathbb{P}(X_{n-1} = k \mid X_0 = i) \\
&= \sum_{k\in\mathbb{Z}} \mathbb{P}(X_n = j \mid X_{n-1} = k)\,\mathbb{P}(X_{n-1} = k \mid X_0 = i) \quad \text{by the Markov property} \\
&= \sum_{k\in\mathbb{Z}} P_{k,j}\, P_{i,k}^{(n-1)} = \sum_{k\in\mathbb{Z}} P_{k,j}\, P_{i,k}^{n-1} \quad \text{by the induction property} \\
&= P_{i,j}^{n}.
\end{aligned}
$$

□

The relationships between the states of the Markov chain generate a classification for these states.

We say that state j is *reached* from state i if there exists $n > 0$ such that $P_{i,j}^{(n)} > 0$. This relationship is denoted $i \longrightarrow j$. If $i \longrightarrow j$ and $j \longrightarrow i$, we say that states i and j *communicate*, denoted $i \longleftrightarrow j$: there exist $n > 0$ and $n' > 0$ such that $P_{i,j}^{(n)} > 0$ and $P_{j,i}^{(n')} > 0$. It is easy to show that this relationship between i and j defines an equivalence relation.

Definition 2.1.8 The equivalence classes of the communication relation $i \longleftrightarrow j$ are called the *classes* of the Markov chain. If there is only one class, the Markov chain is said to be *irreducible*. A class C is said to be *closed* if it is impossible to move from C to the complement of C in one step: $P_{i,j} = 0$ for $i \in C$ and $j \notin C$.

Example 2.1.9 The *base substitution Markovian model* (Allman–Rhodes [3, p. 141], see also [31]) is a 4-state Markov chain describing molecular evolution. Each site in a DNA sequence is one of the 4 bases A, G, C, T (Adenine, Guanine, Cytosine, Thymine), chosen randomly according to the probabilities p_A, p_G, p_C, p_T such that $p_A + p_G + p_C + p_T = 1$. Thus, the vector $P_0 = (p_A, p_G, p_C, p_T)$ describes the ancestral distribution of bases in the DNA sequence. We assume that the sites are independent of each other. Suppose further that these bases can mutate and that only one mutation can occur per generation. The goal is to study the evolution of the Markov chain that describes the state of a site in the DNA sequence with values in $\{A, G, C, T\}$ and initial distribution P_0. The Jukes–Cantor molecular evolution model [46] is a particular case assuming that the transitions describing the mutation probabilities from one base to another in the DNA structure are all equal to a certain number $\frac{\alpha}{3}$, with $0 < \alpha < 1$. Thus the transition matrix is

$$
P = \begin{pmatrix}
1-\alpha & \frac{\alpha}{3} & \frac{\alpha}{3} & \frac{\alpha}{3} \\
\frac{\alpha}{3} & 1-\alpha & \frac{\alpha}{3} & \frac{\alpha}{3} \\
\frac{\alpha}{3} & \frac{\alpha}{3} & 1-\alpha & \frac{\alpha}{3} \\
\frac{\alpha}{3} & \frac{\alpha}{3} & \frac{\alpha}{3} & 1-\alpha
\end{pmatrix}.
$$

The Markov chain with transition matrix P is irreducible. Also note that the sum of the elements of each column is equal to 1 (we already have this property for the rows). In this case, the transition matrix is said to be doubly stochastic. This remains true for all its powers. If the ancestral distribution of bases is uniform, then it will remain uniform. Indeed, we compute, for example, the probability of having a Guanine base at time n. By identifying $\{A,G,C,T\}$ with $\{1,2,3,4\}$, this probability is

$$Q_n^G = \sum_{i=1}^{4} (P_0)_i P_{i,2}^{(n)} = \frac{1}{4}\sum_{i=1}^{4} P_{i,2}^{(n)} = \frac{1}{4}.$$

On the other hand, if the initial distribution is $(\frac{1}{3},\frac{1}{6},\frac{1}{6},\frac{1}{3})$, then at time $n=3$,

$$Q_3^G = \sum_{i=1}^{4} (P_0)_i P_{i,2}^{(3)} = \frac{1}{3}P_{1,2}^{(3)} + \frac{1}{6}P_{2,2}^{(3)} + \frac{1}{6}P_{3,2}^{(3)} + \frac{1}{3}P_{4,2}^{(3)}.$$

One can easily check that $P_{2,2}^{(3)} = (1-\alpha)^3 + \alpha^2(1-\alpha) + 2\frac{\alpha^3}{9}$ and that therefore for $i \neq 2$, $P_{i,2}^{(3)} = \frac{1}{3}(1 - P_{2,2}^{(3)})$. Thus, $Q_3^G = \frac{1}{6}(\frac{5}{3} - \frac{2}{3}P_{2,2}^{(3)})$.

2.2 The Study of Passage Times

2.2.1 Stopping Time and the Strong Markov Property

It is important to know the Markov chain at discrete times n, but some random times will also prove fundamental, particularly those that are linked to the dynamics of the process. Consider a Markov chain $(X_n)_n$. The successive return times of this process to a particular state i are essential. They are defined as follows:

$$\begin{aligned} T_i &= \inf\{n \geq 1\,; X_n = i\};\\ T_i^1 &= T_i;\\ T_i^{k+1} &= \inf\{n > T_i^k\,; X_n = i\}, \qquad \text{for all } k \geq 1, \end{aligned}$$

with the usual convention that the infimum of the empty set is $+\infty$. The sequence $(T_i^k)_k$ describes the successive return times of the chain to state i.

If $X_0 = i$ then T_i is the time of first return to state i and T_j, $j \neq i$, is the time of first passage to state j. The law of T_i is then given by the $f_{i,i}^{(n)} = \mathbb{P}(T_i = n \mid X_0 = i)$, for all $n \geq 1$ (n can possibly be infinite). Note that $f_{i,i}^{(1)} = P_{i,i}$. We will see in the next subsection a method for calculating these laws.

Note that observing $X_0, \ldots, X_n$, that is, the process up to time n, allows us to decide whether T_i equals n or not, or whether it is smaller or larger than n, since for example,

$$\{T_i = n\} = \{X_1 \neq i\} \cap \cdots \cap \{X_{n-1} \neq i\} \cap \{X_n = i\}.$$

In other words, we have

$$\{T_i = n\},\ \{T_i < n\},\ \{T_i > n\}\ \in \mathcal{F}_n.$$

More generally, we will be interested in all the random times that satisfy this property, called stopping times.

Definition 2.2.1 A *stopping time* (relative to the filtration $(\mathcal{F}_n)_n$) is a random variable T with values in $\mathbb{N} \cup \{+\infty\}$ that satisfies, for all $n \in \mathbb{N}$,

$$\{T = n\} \in \mathcal{F}_n.$$

Example 2.2.2 The random times T_i are stopping times. More generally, for any $A \subset \mathbb{Z}$, the hitting time of A, defined by

$$T_A = \inf\{n \geq 0; X_n \in A\},$$

is a stopping time. Indeed,

$$\{T_A = n\} = \{X_0 \notin A\} \cap \cdots \cap \{X_{n-1} \notin A\} \cap \{X_n \in A\} \in \mathcal{F}_n.$$

The k-th passage time into A is also a stopping time. We can show this, for example, by induction on k, noting that

$$\{T_A^k = n\} = \cup_{m=0}^{n-1} \{T_A^{k-1} = m\} \cap \{X_{m+1} \notin A\} \cap \cdots \cap \{X_{n-1} \notin A\} \cap \{X_n \in A\}.$$

On the other hand, the time of last passage in A,

$$L_A = \sup\{n \geq 0; X_n \in A\},$$

is not a stopping time. To know if $L_A \leq n$, we need to know the states of the chain after n.

A Markov chain still satisfies the Markov property if its law is conditioned by the position of the chain at a stopping time. This property is called the strong Markov property and is fundamental for applications.

Theorem 2.2.3 (Strong Markov Property) *Let $(X_n)_n$ be a Markov chain and T a stopping time. Conditionally on X_T on $\{T < \infty\}$, the sequence $(X_{T+n})_{n\geq 1}$ is a Markov chain independent of the trajectory of $(X_n)_n$ until time T. In other words, for any event $A \subset \mathbb{Z}$ such that $A \cap \{T = n\} \in \mathcal{F}_n$ for all n, we have*

$$\begin{aligned}\mathbb{P}(A \cap \{X_{T+1} = i_1, \ldots, X_{T+m} = i_m\} \mid X_T = i, T < \infty) \\ = \mathbb{P}(A \mid X_T = i, T < \infty)\, \mathbb{P}_i(X_1 = i_1, \ldots, X_m = i_m) \qquad (2.2.1)\end{aligned}$$

for all $i_1, \ldots, i_m, i \in \mathbb{Z}$.

Proof It suffices to note that

$$\begin{aligned}\mathbb{P}(A \cap \{T = n\} \cap \{X_{T+1} = i_1, \ldots, X_{T+m} = i_m\} \mid X_T = i) \\ = \mathbb{P}(A \cap \{T = n\} \mid X_T = i)\, \mathbb{P}_i(X_1 = i_1, \ldots, X_m = i_m).\end{aligned}$$

□

2.2.2 Law of the First Return Time to 0 of a Simple Random Walk

Imagine a plankton particle moving in a vertical column. Does it tend to go towards the surface or the bottom, and will it stay there? To answer this, we can study its passage times at a point in the column, depending on the probabilities of this particle going up or down.

In this subsection, we will study the return times to 0 of a simple random walk $(X_n)_n$ (on $\mathbb{Z}$) starting at 0: $X_n = \sum_{k=1}^n Z_k$, and the random variables Z_k are independent and identically distributed. We are interested in the sequence of stopping times when the walk returns to the origin. To describe these random times, it is enough, by the strong Markov property, to consider sums of independent copies of this one. Indeed, if

$$T_0 = \inf\{n \geq 1\,;\, X_n = 0\} \in \mathbb{N}^* \cup \{\infty\}$$

denotes the first return time to 0, where we set $\min\{\emptyset\} = \infty$, and T_0^j denotes the time of the j-th return to 0 for the random walk, we note that on $\{T_0^j < \infty\}$,

$$T_0^{j+1} = T_0^j + \tilde{\tau}_0,$$

where $\tilde{\tau}_0$ is the first hitting time of 0 by the random walk $(X_{T_0^j+n})_{n\geq 0}$. The equality remains true on $\{T_0^j = +\infty\}$. By applying the strong Markov property (2.2.1) to the stopping time T_0^j, we deduce that $\tilde{\tau}_0$ is independent of T_0^j and has the same distribution as T_0.

We want to know the distribution of T_0. Let us compute its generating function. For this, let us introduce, for $n \geq 1$,

$$f(n) = \mathbb{P}(T_0 = n) = \mathbb{P}(X_1 \neq 0, \ldots, X_{n-1} \neq 0, X_n = 0).$$

Warning: it may be that $\mathbb{P}(T_0 = +\infty) > 0$ (if the process never returns to 0), in which case the generating function F of T_0, defined for $s \in [0, 1]$ by

$$F(s) = \mathbb{E}(s^{T_0}) = \sum_{n=1}^{\infty} f(n)\, s^n,$$

satisfies $F(1) = 1 - \mathbb{P}(T_0 = +\infty) = \mathbb{P}(T_0 < +\infty)$.

Proposition 2.2.4 *If $(X_n)_n$ is a simple random walk starting from 0, the generating function of the first return time to 0 is, for all $s \in [0, 1]$, given by*

$$F(s) = 1 - (1 - 4p(1-p)s^2)^{1/2}.$$

Proof We introduce the auxiliary function Q defined for $s \in [0, 1]$ by

$$Q(s) = \sum_{n=0}^{\infty} \mathbb{P}(X_n = 0)\; s^n,$$

which is easier to compute. Let us show that

(i) $Q(s) = 1 + Q(s)\,F(s)$.
(ii) $Q(s) = (1-4p(1-p)s^2)^{-1/2}$.

According to Bayes' formula,

$$\mathbb{P}(X_n = 0) = \sum_{k=1}^{n} \mathbb{P}(X_n = 0 \,|\, T_0 = k)\, \mathbb{P}(T_0 = k).$$

The random variables Z_i are independent and identically distributed and

$$\mathbb{P}(X_n = 0 \,|\, T_0 = k) = \mathbb{P}(X_n - X_k = 0 \,|\, T_0 = k) = \mathbb{P}(X_{n-k} = 0).$$

We deduce that $a_n = \mathbb{P}(X_n = 0)$ is a solution of

$$a_n = \sum_{k=1}^{n} a_{n-k}\, f(k),$$

and $a_0 = 1$. By multiplying by s^n and summing over n, then using Fubini's theorem, we obtain

$$\begin{aligned} Q(s) &= \sum_{n\geq 0} a_n s^n = 1 + \sum_{n\geq 1}\sum_{k=1}^{n} a_{n-k}\, s^{n-k}\, f(k) s^k \\ &= 1 + \sum_{k=1}^{\infty} f(k) s^k \sum_{n=k}^{+\infty} a_{n-k}\, s^{n-k} = 1 + F(s)\, Q(s), \end{aligned}$$

which proves (i).

Let us note that $\frac{X_n+n}{2}$ follows the binomial distribution $\mathcal{B}(n,p)$. Indeed, $\frac{X_n+n}{2} = \sum_{i=1}^{n} \frac{(Z_i+1)}{2}$ and the random variables $\frac{Z_i+1}{2}$ are independent and follow a Bernoulli distribution with parameter p. Thus, $\mathbb{P}(X_n = 0) = \binom{n}{n/2}(p(1-p))^{n/2}$ when n is even and $\mathbb{P}(X_n = 0) = 0$ if n is odd. We then have

$$\begin{aligned} Q(s) &= \sum_{n\geq 0} \binom{2n}{n} (p(1-p))^n s^{2n} = \sum_{n\geq 0} \frac{2^n(2n-1)(2n-3)\cdots 1}{n!}\, (p(1-p)\, s^2)^n \\ &= \sum_{n\geq 0} \frac{(2n-1)(2n-3)\cdots 1}{2^n\, n!}\, (4p(1-p)\, s^2)^n. \end{aligned}$$

We can identify the last term by comparing it to the power series expansion

$$(1+u)^{-\frac{1}{2}} = 1 - \frac{1}{2}u + \frac{(-\frac{1}{2})(-\frac{1}{2}-1)}{2!}u^2 + \cdots + \frac{(-\frac{1}{2})(-\frac{1}{2}-1)\cdots(-\frac{1}{2}-n+1)}{n!}u^n + \cdots$$

(for $|u| < 1$). We deduce (ii), then the value of $F(s)$. □

Corollary 2.2.5 *The probability that the particle returns at least once to the origin is*

$$\mathbb{P}(T_0 < +\infty) = F(1) = 1 - |2p-1| = \begin{cases} 2(1-p) & \text{if } p > 1/2 \\ 2p & \text{if } p < 1/2 \\ 1 & \text{if } p = 1/2. \end{cases}$$

If $p = \frac{1}{2}$, $T_0 < +\infty$ a.s. Moreover, the average time of first return $\mathbb{E}(T_0)$ is infinite.

Proof We have

$$\mathbb{P}(T_0 < +\infty) = \lim_{z \nearrow 1} F(s) = 1 - (1 - 4p(1-p))^{1/2}$$
$$= 1 - |2p-1|$$

according to Proposition 2.2.4. Moreover, when $p = 1/2$, $F(s) = 1 - (1-s^2)^{1/2}$ and

$$\mathbb{E}(T_0) = \lim_{s \nearrow 1} F'(s) = +\infty. \qquad \square$$

Remark 2.2.6 According to Corollary 2.2.5, the simple random walk almost surely returns to its starting point if and only if $p = \frac{1}{2}$. This is consistent with the law of large numbers, which here implies that $\mathbb{P}(\lim_{n\to\infty} \frac{\sum_{i=1}^{n} Z_i}{n} = 2p-1) = 1$.

Note that when $p = \frac{1}{2}$, T_0 is finite almost surely, but the return to 0 is slow since the mean return time to 0 is infinite. By the strong Markov property, we deduce that the random walk will almost surely return an infinite number of times to 0. This property, true for point 0, can be applied to any point of the random walk by changing the origin. In this case, we say that the random walk is *recurrent*. In the opposite case (here under the assumption $p \neq \frac{1}{2}$), we will say that the random walk is *transient*. A general definition of these notions will be given in the following section.

Let us note that the properties of transience or recurrence of the symmetric simple random walk are linked to the dimension of the space. We will not prove the following theorem.

Theorem 2.2.7 (Pólya's theorem) *(See for example [38, pp. 114–115].) The symmetric simple random walk is recurrent if the dimension d is equal to* 1 *or* 2. *If $d \geq 3$, the random walk is transient.*

In what follows, we will only consider dimension one.

2.3 Recurrence and Transience – Ergodic Theorems

We will now give a precise definition of recurrence and transience for a Markov chain and deduce certain long-term behaviors. These results are part of an important probabilistic theory that is not covered in this book. We refer, for example, to the books [23], [38] or [32] for more details.

We consider a Markov chain $(X_n)_n$ and for each state $i \in \mathbb{Z}$, we define the number N_i of visits to point i. Thus,

$$N_i = \sum_{n \geq 0} \mathbb{1}_{\{X_n = i\}}.$$

Proposition 2.3.1 *The random variable N_i follows, under $\mathbb{P}_i$, a geometric law on $\mathbb{N}^*$ of parameter $\mathbb{P}_i(T_i < +\infty)$: for all $k \in \mathbb{N}^*$,*

$$\mathbb{P}_i(N_i = k) = \mathbb{P}_i(T_i < +\infty)^{k-1}(1 - \mathbb{P}_i(T_i < +\infty)).$$

If $\mathbb{P}_i(T_i < +\infty) < 1$, the average number of visits to state i is then

$$\mathbb{E}_i(N_i) = \frac{1}{1 - \mathbb{P}_i(T_i < +\infty)} < +\infty.$$

Proof Note that if the Markov chain starts from i, necessarily $N_i \geq 1$. The strong Markov property (Theorem 2.2.3) allows us to show that for all $k \geq 0$,

$$\mathbb{P}_i(N_i \geq k) = \mathbb{P}_i(T_i < +\infty)^{k-1}.$$

Indeed, $\mathbb{P}_i(N_i \geq 1) = 1$ and $\mathbb{P}_i(N_i \geq 2) = \mathbb{P}_i(T_i < +\infty)$ and for $k \geq 3$,

$$\mathbb{P}_i(N_i \geq k) = \mathbb{P}_i(N_i \geq k, T_i < +\infty) = \mathbb{P}_i(N_i \geq k-1)\mathbb{P}_i(T_i < +\infty).$$

We easily deduce that

$$\mathbb{E}_i(N_i) = \sum_{k \geq 0} \mathbb{P}_i(N_i > k) = \frac{1}{1 - \mathbb{P}_i(T_i < +\infty)}.$$ □

We can thus observe two opposite behaviors that lead to the following definition.

Definition 2.3.2 A state $i \in \mathbb{Z}$ is said to be *transient* if and only if

$$\mathbb{P}_i(T_i < +\infty) < 1 \Longleftrightarrow \mathbb{P}_i(N_i = +\infty) = 0 \Longleftrightarrow \mathbb{E}_i(N_i) < +\infty.$$

A state i is said to be *recurrent* if and only if

$$\mathbb{P}_i(T_i < +\infty) = 1 \Longleftrightarrow \mathbb{P}_i(N_i = +\infty) = 1 \Longleftrightarrow \mathbb{E}_i(N_i) = +\infty.$$

An integer i is therefore a recurrent state if the Markov chain starting from state i returns to i in finite time with probability 1, that is, if $\sum_{n=1}^{\infty} f_{ii}^{(n)} = 1$, where $f_{i,i}^{(n)} = \mathbb{P}(T_i = n \mid X_0 = i)$. The state i is transient otherwise.

Proposition 2.3.3 *If $i \longleftrightarrow j$, then i and j are simultaneously transient or recurrent. If the chain is irreducible, the states are either all recurrent or all transient and the* *chain is then said to be* recurrent *or* transient, *respectively. In the case where all states are recurrent, the chain is said to be* irreducible recurrent.

Proof Suppose that i is recurrent. Then the successive return times to i, denoted T_i^k, are finite $\mathbb{P}_i$-almost surely. Let Y_k be the number of visits to j between T_i^{k-1} and T_i^k. The strong Markov property implies that the random variables Y_k are independent and identically distributed under $\mathbb{P}_i$ and, by hypothesis, $\mathbb{P}_i(Y_1 = 0) < 1$ (since $i \longrightarrow j$). In particular, for all n,

$$\begin{aligned}\mathbb{P}_i(\cup_{k\geq 0}\{Y_{k+1} = Y_{k+2} = \cdots = 0\}) &= \lim_k \mathbb{P}_i(Y_{k+1} = Y_{k+2} = \cdots = 0)\\ = \mathbb{P}_i(\forall k, Y_k = 0) \leq \mathbb{P}_i(Y_1 = \cdots = Y_n = 0) &= \mathbb{P}_i(Y_1 = 0)^n,\end{aligned}$$

from which $\mathbb{P}_i(\cup_{k\geq 0}\{Y_{k+1} = Y_{k+2} = \cdots = 0\}) = 0$. Let $N_j = \sum_{k\geq 1} Y_k$ be the number of visits to j. Then

$$\mathbb{P}_i(N_j < +\infty) = \mathbb{P}_i(\cup_{k\geq 0}\{Y_{k+1} = Y_{k+2} = \cdots = 0\}) = 0,$$

and $\mathbb{P}_i(N_j = +\infty) = 1$. The Markov property also implies that

$$\mathbb{P}_i(N_j = +\infty) = \mathbb{P}_i(T_j < +\infty)\,\mathbb{P}_j(N_j = +\infty). \tag{2.3.1}$$

From this, we deduce that $\mathbb{P}_i(T_j < +\infty) = 1$ and $\mathbb{P}_j(N_j = +\infty) = 1$ and then that state j is recurrent. □

Let us note that another proof of Proposition 2.3.3 is given in Exercise 2.6.2.

The communication relation $i \longleftrightarrow j$ is an equivalence relation. This means that the states can be grouped into communicating classes, with each class containing all the elements that communicate with each other.

We will now study the long-term behavior of an irreducible and recurrent Markov chain.

Definition 2.3.4 A *stationary probability measure* for a Markov chain with transition matrix $(P_{i,j})_{i,j\in\mathbb{N}}$ is a probability $\pi = (\pi_i)_{i\in\mathbb{Z}}$ such that

$$\pi P = \pi \iff \sum_i \pi_i P_{i,j} = \pi_j,\ \forall j \in \mathbb{Z}. \tag{2.3.2}$$

This probability measure represents a distribution equilibrium of the chain, in the sense that if the initial condition X_0 has the distribution π, then at each time n, the random variable X_n also has the distribution π. This measure is also called the *invariant probability measure* of X. The equation (2.3.2), rewritten in the form

$$\sum_{i\neq j} \pi_i P_{i,j} = \pi_j(1 - P_{j,j}),$$

can be seen as a balance equalizing what leaves state j and what arrives in j. The search for an invariant probability measure reduces to a linear algebra problem, since the vector π is a left eigenvector of the matrix P associated with the eigenvalue 1. In the case of finite dimension, the existence and uniqueness of such a vector are linked to the Perron–Frobenius theorem.

Indeed, in the particular case of a finite state space, we have the following theorem.

Theorem 2.3.5 *An irreducible Markov chain with a finite state space admits at least one stationary probability measure π.*

Proof Suppose that the state space is $\{1, \ldots, N\}$. The matrix P is such that the sum of the elements of each of its rows is equal to 1 and each of its coefficients is positive. Thus, it is immediate that 1 is an eigenvalue associated with the right eigenvector whose coordinates are all equal to 1. Moreover, if λ is an eigenvalue of P and v is a right eigenvector associated with λ such that $|v_{i_0}| = \max_{i=1,\ldots,N} |v_i|$, we have

$$|\lambda||v_{i_0}| = |\sum_j P_{i_0,j}\, v_j| \leq \max_{i=1,\ldots,N} |v_i| = |v_{i_0}|,$$

from which $|\lambda| \leq 1$. The eigenvalue 1 is therefore an eigenvalue greater in modulus than any other eigenvalue. As the Markov chain is irreducible, the Perron–Frobenius theorem applies (see Chapter 3, Theorem 3.7.7). We deduce that the eigenvalue 1 is associated with a left eigenvector π which is a stationary probability measure of the Markov chain. □

More generally, we give below conditions ensuring the existence and uniqueness of a stationary probability. We will say that an irreducible Markov chain is *positive recurrent* if it is irreducible and if at least one of the states i is recurrent and satisfies $\mathbb{E}_i(T_i) < +\infty$.

Theorem 2.3.6

(1) *If a Markov chain $(X_n)_n$ is irreducible with transition matrix $(P_{i,j})$, positively recurrent such that for a state i, $\mathbb{E}_i(T_i) < +\infty$, then there exists a stationary probability measure π such that $\pi_j > 0$ for all j and such that*

$$\pi_i = \frac{1}{\mathbb{E}_i(T_i)} > 0.$$

(2) *Under the assumptions of (1), every state j is recurrent and also satisfies $\mathbb{E}_j(T_j) < +\infty$. We deduce the existence and uniqueness of a stationary probability measure π satisfying for all j that $\pi_j = \dfrac{1}{\mathbb{E}_j(T_j)} > 0$.*

(3) *Conversely, if there exists a unique invariant probability measure π strictly positive in each state, then all states are positively recurrent and $\mathbb{E}_i(T_i) = \dfrac{1}{\pi_i}$.*

Proof (1) Recall that if the chain is recurrent, the stopping time T_i is finite almost surely. We further assume that $\mathbb{E}_i(T_i) < +\infty$.

Let's introduce m^i_j, the average number of visits of the chain to state j between two passage times to i. We have $m^i_i = 1$ and by the Markov property, for $j \neq i$,

$$\begin{aligned}
m^i_j &= \mathbb{E}_i\Big(\sum_{n=1}^{T_i} \mathbb{1}_{X_n=j}\Big) = \mathbb{P}_i(X_1 = j, T_i \geq 1) + \sum_{n\geq 2} \mathbb{P}_i(X_n = j, T_i \geq n)\\
&= P_{i,j} + \sum_{i\neq k}\sum_{n\geq 2} \mathbb{P}_i(X_{n-1} = k, T_i \geq n-1, X_n = j)\\
&= P_{i,j} + \sum_{k\neq i}\Big(\sum_{n\geq 2} \mathbb{P}_i(X_{n-1} = k, T_i \geq n-1)\Big)P_{k,j} = (m^i.P)_j.
\end{aligned}$$

Thus, $m^i.P = m^i$ and the measure m^i is an invariant measure. Let us show that this measure is strictly positive at each point and has finite mass.

The chain is irreducible and therefore there exists $n > 0$ such that $P^{(n)}_{i,j} > 0$. Thus, since $m^i_i = 1$, we have

$$0 < P^{(n)}_{i,j} = m^i_i P^{(n)}_{i,j} \leq (m^i.P)_j = m^i_j.$$

Moreover, by Fubini's Theorem, we have $\sum_{j\in\mathbb{Z}} m^i_j = \mathbb{E}_i(T_i)$, which is finite by hypothesis.

We have therefore shown the existence of an invariant probability measure π defined by $\pi_j = \frac{m^i_j}{\mathbb{E}_i(T_i)}$. In particular, $\pi_i = \frac{1}{\mathbb{E}_i(T_i)}$.

(2) Let ν be another invariant measure, that is, a (positive) measure on $\mathbb{Z}$ satisfying $\nu.P = \nu$. We can assume without loss of generality that $\nu_i = 1$. Then, for all j,

$$\begin{aligned}
\nu_j &= P_{i,j} + \sum_{k_1\neq i} \nu_{k_1}\, P_{k_1,j}\\
&= P_{i,j} + \sum_{k_1\neq i} P_{i,k_1}\, P_{k_1,j} + \sum_{k_1,k_2\neq i} \nu_{k_2} P_{k_2,k_1}\, P_{k_1,j}\\
&\geq \sum_{n=1}^{\infty} \sum_{k_1,\cdots,k_n\neq i} P_{i,k_n} P_{k_n,k_{n-1}} \cdots P_{k_1,j} + P_{i,j},
\end{aligned}$$

from which we deduce

$$\nu_j \geq \sum_{n=0}^{\infty} \mathbb{P}_i(X_{n+1} = j, T_i \geq n+1) = m^i_j.$$

Thus, the measure $\mu = \nu - m^i$ is a positive, stationary measure and satisfies $\mu_i = 0$.

Let us then introduce j and n such that $P^{(n)}_{j,i} > 0$. We have

$$0 = \mu_i = \sum_k \mu_k P^{(n)}_{k,i} \geq \mu_j P^{(n)}_{j,i}.$$

This implies that necessarily $\mu_j = 0$, and this for all j. We have therefore shown the uniqueness of an invariant measure such that $\nu_i = 1$. Thus, any invariant measure is proportional to m^i. In particular, this is true for the measures m^j, for all $j \in \mathbb{Z}$. We deduce that the mass of such a measure is finite and therefore, for all $j \in \mathbb{Z}$, $\mathbb{E}_j(T_j) < +\infty$. Furthermore, we deduce the uniqueness of the invariant probability measure π such that for all j, $\pi_j = \frac{1}{\mathbb{E}_j(T_j)} > 0$.

(3) Conversely, let π be a stationary probability measure strictly positive in each state. Moreover, suppose that a state j is transient. Then $\sum_{n\geq 0} P^{(n)}_{j,j} = \mathbb{E}_j(N_j) < +\infty$, from which we deduce that $\lim_{n\to\infty} P^{(n)}_{j,j} = 0$. We can show as for (2.3.1) that $\mathbb{E}_i(N_j) = \mathbb{P}_i(T_j < +\infty)\,\mathbb{E}_j(N_j)$. It follows that $\mathbb{E}_i(N_j) < \infty$ and then that for all i, $\lim_{n\to\infty} P^{(n)}_{i,j} = 0$. Thus, $\pi_j = \sum_i \pi_i P^{(n)}_{i,j} = 0$ by the dominated convergence theorem, which is contradictory. We therefore deduce that all states are recurrent. We saw in the first part of the proof that in this case, the measure m^i is the unique stationary measure such that $m^i_i = 1$. We therefore have $m^i = \frac{\pi}{\pi_i}$ and $\mathbb{E}_i(T_i) = \sum_j m^i_j = \frac{1}{\pi_i}\sum_j \pi_j = \frac{1}{\pi_i} < +\infty$. Thus, every state i is positive recurrent. □

Proposition 2.3.7 *A Markov chain with values in a finite and irreducible state space is positive recurrent. Theorem 2.3.6 then gives us the value of the unique invariant measure whose existence had been proved in Theorem 2.3.5.*

Proof Indeed, suppose that the state space is a finite set F. It is then immediate that $\sum_{i\in F} N_i = +\infty$ and $\sum_{i\in F} \mathbb{E}(N_i) = +\infty$. As F is finite, there exists i such that $\mathbb{E}(N_i) = +\infty$. But the Markov property implies that $\mathbb{E}(N_i) = \mathbb{P}(T_i < +\infty)\,\mathbb{E}_i(N_i)$ and therefore that i is recurrent. We can then define the associated invariant measure m^i. Since F is finite, $\mathbb{E}_i(T_i) = \sum_{j\in F} m^i_j$ is finite and i is positive recurrent. □

We wish to study the long-term behavior of the Markov chain by studying the limit, as n tends to infinity, of $P^{(n)}_{i,j} = \mathbb{P}_i(X_n = j)$. However, we must ensure that this limit exists, which is not always the case. Suppose, for example, that the random walk takes its values in $\{1, 2\}$ with the transition probability measures $P_{1,1} = P_{2,2} = 0$ and $P_{1,2} = P_{2,1} = 1$. It is immediate that for all $n \in \mathbb{N}$,

$$\mathbb{P}_1(X_n = 1) = 1 - \mathbb{P}_1(X_n = 2) = \frac{1 + (-1)^n}{2},$$

and $P^{(n)}_{1,1}$ cannot have a limit as n tends to infinity.

To overcome this difficulty, we will consider irreducible Markov chains for which two states can be connected with positive probability as soon as the time is sufficiently large. More precisely, we introduce the following definition.

Definition 2.3.8 The Markov chain is said to be *aperiodic* if for all states $i, j \in \mathbb{Z}$, there exists $n(i, j) \in \mathbb{N}^*$ such that for all $n \geq n(i, j)$,

$$\mathbb{P}_i(X_n = j) = P^{(n)}_{i,j} > 0.$$

Note that if the Markov chain is irreducible and if there exists a state i such that $P^{(n)}_{i,i} > 0$ for $n \geq n(i)$ large enough, then the chain is aperiodic. Indeed, let two states

j and k be given. Since the chain is irreducible, there exist n_1 and n_2 such that $P^{(n_1)}_{j,i} > 0$ and $P^{(n_2)}_{i,k} > 0$. But then, for $n \geq n_1 + n(i) + n_2$, we have

$$P^{(n)}_{j,k} \geq P^{(n_1)}_{j,i} P^{(n-n_1-n_2)}_{i,i} P^{(n_2)}_{i,k} > 0.$$

We also note that there is another definition of the aperiodicity of a Markov chain, equivalent to the one we have given, which is developed in Exercise 2.6.5.

We observed in the proof of Theorem 2.3.6 (3) that if state j is transient, then as n tends to infinity, the limit of $P^{(n)}_{i,j}$ is zero. In order to obtain a non-trivial long-time behavior of the Markov chain, we shall therefore assume that it is irreducible, aperiodic and positively recurrent. We will prove the following convergence theorem, called the strong ergodic theorem.

Theorem 2.3.9 *Let $(X_n)_n$ be an irreducible, positively recurrent and aperiodic Markov chain with transition matrix $(P_{i,j})$. Then for all $i, j \in \mathbb{N}$,*

$$\lim_{n\to\infty} P^{(n)}_{i,j} = \lim_{n\to\infty} \mathbb{P}_i(X_n = j) = \frac{1}{\mathbb{E}_j(T_j)}.$$

Proof We know from Theorem 2.3.6 that the chain X has a unique stationary probability measure π and that $\pi_j = \frac{1}{\mathbb{E}_j(T_j)}$. The proof of the result uses a coupling technique. We introduce two independent chains X^1 and X^2 with the same transition matrix as the chain X. The chain X^1 starts from i and X^2_0 has distribution π. Let $Y = (X^1, X^2) \in \mathbb{Z}^2$. We will show in the next Lemma 2.3.10 that the process Y is an irreducible, positively recurrent and aperiodic Markov chain. Let $T = \inf\{n \geq 0, X^1_n = X^2_n\}$. Although the two chains do not have the same initial condition, they will have the same distribution after time T. Note that

$$\begin{aligned}
\mathbb{P}(X^1_n = j) &= \mathbb{P}(X^1_n = j, T > n) + \mathbb{P}(X^1_n = j, T \leq n) \\
&= \mathbb{P}(X^1_n = j, T > n) + \mathbb{P}(X^2_n = j, T \leq n) \\
&\leq \mathbb{P}(T > n) + \mathbb{P}(X^2_n = j).
\end{aligned}$$

By reversing the roles of X^1 and X^2 we finally obtain that

$$|\mathbb{P}(X^1_n = j) - \mathbb{P}(X^2_n = j)| \leq \mathbb{P}(T > n).$$

Furthermore, we know that $\mathbb{P}(X^2_n = j) = \pi_j$. The proof of the theorem will be immediate once we show that $\mathbb{P}(T > n)$ tends to 0 as n tends to infinity. Since Y is positively recurrent, T, which is the time of reaching the diagonal by Y, is finite almost surely. This completes the proof. □

Lemma 2.3.10 *If X^1 and X^2 are two independent Markov chains with the same transition matrix P, irreducible, positively recurrent and aperiodic, the same is true for $Y = (X^1, X^2)$.*

Proof The process Y is clearly a Markov chain. Its transition matrix $(P \otimes P)$ is defined by

$$(P \otimes P)_{(i,k),(j,l)} = P_{i,j} P_{k,l}$$

and $(P \otimes P)^{(n)}_{(i,k),(j,l)} = \mathbb{P}_{(i,k)}(Y_n = (j,l)) = P^{(n)}_{i,j} P^{(n)}_{k,l}$. Let us show that it is irreducible, that is, for all i, j, k, l, there exists n such that $(P \otimes P)^{(n)}_{(i,k),(j,l)} > 0$. We know that there exist n_1 and n_2 such that $P^{(n_1)}_{i,j} > 0$ and $P^{(n_2)}_{k,l} > 0$ since X is irreducible. We also know that X is aperiodic. In this case, for any state j, there exists $n(j)$ such that $P^{(n)}_{j,j} > 0$ for all $n \geq n(j)$. Let us take $n \geq n(j) + n(l) + n_1 + n_2$. We have $P^{(n)}_{i,j} \geq P^{(n_1)}_{i,j} P^{(n(j)+n(l)+n_2)}_{j,j} > 0$, and similarly $P^{(n)}_{k,l} > 0$. Thus $(P \otimes P)^{(n)}_{(i,k),(j,l)} > 0$. Since X is positively recurrent, it admits a unique stationary probability measure π and it is easy to see that $\pi \otimes \pi$ is the unique stationary probability measure for Y. By Theorem 2.3.6 (3), we deduce that Y is positively recurrent. □

Example 2.3.11 Consider a birth and death Markov chain X with values in $\{0, \ldots, N\}$, such that

$$\begin{aligned} P_{i,i+1} &= b(i), \\ P_{i,i-1} &= d(i), \\ P_{i,i} &= 1 - b(i) - d(i), \\ d(0) &= b(N) = 0. \end{aligned}$$

Suppose that for all $i \in \{0, \ldots, N-1\}$, $b(i) > 0$ and that for all $i \in \{1, \ldots, N\}$, $d(i) > 0$. (The condition $b(0) > 0$ can model a form of immigration when the population goes extinct.) The chain is irreducible and has a finite state space. It is therefore positively recurrent and admits a unique invariant probability measure, a solution of $\pi P = \pi$, which is written here: for all $i < N$,

$$\begin{aligned} b(i-1)\pi_{i-1} + d(i+1)\pi_{i+1} - \pi(i)(b(i) + d(i)) &= 0; \\ b(N-1)\pi_{N-1} - d(N)\pi_N &= 0. \end{aligned}$$

In particular, this gives for $i = N$ that $\frac{\pi(N)}{\pi(N-1)} = \frac{b(N-1)}{d(N)}$. Step by step, we can show that for $i \in \{1, \ldots, N\}$,

$$\pi(i) = \frac{b(0)b(1)\cdots b(i-1)}{d(1)d(2)\cdots d(i)} \pi(0),$$

with

$$\pi(0) = \left(1 + \sum_{i=1}^{N} \frac{b(0)b(1)\cdots b(i-1)}{d(1)d(2)\cdots d(i)}\right)^{-1}.$$

This last expression is obtained by using that $\sum_{i=0}^{N} \pi(i) = 1$.

2.4 Absorbed or Reflected Random Walks

2.4.1 Absorbing Barriers

Let us consider a random walk on $\mathbb{Z}$ which moves between two absorbing barriers at points 0 and $a \in \mathbb{N}^*$, where each step is +1 with probability p and −1 with probability $1-p$. The walk stops as soon as it reaches 0 or a. Therefore, with the previous notation, we have $P_{0,0} = P_{a,a} = 1$. A random walk absorbed at 0 and a can model any capture situation, such as the dynamics of a fish in a river with two fishing nets at positions 0 and a or that of a plankton particle moving vertically when the sea floor is covered with mussels (plankton predators) and the surface is covered with a pollution slick.

We want to know with what probability the random walk will be absorbed, and in this case, what will be the probability of being absorbed in each of the traps.

As before, let us denote by $\mathbb{P}_k$ the probability measure conditioned on the fact that $X_0 = k$ and by O, the event "the walk is absorbed at 0". Thus, $O = \{T_0 < \infty\}$. Let us determine the probability $\mu_k = \mathbb{P}_k(O)$ of reaching 0 from an initial position k. We immediately have

$$\mathbb{P}_k(O) = \mathbb{P}_k(O \mid Z_1 = 1)\, \mathbb{P}(Z_1 = 1) + \mathbb{P}_k(O \mid Z_1 = -1)\, \mathbb{P}(Z_1 = -1),$$

from which we obtain

$$p\,\mu_{k+1} + (1-p)\,\mu_{k-1} - \mu_k = 0 \tag{2.4.1}$$

for $k \in \{1, \ldots, a-1\}$, with the boundary conditions $\mu_0 = 1$, $\mu_a = 0$. Equation (2.4.1) is a linear equation, for which we first look for solutions of the form $\mu_k = r^k$. The number r must satisfy the characteristic equation

$$p\,r^2 - r + (1-p) = 0,$$

whose solutions are $r_1 = 1$, $r_2 = (1-p)/p$. Two cases then arise.

- $p \neq 1/2$ (asymmetric walk). The two roots are different, the solutions of (2.4.1) are of the form $\mu_k = \alpha\, r_1^k + \beta\, r_2^k$ and α and β are determined by the boundary conditions. We obtain

$$\mathbb{P}_k(O) = \frac{\left(\frac{1-p}{p}\right)^a - \left(\frac{1-p}{p}\right)^k}{\left(\frac{1-p}{p}\right)^a - 1}, \quad 0 \le k < a. \tag{2.4.2}$$

  Proceeding in the same way for the event A: "the random walk reaches a before 0", we find that $\mathbb{P}_k(A)$ satisfies (2.4.1), with the boundary conditions $\mu_0 = 0$, $\mu_a = 1$. The computation then gives that $\mathbb{P}_k(A) = 1 - \mathbb{P}_k(O)$: the walker is trapped in a finite time, with probability 1.

- $p = 1/2$ (symmetric walk). In this case, $r_1 = r_2 = 1$ and the solutions of (2.4.1) are of the form $\mu_k = \alpha + \beta k$. Taking into account the boundary conditions, we obtain

$$\mathbb{P}_k(O) = 1 - \frac{k}{a}, \qquad \mathbb{P}_k(A) = \frac{k}{a}. \tag{2.4.3}$$

Here again, the walker ends up being trapped (almost surely).

Note that if $p > \frac{1}{2}$ and if a is large enough, then $\left(\frac{1-p}{p}\right)^a$ is negligible, and $\mathbb{P}_k(O) \simeq \left(\frac{1-p}{p}\right)^k$ therefore depends exponentially on k. For example, suppose that $p = 0.505$ and that $a = 1000$. Then $\mathbb{P}_2(O) \simeq 0.96$ and $\mathbb{P}_{500}(O) \simeq 4.5 \times 10^{-5}$.

Now let us compute the average duration before absorption.

We define the average number of steps before absorption for the process starting from k, given by $V_k = \mathbb{E}_k(T)$, where T is the time to reach one of the barriers,

$$T = \min\{n \geq 0;\, X_n = 0 \text{ or } a\} = T_0 \wedge T_a.$$

By the Markov property, by conditioning on the behavior of the chain at time 1, we obtain that for $1 \leq k \leq a - 1$,

$$\mathbb{E}_k(T) = 1 + \mathbb{P}(Z_1 = 1)\,\mathbb{E}_{k+1}(T) + \mathbb{P}(Z_1 = -1)\,\mathbb{E}_{k-1}(T).$$

This leads us to the linear recurrence equation

$$p\,V_{k+1} + (1-p)V_{k-1} - V_k = -1$$

for $1 \leq k \leq a - 1$, with $V_0 = V_a = 0$. This equation is solved by finding the general solutions and a particular solution. (The characteristic equation is the same as in the previous computation.) Finally, for $0 \leq k \leq a$,

$$\mathbb{E}_k(T_0 \wedge T_a) = \begin{cases} \frac{1}{1-2p}\left(k - a\,\frac{1-\left(\frac{1-p}{p}\right)^k}{1-\left(\frac{1-p}{p}\right)^a}\right) & \text{if } p \neq 1/2, \\ k(a-k) & \text{if } p = 1/2. \end{cases} \tag{2.4.4}$$

For example, if $p = 0.5$ and $a = 10$, then $\mathbb{E}_1(T_0 \wedge T_{10}) = 9$ and $\mathbb{E}_5(T_0 \wedge T_{10}) = 25$. If $a = 1000$, we have $\mathbb{E}_1(T_0 \wedge T_{1000}) = 999$ and $\mathbb{E}_{500}(T_0 \wedge T_{1000}) = 250000$.

Random walk absorbed at 0

Consider a random walk (still moving by +1 with probability p and -1 with probability $1 - p$), absorbed only at 0. To know if it will reach 0 in finite time, we will let a tend to infinity in (2.4.2) and (2.4.3). The events $\{T_0 < T_a\}$ increase towards

$\{T_0 < +\infty\}$. If $p > 1/2$, the probability of reaching 0 starting from k is $\left(\frac{1-p}{p}\right)^k$. On the other hand, if $p \leq 1/2$, this probability is 1 for all k.

Link with the extinction probability of a population

As we saw in the introduction, this model is suitable for describing the dynamics of a population. Imagine that X_n is the size of a population at time n. At each time unit, it grows by one individual (birth) with probability p and decreases by one individual (death) with probability $1-p$. Reaching 0 signifies the extinction of the population. If there is no immigration, once the population is extinct, it cannot be reborn and the process X is therefore a random walk with values in $\mathbb{N}$ absorbed at 0. The time to reach 0 is then the extinction time. We have seen above that if $1-p < p \Leftrightarrow p > 1/2$, the extinction probability of a population with k individuals at time 0 is $\left(\frac{1-p}{p}\right)^k$. On the other hand, if $p \leq 1/2$, this probability is 1 for all k. The interpretation of this result for a population dynamics is clear: if the birth probability p is strictly greater than the death probability $1-p$, there is a positive probability of not going extinct, which depends on the initial size of the population, but in the opposite case, the probability of extinction is 1. Let us note that by making a tend to $+\infty$ in (2.4.4), we obtain that in the case of almost-sure extinction,

$$\mathbb{E}_k(T_0) = \begin{cases} \frac{k}{1-2p} & \text{if } p < 1/2, \\ +\infty & \text{if } p = 1/2. \end{cases}$$

In the case where $p > 1/2$ and on the event where there is no extinction, the law of large numbers shows us that the size of the population tends towards infinity almost surely.

2.4.2 Reflecting Barriers

Let us now assume that points 0 and a are not completely absorbing and that, with a strictly positive probability, the random walk can instantly go back in the other direction. If this probability is 1, we will say that the random walk is reflected at this point. In the general case, we will assume that if the random walk is at 0, it stays there with probability $1-p$ and goes to 1 with probability p, and similarly, if it reaches a, it stays there with probability p and goes to $a-1$ with probability $1-p$. In this case, we can imagine that the random walk moves indefinitely between the barriers with positive probability. We want to know how it stabilizes over a long period of time. This random walk only takes a finite number of values and it is easy to show that it is irreducible and aperiodic. By Theorems 2.3.5 and 2.3.9, we know that there exists a unique stationary probability $\pi = (\pi_i)_{i\in\{0,\ldots,a\}}$ such that the law of X_n converges to π as n tends to infinity (for any initial condition). The probability π satisfies the following equation

$$\pi_i = p\pi_{i-1} + (1-p)\pi_{i+1}, \qquad \text{for } 1 \le i \le a-1$$

with the boundary conditions

$$\pi_0 = (1-p)\pi_0 + (1-p)\pi_1;$$
$$\pi_a = p\pi_{a-1} + p\pi_a.$$

This equation is easily solved and we obtain that

$$\pi_i = \left(\frac{p}{1-p}\right)^i \frac{1-\left(\frac{p}{1-p}\right)}{1-\left(\frac{p}{1-p}\right)^{a+1}}, \qquad \forall i = 0, \dots, a.$$

If $p = \frac{1}{2}$, then $\pi_i = \frac{1}{a+1}$ and the invariant distribution is uniform over $\{0, \dots, a\}$, which is a very intuitive result. If, on the other hand, $1-p > p \Leftrightarrow p < 1/2$ (respectively $1-p < p \Leftrightarrow p > 1/2$), then π_i decreases geometrically from the barrier 0 (respectively from the barrier a).

2.5 Discrete Time Martingales

If we consider a random walk $(X_n)_n$ defined by $X_n = X_0 + \sum_{i=1}^n Z_i$, where X_0 is integrable and the Z_i are independent, equidistributed and centered, we see that the expectation of X_n remains constant equal to $\mathbb{E}(X_0)$. We also see that if, as before, $\mathcal{F}_n = \sigma(X_0, \dots, X_n)$ denotes the σ-field of the process up to time n, then

$$\mathbb{E}(X_{n+1} \mid \mathcal{F}_n) = X_n.$$

The expectation of a future increment, conditionally on the past of the process, is zero. The random fluctuations created by these increments will not be visible at the macroscopic level. In this section, we will formalize this property. More details can be found in Neveu [64], Karlin–Taylor [49] and in Jacod–Protter [44].

We consider a probability space equipped with a filtration, namely a sequence $\mathcal{F}_0 \subset \mathcal{F}_1 \subset \cdots$ of nested σ-fields. Typically $\mathcal{F}_n = \sigma(X_0, \dots, X_n)$, where $(X_0, X_1, \dots)$ is a sequence of random variables. We can easily generalize the notion of stopping time to a general filtration $(\mathcal{F}_n)_n$.

Definition 2.5.1 (i) A random variable T with values in $\mathbb{N} \cup \{+\infty\}$ is an *$\mathcal{F}_n$-stopping time* if $\{T = n\} \in \mathcal{F}_n$ for all n.

(ii) For T an $\mathcal{F}_n$-stopping time, the σ-field

$$\mathcal{F}_T = \{\text{events } A \text{ such that } A \cap \{T = n\} \in \mathcal{F}_n\}$$

is the *σ-field of events prior to* T. It represents the information available at the random time T.

Let us now define what a martingale is.

Definition 2.5.2 A sequence $(X_0, X_1, \ldots)$ of random variables is a *martingale* for the filtration $(\mathcal{F}_n)_n$ if for all n, X_n is $\mathcal{F}_n$-measurable and integrable, and satisfies

$$\mathbb{E}(X_{n+1}|\mathcal{F}_n) = X_n.$$

Note that the average of a martingale is constant over time:

$$\mathbb{E}(X_n) = \mathbb{E}(X_0), \quad \text{for all } n \in \mathbb{N}. \tag{2.5.1}$$

The proof is immediate. Indeed, $\mathbb{E}(X_{n+1}) = \mathbb{E}(\mathbb{E}(X_{n+1} \,|\, \mathcal{F}_n)) = \mathbb{E}(X_n)$.

A process $(X_n)_n$ such that X_n is $\mathcal{F}_n$-measurable for all n is called a *process adapted to the filtration* $(\mathcal{F}_n)_n$.

Definition 2.5.3 An $(\mathcal{F}_n)_n$-adapted process $(X_n)_n$ is a *supermartingale* for the filtration $(\mathcal{F}_n)$ if the variables X_n are integrable and satisfy, for all n, $\mathbb{E}(X_{n+1}|\mathcal{F}_n) \le X_n$. Similarly, the process is a *submartingale* if, for all n, $\mathbb{E}(X_{n+1}|\mathcal{F}_n) \ge X_n$.

Such examples can be obtained by considering concave or convex functions of martingales, as is easy to prove using Jensen's inequality for conditional expectations. For example, if $(M_n)_n$ is a martingale, the process $(\exp(-M_n))_n$ is a submartingale.

We will first show that equality (2.5.1) remains true at a stopping time T, under certain assumptions, either of boundedness of T, or of domination of the martingale. This property will give us a powerful computation tool, as seen later.

Notation: We denote by $n \wedge T$ the stopping time $\min(n, T)$ and the stopped process $X_{n\wedge T}$ is the random variable

$$\begin{aligned} X_{n\wedge T} : \Omega &\longrightarrow \mathbb{R} \\ \omega &\longmapsto X_{n\wedge T(\omega)}(\omega). \end{aligned}$$

Theorem 2.5.4 (Stopping Theorem) *Consider a martingale $(X_n)_n$ for the filtration $(\mathcal{F}_n)_n$ and let T be a stopping time with respect to the same filtration.*

(i) *The stopped process $(X_{n\wedge T})_{n\in\mathbb{N}}$ is a martingale.*
(ii) *If there exists $N \in \mathbb{N}$ such that $\mathbb{P}(T \le N) = 1$, then $\mathbb{E}(X_T) = \mathbb{E}(X_0)$.*
(iii) *If $\mathbb{P}(T < +\infty) = 1$ and there exists Y such that $\mathbb{E}(Y) < +\infty$ and $|X_{n\wedge T}| \le Y$ for all $n \in \mathbb{N}$ (domination condition), then $\mathbb{E}(X_T) = \mathbb{E}(X_0)$.*

Be careful, in many situations $\mathbb{E}(X_T) \neq \mathbb{E}(X_0)$: these are cases where the assumptions of the theorem are not met. For example, consider a simple random walk on $\mathbb{N}$ absorbed at 0 starting from $X_0 = 1$. Suppose that $\mathbb{P}(Z_i = 1) = 1/2 = \mathbb{P}(Z_i = -1)$. We have seen that in this case $T_0 < \infty$ almost surely (see Corollary 2.2.5). However, $X_{T_0} = 0$ and its expectation is zero while $\mathbb{E}(X_0) = 1$. The domination condition is therefore not met. Indeed, the random walk can reach any integer with positive probability. Moreover, we have seen that $\mathbb{E}(T_0) = +\infty$ and T_0 cannot be bounded.

Proof (i) Let us denote by M_n the random variable $X_{n\wedge T}$, which is $\mathcal{F}_n$-measurable and integrable as soon as X_n is.

We have $M_n = X_n \mathbb{1}_{T>n-1} + X_T \mathbb{1}_{T\le n-1}$. As

$$X_T \mathbb{1}_{T\le n-1} = X_1 \mathbb{1}_{T=1} + \cdots + X_{n-1}\mathbb{1}_{T=n-1},$$

with $\{T=k\} \in \mathcal{F}_k \subset \mathcal{F}_{n-1}$ for $k \le n-1$, the random variable $X_T \mathbb{1}_{T\le n-1}$ is $\mathcal{F}_{n-1}$-measurable. We deduce that

$$\begin{aligned}
\mathbb{E}(M_n \mid \mathcal{F}_{n-1}) &= \mathbb{E}\big(X_n \mathbb{1}_{\underbrace{\{T>n-1\}}_{\in \mathcal{F}_{n-1}}} \mid \mathcal{F}_{n-1}\big) + \mathbb{E}\big(X_T \mathbb{1}_{T\le n-1} \mid \mathcal{F}_{n-1}\big) \\
&= \mathbb{1}_{T>n-1} \underbrace{\mathbf{E}(X_n \mid \mathcal{F}_{n-1})}_{\substack{= X_{n-1} \text{ because} \\ (X_n) \text{ is a martingale}}} + X_T \mathbb{1}_{T\le n-1} \\
&= X_{n-1}\mathbb{1}_{T>n-1} + X_T \mathbb{1}_{T\le n-1} = X_{(n-1)\wedge T} = M_{n-1}.
\end{aligned}$$

So (M_n) is a martingale.

(ii) The process $(X_{n\wedge T})$ is a martingale, so at time N

$$\mathbb{E}(X_T) = \mathbb{E}(X_{N\wedge T}) = \mathbb{E}(X_{0\wedge T}) = \mathbb{E}(X_0).$$

(iii) The process $(X_{n\wedge T})$ is a martingale, so $\mathbb{E}(X_{n\wedge T}) = \mathbb{E}(X_0)$. Let n tend to infinity in the previous equality. Since T is finite almost surely, $X_{n\wedge T}$ converges almost surely to X_T as n tends to infinity. Moreover, the domination assumption allows us to apply the dominated convergence theorem and $\mathbb{E}(X_{n\wedge T})$ tends to $\mathbb{E}(X_T)$. We deduce that $\mathbb{E}(X_T) = \mathbb{E}(X_0)$. □

A fundamental property of martingales is their asymptotic behavior as n tends to infinity, under a stronger integrability assumption. This often justifies the identification of martingales as soon as we want to show a long-term convergence result. We do not give here the weakest assumptions allowing this convergence, but we state the theorem most often used in applications, for which the assumption is easy to state and often easy to verify. For more general statements, we refer to

Theorem 2.5.5 (Martingale convergence theorem) *Consider a uniformly square-integrable martingale $(M_n)_n$, i.e. a martingale satisfying*

$$\sup_n \mathbb{E}(M_n^2) < +\infty.$$

Then $(M_n)_n$ converges almost surely as n tends to infinity, towards an integrable limit M_∞ satisfying $M_n = \mathbb{E}(M_\infty | \mathcal{F}_n)$ for all $n \in \mathbb{N}$.

The proof of Theorem 2.5.5 relies on the martingale inequalities stated below, which are essential to obtain trajectorial estimates. These inequalities are often called the Maximal Lemma (i) and Doob's inequality (ii).

Proposition 2.5.6 *Let $(M_n)_n$ be an $(\mathcal{F}_n)$-martingale.*

(i) *Define $M_n^* = \sup_{0\le k\le n} |M_k|$. Then for $a > 0$,*

$$a\,\mathbb{P}(M_n^* > a) \le \mathbb{E}(|M_n|\mathbb{1}_{M_n^*>a}). \tag{2.5.2}$$

(ii) *Suppose that $\sup_n \mathbb{E}(M_n^2) < +\infty$. Then we have Doob's inequality:*

$$\mathbb{E}((M_n^*)^2) = \mathbb{E}(\sup_{k\le n} |M_k|^2) \le 4\,\mathbb{E}(M_n^2). \tag{2.5.3}$$

Proof of Proposition 2.5.6 (i) By Jensen's inequality, we first note that $|M_n| \le \mathbb{E}(|M_{n+1}||\mathcal{F}_n)$. Consider the stopping time $T = \inf\{k \ge 0, |M_k| > a\} \wedge n$ bounded by n. We have the inclusions $\{M_n^* \le a\} \subset \{T = n\}$ and $\{M_n^* > a\} \subset \{|M_T| > a\}$. Thus,

$$\mathbb{E}(|M_T|) = \mathbb{E}(|M_T|\mathbb{1}_{\{M_n^*\le a\}}) + \mathbb{E}(|M_T|\mathbb{1}_{\{M_n^*>a\}}) \ge \mathbb{E}(|M_n|\mathbb{1}_{\{M_n^*\le a\}}) + a\,\mathbb{P}(M_n^* > a).$$

Since T is bounded, an argument similar to that of the proof of Theorem 2.5.4 (ii) implies that $\mathbb{E}(|M_n|) \ge \mathbb{E}(|M_T|)$. Finally

$$\mathbb{E}(|M_n|) \ge \mathbb{E}(|M_n|\mathbb{1}_{\{M_n^*\le a\}}) + a\,\mathbb{P}(M_n^* > a),$$

and we deduce the result.

(ii) Suppose that $\sup_n \mathbb{E}((M_n)^2) < +\infty$. By integrating (2.5.2) with respect to $a > 0$, we obtain that $\int_0^{+\infty} a\,\mathbb{P}(M_n^* > a)\,\mathrm{d}a \le \int_0^{+\infty} \mathbb{E}(|M_n|\mathbb{1}_{M_n^*>a})\,da$. But by Fubini's theorem,

$$\int_0^{+\infty} a\,\mathbb{P}(M_n^* > a)\,\mathrm{d}a = \mathbb{E}\left(\int_0^{M_n^*} a\,\mathrm{d}a\right) = \frac{1}{2}\mathbb{E}((M_n^*)^2).$$

Using Fubini's theorem and Schwartz's inequality, we deduce that

$$\frac{1}{2}\mathbb{E}((M_n^*)^2) \le \mathbb{E}(M_n^2)^{1/2}\mathbb{E}((M_n^*)^2)^{1/2}$$

and (2.5.3) follows. □

Proof of Theorem 2.5.5 Let us introduce the increments $\Delta_n = M_n - M_{n-1}$. Note first that by the martingale property, $\mathbb{E}(\Delta_m\Delta_n) = 0$ for $m \ne n$. Indeed, assuming for example that $m > n$, we have

$$\begin{aligned}&\mathbb{E}(\mathbb{E}((M_m - M_{m-1})(M_n - M_{n-1})|\mathcal{F}_{m-1}))\\ &\quad = \mathbb{E}((\mathbb{E}(M_m|\mathcal{F}_{m-1}) - M_{m-1})(M_n - M_{n-1})) = 0.\end{aligned}$$

We can deduce that $\mathbb{E}(M_n^2) = \sum_{k=0}^n \mathbb{E}(\Delta_k^2)$. Thus, the boundedness assumption in $\mathbb{L}^2$ is equivalent to the convergence of the series with general term $\mathbb{E}(\Delta_k^2)$. In particular, for all integers n and m, $\mathbb{E}((M_{n+m} - M_n)^2) = \sum_{k=n}^{n+m} \mathbb{E}(\Delta_k^2)$, and the sequence $(M_n)_n$ is Cauchy in $\mathbb{L}^2$. It therefore converges in $\mathbb{L}^2$ to a square-integrable random variable M_∞. Moreover, the martingale property and Schwartz's inequality imply that for all $A \in \mathcal{F}_n$ and for all m and n,

$$|\mathbb{E}((M_n - M_\infty)\mathbb{1}_A)| = |\mathbb{E}((M_{n+m} - M_\infty)\mathbb{1}_A)| \le \mathbb{E}(|M_{n+m} - M_\infty|^2)^{1/2},$$

which tends to 0 as m tends to infinity. Thus, $M_n = \mathbb{E}(M_\infty \,|\, \mathcal{F}_n)$. Let us show that the convergence also occurs almost surely. By applying the Bienaymé–Chebyshev inequality, we obtain that for $\varepsilon > 0$,

$$\begin{aligned}\varepsilon^2\,\mathbb{P}(\sup_{k\ge n}|M_k - M_\infty| > \varepsilon) &\le \mathbb{E}\Big(\sup_{k\ge n}|M_k - M_\infty|^2\Big)\\ &\le 2\left(\mathbb{E}\Big(|M_n - M_\infty|^2\Big) + \mathbb{E}\Big(\sup_{k\ge n}|M_k - M_n|^2\Big)\right).\end{aligned}$$

The monotone convergence theorem implies that

$$\mathbb{E}\Big(\sup_{k\ge n}|M_k - M_n|^2\Big) = \lim_{N\to\infty}\mathbb{E}\Big(\sup_{n\le k\le N}|M_k - M_n|^2\Big).$$

We can then apply Doob's inequality (2.5.3) to the martingale $(M_k - M_n)_{k\ge n}$ to deduce that

$$\begin{aligned}\mathbb{E}\Big(\sup_{k\ge n}|M_k - M_n|^2\Big) &\le \lim_{N\to\infty} 4\,\mathbb{E}\Big(|M_N - M_n|^2\Big)\\ &= 4\,\mathbb{E}\Big(|M_\infty - M_n|^2\Big).\end{aligned}$$

This implies that for all $\varepsilon > 0$, $\mathbb{P}(\sup_{k\ge n}|M_k - M_\infty| > \varepsilon)$ tends to 0 as n tends to infinity. The sets $\{\sup_{k\ge n}|M_k - M_\infty| > \varepsilon\}$ are decreasing in n and their intersection equals $\limsup_k \{|M_k - M_\infty| > \varepsilon\}$. We deduce that this set has null probability, which implies that the sequence (M_n) converges almost surely to M_∞. □

2.6 Exercises

Exercise 2.6.1 (A tree growth model)
Consider a Markov chain model for the life cycle of trees. Suppose there are 3 stages of growth: young tree, mature tree, and old tree. When a tree is old, it is replaced by a young tree with probability $p > 0$. Moreover, knowing that the transition times are periods of 8 years, the matrix of the chain is given by

$$P = \begin{pmatrix} 1/4 & 3/4 & 0 \\ 0 & 1/2 & 1/2 \\ p & 0 & 1-p \end{pmatrix}.$$

Show that the chain is irreducible and aperiodic. Compute its stationary probability measure.

Deduce the average recurrence times for each state, namely the time it will take for a tree of a certain stage to be replaced by a tree at the same stage. What are these times for $p = 7/10$?

Exercise 2.6.2

1. Consider an integer-valued Markov chain with transition matrix $(P_{i,j})$. Let $F_{i,i}(s)$ denote the generating function of the first return time T_i to state i given that $X_0 = i$ and $f_{i,i}^{(n)}$ the probability that this time is equal to n. Show that

$$P_{i,i}^{(n)} = \sum_{k=1}^{n} f_{i,i}^{(k)} P_{i,i}^{(n-k)}.$$

   Deduce that for any real number s with $|s| < 1$,

$$P_{i,i}(s) = \frac{1}{1 - F_{i,i}(s)}.$$

2. Show that state i is recurrent if and only if $\sum_{n=0}^{\infty} P_{i,i}^{(n)} = \infty$.
3. Deduce that if $i \longleftrightarrow j$, then i is recurrent if and only if j is recurrent.

Exercise 2.6.3 (A reflected random walk)
Consider a sequence of independent random variables $(Z_n)_n$ with values in $\{-1, 1\}$ and such that $\mathbb{P}(Y_1 = 1) = p$, $0 < p < 1$. Define the Markov chain reflected at 0 by $X_0 \in \mathbb{N}$ and by the recursive formula

$$X_{n+1} = X_n + Z_{n+1} \mathbb{1}_{X_n > 0} + \mathbb{1}_{X_n = 0}.$$

1. Give the transition matrix of the chain. Deduce that it is irreducible.
2. Consider also the random walk (Y_n) defined by $Y_n = X_0 + \sum_{k=1}^{n} Z_k$. Show that almost surely,

$$X_n \geq Y_n, \quad \forall n.$$

   What can be deduced about the behavior of $(X_n)_n$ when $p > 1/2$?

3. Let T_0 be the time it takes for the chain $(X_n)_n$ to reach 0. Show that on the random event $\{T_0 \geq n\}$, we have $X_n = Y_n$. Deduce that if $p \leq 1/2$, state 1 is recurrent for $(X_n)_n$. What can be deduced for this chain?
4. For $p \leq 1/2$, does the chain $(X_n)_n$ have an invariant probability measure? Comment on the results obtained.

Exercise 2.6.4 (Modeling of waiting lines)
A queue forms at a counter. At each time n, a customer arrives with probability p and no customer arrives with probability $1-p$ $(0 < p < 1)$. When at least one customer is waiting, at each moment either a customer is served over a unit of time and leaves the system at time $n+1$ with probability q, $0 < q < 1$, or no one leaves the system with probability $1-q$. All the events thus described are independent of each other. We denote by X_n the number of customers present in the queue at time n.

1. Show that the sequence $(X_n)_n$ is an irreducible Markov chain and give the transition matrix.
2. Give a necessary and sufficient condition on p and q for the chain $(X_n)_n$ to have a unique stationary probability, denoted by π. Compute in this case $\mathbb{E}_\pi(X_n)$.
3. Assume that this condition is satisfied. As customers are served in the order of arrival, what is the average waiting time of a customer knowing that the system is initialized with its stationary probability measure?
4. Now suppose that Y_{n+1} customers can arrive between times n and $n+1$, the random variables Y_n being independent and equidistributed and $0 < \mathbb{P}(Y_1 = 0) < 1$. Denote their generating function by φ. Then give the transition matrix of the chain $(X_n)_n$. Calling ψ_n the generating function of X_n, give a recursive relation between ψ_{n+1}, φ and ψ_n. Deduce that there exists a unique stationary probability measure for the chain $(X_n)_n$ if and only if $\mathbb{E}(Y_1) < q$. Comment on this result.

Exercise 2.6.5 (Another definition of aperiodicity)
There is another definition of aperiodicity than the one given by Definition 2.3.8.

We define the period $d(i)$ of a state i as the greatest common divisor of all integers $n \geq 1$ such that $P^{(n)}_{i,i} > 0$. If the state has a period $d(i) > 1$, it is said to be periodic with period $d(i)$. If the period is equal to 1, the state i is said to be aperiodic.

If $P^{(n)}_{i,i} = 0$ for all n, we define by extension $d(i) = 0$.

1. Show that periodicity is a class property, that is, if $i \longleftrightarrow j$, then $d(i) = d(j)$.

For this, we will show that $d(j)$ divides any number k such that $P^{(k)}_{i,i} > 0$.

2. Show that if a and b are two coprime natural numbers, the set

$$I = \{au + bv; u, v \in \mathbb{N}\}$$

contains all natural numbers, except perhaps a finite number of them.

3. Deduce that if $d(i) = 1$, there exists $n(i)$ such that $P^{(n)}_{i,i} > 0$ for all $n \geq n(i)$. We can study the set $I(i) = \{n \geq 1, P^{(n)}_{i,i} > 0\}$.

The converse of the previous assertion is immediate. We have therefore shown that the random walk is aperiodic if and only if all its elements are aperiodic.

Chapter 3
Population Dynamics in Discrete Time

The randomness of events will constantly disturb the slow but steady course of nature, often delaying it, sometimes accelerating it.

Condorcet (1743–1794),
Esquisse d'un tableau historique des progrès de l'esprit humain

Birth and death processes and branching processes are fundamental models in population dynamics. We wish to study here, through certain probabilistic models, the variations, growth or extinction of a population. Many areas of biology are concerned with these models: population biology, cell kinetics, bacterial growth, DNA replication. Some questions, for example, concern the frequency of extinctions of families, local populations or species. Are these extinctions the consequences of disasters or environmental changes external to populations or are they linked to intrinsic behaviors of populations? If this is the case, how is this compatible with the famous Malthusian exponential growth law in the case of significant resources? Is there a dichotomy between exponential growth and extinction or are there slower forms of growth? How does the population evolve if it persists over a long time? Does it stabilize?

In the following, the term population shall very generally designate a set of living individuals, and an individual can refer to any type of living being, as diverse as bacteria, plants, humans or elephants.

We shall first study certain discrete time models, mainly describing genealogy dynamics. We will recall in particular the main properties satisfied by the Galton–Watson process (also called the Bienaymé–Galton–Watson process) and will develop the study of quasi-stationary distributions.

There is a large literature concerning these processes and our bibliography is not exhaustive. We will refer, for example, to the books by Haccou, Jagers & Vatutin [39], Allen [2], and Athreya & Ney [5].

S. Méléard, *Random Models in Biology, Ecology and Evolution*,
Texts in Applied Mathematics 87, https://doi.org/10.1007/978-3-662-73483-4_3

3.1 Birth and Death Markov Processes

In discrete time models, time is generally expressed in number of generations or it describes a periodic aspect of the reproduction law (seasons, years,...). It can also, in certain cases, simply correspond to a discretization of continuous time. The variable X_n is the size of the population at generation n. It is therefore valued in $\mathbb{N}$.

We will study the following general model. If we assume that at a time n, the population is of size $i \in \mathbb{N}$ ($X_n = i$), then each living individual at generation n

- dies when moving from generation n to generation $n+1$,
- gives birth to a random number of descendants, independently of other individuals. We also assume that the law of the variable "number of descendants" does not depend on n or on the individual of the n-th generation from which the descendants are issued, but can possibly depend on the size i of the population, thus modeling the influence of the number of living individuals on reproduction. This can be, for example, through a decrease in resources and therefore less energy for reproduction, or on the contrary through a stimulation favoring reproduction. This law will be denoted by $Q^i = (q^i_l, l \in \mathbb{N})$. Thus, q^i_l is the probability that an individual from a population of size i gives birth, in the next generation, to l individuals.
- In addition, there is possibly immigration of a random number of individuals, independently of the birth and death process, with law $\eta^i = (\eta^i_l, l \in \mathbb{N})$. Thus η^i_l denotes the probability that a number l of migrants join a population of size i in the next generation. Once the migrants have arrived, they reproduce and die like the other individuals of the population.

Construction of the chain

On a measurable space $(\Omega, \mathcal{A})$, let us give a family of integer-valued random variables

$$(X_0, (Y_{n,i,k})_{n,i,k\geq 0}, (Z_{n+1,i})_{n,i\geq 0}). \tag{3.1.1}$$

The random variable X_0 is the number of individuals at time 0, $Y_{n,i,k}$ the number of children of the k-th individual of the n-th generation when the total number of individuals of this generation is i, and $Z_{n+1,i}$ the number of migrants arriving at the $(n+1)$-th generation, under the same assumption $X_n = i$.

Definition 3.1.1 For each probability measure μ on $\mathbb{N}$, we denote by $\mathbb{P}_\mu$ the unique probability measure on the space $(\Omega, \mathcal{A})$ under which the random variables (3.1.1) are independent and such that

(i) X_0 has law μ.
(ii) For each n, i, k, the random variable $Y_{n,i,k}$ has law Q^i.
(iii) For each n, i, the random variable $Z_{n+1,i}$ has law η^i.

Starting from X_0, we thus define the random variables X_n by induction:

$$X_{n+1} = \sum_{k=1}^{i} Y_{n,i,k} + Z_{n+1,i} \quad \text{if} \quad X_n = i, \ \text{ for } i \in \mathbb{N}, \tag{3.1.2}$$

with the convention that an "empty" sum equals 0.

We denote by $\mathbb{P}_i$ the probability $\mathbb{P}_\mu$ when μ is the Dirac mass at i, that is, when $X_0 = i$.

The random variable X_{n+1} is a function of X_n and of the variables

$$(Y_{n,i,k}, Z_{n+1,i}, i, k \geq 0).$$

Thanks to the independence properties, it is easy to show that $(X_n)_n$ is a Markov chain. Its transition probabilities are given for all i, j by

$$P_{i,j} = \mathbb{P}(X_{n+1} = j | X_n = i) = \sum_{k_1,k_2,\ldots,k_i,r \in \mathbb{N}, k_1+\cdots+k_i+r=j} q^i_{k_1} q^i_{k_2} \cdots q^i_{k_i} \eta^i_r. \tag{3.1.3}$$

Definition 3.1.2 1) When the probability measures Q^i and η^i are independent of i, the population dynamics is said to be *density-independent*. Otherwise, it is said to be *density-dependent*.

2) There is no immigration when $\eta^i_0 = 1$ for all i. The population is then an *isolated population*.

We will first study density-independent population models without immigration, when the reproduction law does not depend on the number of individuals present. Such a model can only be realistic in situations of infinite resources. Indeed, in such a case, the individuals do not exert pressure on each other and each can reproduce freely and independently of others. On the other hand, if resources are limited, individuals could fight to survive and the number of individuals in the population may have an effect on the reproduction law. This density-dependence introduces non-linearity into the mathematical models. We will study this type of model in Chapter 5, in the context of continuous time.

The birth and death Markov chain can also be elaborated to describe the dynamics of various interacting subpopulations or to include mutation. Let us give two examples of multi-type models with interaction.

Example 3.1.3 (Two interacting populations) We consider a population made up of individuals of two types $a = 1$ or $a = 2$. There is no immigration. The population at generation n is composed of $X^{(1)}_n$ individuals of type 1, and $X^{(2)}_n$ individuals of type 2. The reproduction law of a type 1 individual (resp. type 2), in a population of i_1 individuals of type 1 and i_2 individuals of type 2, is the probability $Q^{i_1,i_2}_1 = (Q^{i_1,i_2}_{1,l})_{l \in \mathbb{N}}$ (resp. $Q^{i_1,i_2}_2 = (Q^{i_1,i_2}_{2,l})_{l \in \mathbb{N}}$). In this model, type 1 individuals generate descendants of type 1, but the reproduction law depends on the composition of the population in types 1 and 2. This can be an example of a predator-prey or host-parasite model.

To rigorously define this dynamics, we consider a sequence of random variables $(Y^a_{n,i_1,i_2,k}; n, i_1, i_2, k \in \mathbb{N}, a = 1 \text{ or } 2)$ independent of each other and of law $Q_a^{i_1,i_2}$ and we denote by $X_0^{(1)}$ and $X_0^{(2)}$ the initial numbers of individuals respectively of type 1 and type 2. We then define, for $a = 1$ or $a = 2$,

$$X^{(a)}_{n+1} = \sum_{k=1}^{i_a} Y^a_{n,i_1,i_2,k} \quad \text{if } X^{(1)}_n = i_1,\ X^{(2)}_n = i_2, \ \text{ for all } i_1, i_2 \in \mathbb{N}.$$

The process $(X^{(1)}_n, X^{(2)}_n)$ is a Markov chain with values in $\mathbb{N}^2$, with transition matrix

$$\begin{aligned} P_{(i_1,i_2),(j_1,j_2)} &= \mathbb{P}(X^{(1)}_{n+1} = j_1, X^{(2)}_{n+1} = j_2 | X^{(1)}_n = i_1, X^{(2)}_n = i_2) \\ &= \sum_{k_1+\cdots+k_{i_1}=j_1; r_1+\cdots+r_{i_2}=j_2} \prod_{l=1}^{i_1} Q^{i_1,i_2}_{1,k_l} \prod_{m=1}^{i_2} Q^{i_1,i_2}_{2,r_m}. \end{aligned}$$

Example 3.1.4 (Isolated population with mutation) The situation is similar to that of Example 3.1.3, but each descendant of a type 1 individual (resp. type 2) can "mutate" and thus becomes type 2 (resp. type 1) with probability α_1 (resp. α_2), the mutations being independent of reproduction. This model can be used in epidemiology to represent a population of healthy individuals (type 1) and infected individuals (type 2). The chain $(X^{(1)}_n, X^{(2)}_n)$ is still Markovian, with transition matrix given by

$$P_{(i_1,i_2),(j_1,j_2)} = \\ \sum_{A(i_1,i_2,j_1,j_2)} \prod_{l=1}^{i_1} Q^{i_1,i_2}_{1,k_l} \binom{k_l}{k'_l} \alpha_1^{k'_l} (1-\alpha_1)^{k_l - k'_l} \prod_{m=1}^{i_2} Q^{i_1,i_2}_{1,r_m} \binom{r_m}{r'_m} \alpha_2^{r'_m} (1-\alpha_2)^{r_m - r'_m},$$

where $A(i_1, i_2, j_1, j_2)$ is the set of families of integers

$$((k_l, k'_l, r_m, r'_m),\ l = 1, \ldots, i_1;\ m = 1, \ldots, i_2)$$

such that $0 \le k'_l \le k_l$ and $0 \le r'_m \le r_m$, with

$$\begin{aligned} j_1 &= (k_1 - k'_1) + \cdots + (k_{i_1} - k'_{i_1}) + r'_1 + \cdots + r'_{i_2} \text{ and} \\ j_2 &= k'_1 + \cdots + k'_{i_1} + (r_1 - r'_1) + \cdots + (r_{i_2} - r'_{i_2}). \end{aligned}$$

The problems naturally arising from this type of model concern the behavior of the chain $(X_n)_n$ as n tends to infinity. For example, is there extinction of the entire population ($X_n \to 0$) or of a subpopulation of a particular type, or does a stationary regime appear (X_n converges in a certain sense), or is there an "explosion" of the population ($X_n \to +\infty$)? If $(X_n)_n$ tends to infinity, can we find a suitable normalization, a sequence a_n tending to 0, such that $a_n X_n$ converges? This will give an order of magnitude of the population size and the speed of convergence. More complex problems may arise. For example, what happens when we study the behavior of a population conditionally on non-extinction at time n? This is related to the notion of quasi-stationarity distribution, which gives a mathematical sense to quasi-stability before population extinction.

We will not be able to address these problems in all their generality. We will first develop them in the simplest framework of the Galton–Watson process.

3.2 The Galton–Watson or Bienaymé–Galton–Watson Process

3.2.1 Definition

The simplest birth and death Markov chain is the *Galton–Watson* or *Bienaymé–Galton–Watson process*, which we will abbreviate as BGW. It is a Markov chain without immigration and density-independent. The reproduction law is a probability measure on $\mathbb{N}$ independent of i,

$$Q = (q_l, l \in \mathbb{N}),$$

and the immigration probability measure η^i is the Dirac mass at 0. We will study the dynamics of the number X_n of individuals in this population over successive generations $n = 0, 1, 2, \ldots$ assuming that each of the X_n individuals of the n-th generation generates a random number $Y_{n,k}$ of children $(1 \le k \le X_n)$ so that

$$X_{n+1} = \sum_{k=1}^{X_n} Y_{n,k} \qquad (n \ge 0). \tag{3.2.1}$$

The random variables $(Y_{n,k}, n \ge 0, k \ge 1)$ are assumed to be independent of each other and have the distribution Q of a generic variable Y. This law is characterized by its generating function $g(s) = \mathbb{E}(s^Y) = \sum_k q_k s^k$ for all $s \in [0, 1]$. In the following, we will exclude the case where $g(s) \equiv s$.

Figure 3.1 represents the individuals of generations 0 to 3, when $X_0 = 1$, $Y_{0,1} = 2$, $Y_{1,1} = 3$, $Y_{1,2} = 1$, $Y_{2,1} = 2$, $Y_{2,2} = 0, Y_{2,3} = 1$, $Y_{2,4} = 3$. This figure is a realization of a random tree. In more complicated situations, it is possible to structure this tree by adding a mark to each branch, such as the type of the individual and including possible mutations of this type, the age of the individual or its position in space. On this figure, we observe the dynamics of the population in the physical sense of time (from left to right). If, on the other hand, we consider it from right to left, we observe the ancestral lines of the individuals living at time 3. These two aspects essentially correspond to Chapters 4, 5 and 6 of this book. In Chapters 4 and 5, we will study the dynamics of populations and in Chapter 6, we will seek to understand the genealogy of a sample of individuals.

The Bienaymé–Galton–Watson process is the simplest prototype of a branching process, defined for discrete times and a discrete state space. This model was motivated by the study of the survival of French noble families (Bienaymé, 1845) and that

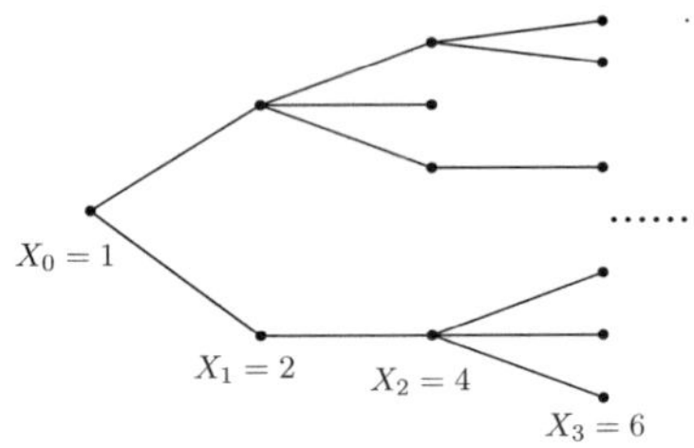

Fig. 3.1 The genealogical tree of the branching process.

of English aristocratic families (Galton, 1873), and more specifically by the study of the non-extinction of male descendants who alone transmitted the name (see Bacaer [6]): N adult men of a nation who carry different family names set out to colonize a country. It is assumed that at each generation, the proportion of men who have l boys is q_l, $l \in \mathbb{N}$. What can we say about the evolution of the number of men carrying this or that name, over time?

Many population dynamics phenomena can be modeled by such a process, whose description is simple, but which nevertheless presents non-trivial behavior. The model can be used to describe (for example) the growth of a cell population, epidemic propagation, or the replication of a DNA molecule in genetics. Consider, for example (see Kimmel–Axelrod [51], Delmas–Jourdain [31]), a DNA macromolecule consisting of a chain of N nucleotides. In one unit of time, this chain is replicated, each nucleotide being copied correctly with probability p and independently of the other nucleotides. At the end of the replication, the molecule is destroyed with probability q or gives birth to two molecules with probability $1-q$. The probability of disappearance of the population of correct macro-molecules is then equal to the probability of extinction of a family name in the case where $q_0 = q$ (destruction), $q_1 = (1-q)(1-p^N)$ (non-destruction but incorrect replication), $q_2 = (1-q)p^N$ (non-destruction and correct replication), and $q_k = 0$ for $k \geq 3$. It is easy to compute the expectation of the reproduction law:

$$\mathbb{E}(Y) = (1-q)(1-p^N) + 2(1-q)p^N = (1-q)(1+p^N).$$

3.2.2 Elementary Results

An immediate but fundamental property is that 0 is an absorbing state: if $X_n = 0$, then $X_{n+1} = 0$. The time to reach 0, that is,

$$T_0 = \inf\{n \geq 0, X_n = 0\},$$

is called the *extinction time* and satisfies $X_n = 0$ if and only if $n \geq T_0$. It is a stopping time for the filtration $(\mathcal{F}_n)_n$ where $\mathcal{F}_n = \sigma(X_k, k \leq n)$.

From a biological point of view, it is important to know the law of the extinction time, and in particular the extinction probability starting from a number i of individuals, for $i \geq 1$, defined by

$$p_i = \mathbb{P}_i(T_0 < +\infty).$$

Of course, $p_0 = 1$.

Proposition 3.2.1 *The BGW process $(X_n)_n$ has the branching property: its law (as a process) under $\mathbb{P}_i$ is the same as the law of*

$$(Z_n^1 + \cdots + Z_n^i)_n,$$

where the $((Z_n^j)_n, 1 \leq j \leq i)$ are Markov chains independent of each other and all have the same law as the process $(X_n)_n$ under $\mathbb{P}_1$.

The branching property for a Markov chain of type (3.1.2) *is characteristic of BGW chains.*

Remark 3.2.2 The branching property allows the study of the law $\mathbb{P}_i$ of a BGW process originating from i individuals to be reduced to the law $\mathbb{P}_1$ of such a process originating from a single ancestor.

Proof By (3.2.1) and $X_0 = i$, we can write $X_n = \sum_{j=1}^{i} Z_n^j$, where Z_n^j denotes the number of individuals living at time n, and descending from the j-th initial individual. By hypothesis, the processes $((Z_n^j)_n, 1 \leq j \leq i)$ are independent.

Conversely, the branching property for a Markov chain of type (3.1.2) implies that the law $(P_{i,j}, j \in \mathbb{N})$ is the i-th power of convolution of the law $(P_{1,j}, j \in \mathbb{N})$. In other words, we have (3.1.3) with $q_j^i = P_{1,j}$ and $\eta_0^i = 1$. The chain is therefore BGW. □

Since the distribution $(P_{i,j}^{(n)})_{j\in\mathbb{N}}$ is the i-th power of convolution of the distributions $(P_{1,j}^{(n)})_{j\in\mathbb{N}}$, we can more easily compute certain quantities associated with a population starting from i individuals. For example, the probability of extinction before time n is

$$\mathbb{P}_i(T_0 \leq n) = \mathbb{P}_1(T_0 \leq n)^i.$$

To prove this, it is enough to note that the extinction time of $Z_n^1 + \cdots + Z_n^i$ is the maximum of the extinction times of Z_n^j, $1 \leq j \leq i$, which are independent of each other.

Certain reproduction laws lead to trivial asymptotic behaviors:

- If $q_1 = 1$, we have $X_n = X_0$ and the population size does not change.
- If $q_1 < 1$ and if $q_0 + q_1 = 1$, we have $X_{n+1} \leq X_n$. The population size decreases and $T_0 < +\infty$ $\mathbb{P}_1$-a.s. If $q_0 = 1$, $T_0 = 1$ $\mathbb{P}_1$-a.s. and if $q_0 < 1$, $\mathbb{P}_1(T_0 = k) = q_0 q_1^{k-1}$. ($T_0$ follows a geometric distribution.)
- If $q_1 < 1$ and $q_0 = 0$, then $X_{n+1} \geq X_n$ and the population size increases. We have $T_0 = +\infty$ $\mathbb{P}_1$-a.s.

In the following, we will eliminate these three cases by assuming that

$$q_0 > 0, \qquad q_0 + q_1 < 1. \tag{3.2.2}$$

Proposition 3.2.3 *The state* 0 *is absorbing and under* (3.2.2), *the other states are transient. If in addition d is the greatest common divisor of the $i \geq 1$ such that $q_i > 0$, the communicating classes are the class $C = d\mathbb{N}^* = \{nd, n \geq 1\}$, and all the singletons $\{i\}$ with $i \notin C$.*

Proof According to (3.2.1), if $i \geq 1$, we have $P_{i,0} \geq (q_0)^i > 0$, so i leads to 0, while 0 does not lead to $i \geq 1$ since 0 is absorbing. The state i cannot therefore be recurrent. The random variables $Y_{n,k}$ almost surely take their values in the set $\{i, q_i > 0\}$, which is contained in $C \cup \{0\}$, so that for all $n \geq 1$, X_n belongs to $C \cup \{0\}$ almost surely. Consequently, a state i can only lead to 0 and the points of C and every singleton not contained in C is a communicating class. □

From the point of view of the ergodic properties or the possible stationarity of the chain $(X_n)_n$, everything is covered by the previous proposition: there is only one recurrent class $\{0\}$ and only one invariant probability measure, i.e. the Dirac mass at 0. The chain is stationary if and only if it starts from 0 and then always stays at 0. Otherwise, due to the fact that all non-zero states are transient, the chain only has two possible behaviors, either it converges to 0, or it converges to + infinity. From the applications point of view, however, we will be able to answer a number of questions concerning the computation of extinction probabilities p_i, the law of extinction time T_0, the explosion speed if $(X_n)_n$ tends towards infinity, etc.

3.2.3 Behavior at Infinity

For BGW chains, we are able to describe the almost sure behavior of the chain as n tends to infinity. Let us denote by m and m_2 the first two moments of the reproduction law Q (which may possibly be infinite)

$$m = \sum_{k=0}^{\infty} k q_k; \; m_2 = \sum_{k=0}^{\infty} k^2 q_k.$$

Also consider the generating function g of Q defined for $s \in [0, 1]$ by

$$g(s) = \sum_{k=0}^{\infty} q_k s^k = \mathbb{E}(s^Y),$$

where $Y \in \mathbb{N}$ is a random variable with distribution Q (the number of children of an individual). The function g, increasing on $[0, 1]$, plays an important role in the study of the process. By differentiating under the expectation and using the dominated convergence theorem, it is easy to show that g is convex. Moreover, it is of class C^∞

on $[0,1[$ and is once (resp. twice) left-differentiable at $s=1$ if and only if $m<+\infty$ (resp. $m_2<+\infty$), and in this case,

$$m=g'(1); \quad m_2-m=g''(1).$$

For $n\geq 1$, let us denote by g_n the n-th iterate of g, that is $g\circ g\circ\cdots\circ g$, n times. Thus $g_{n+1}=g\circ g_n=g_n\circ g$. Let us extend the definition by $g_0(s)=s$. Finally, define

$$s_0=\inf\{s\in[0,1], g(s)=s\}. \tag{3.2.3}$$

Note that $s_0>0$ because $q_0>0$ and $s_0\leq 1$ because $g(1)=1$.

The study of the asymptotic behavior of the sequence $(g_n)_n$ comes down to an inductive argument, developed in the following lemma.

Lemma 3.2.4

a) *If $g'(1)\leq 1$, then $s_0=1$ and $g_n(s)$ tends to 1 as n tends to infinity, for all $s\in[0,1]$.*
b) *If $g'(1)>1$, the equation $g(v)=v$ has a unique solution s_0 in $]0,1[$ and $g_n(s)$ tends to s_0 (respectively decreases to s_0) as n tends to infinity, for all $s\in[0,s_0]$ (respectively all $s\in[s_0,1[$).*

Proof We refer to Figure 3.2 for this proof. The function g from $[0,1]$ into itself is increasing and strictly convex. Moreover, $g(1)=1$. As we have excluded the case $g(s)\equiv s$, the curve g either does not cut the diagonal of the square $[0,1]^2$ or it cuts it at a point distinct from $(1,1)$, depending on whether $g'(1)\leq 1$ or whether $g'(1)>1$. Thus, depending on the case, the fixed point equation $g(v)=v$ has no solution or has a unique solution in $]0,1[$, that is, either $s_0=1$ or $s_0<1$.

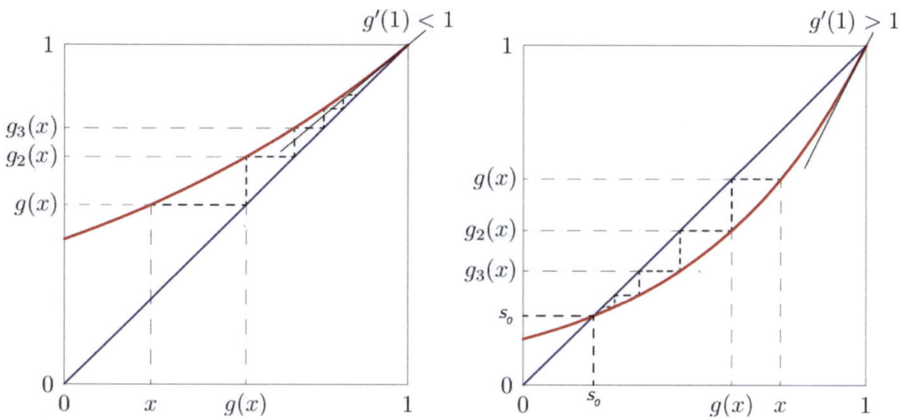

Fig. 3.2 The function g and its iterations in the two cases $m\leq 1$ and $m>1$.

a) When $g'(1)\leq 1$, we have $s\leq g(s)$ and therefore $g_n(s)\leq g_{n+1}(s)$ (since $g_{n+1}=g\circ g_n$) for all s. The limit $\lim_n g_n(s)$ is less than 1 and a solution of $g(u)=u$. It can then only be 1.

b) Similarly, if $g'(1) > 1$, we have $s \le g(s) \le s_0$ or $s \ge g(s) \ge s_0$ depending on whether $s \le s_0$ or $s \ge s_0$; it follows that $g_n(s)$ increases (resp. decreases) with n depending on the case. The limit $\lim_n g_n(s)$, which is a solution of $g(u) = u$ and strictly less than 1 for $s \ne 1$, is then necessarily equal to s_0. □

We also have the following properties.

Lemma 3.2.5

1) *For all* $s \in [0, 1[$, $1 - g_n(s) \le m^n(1 - s)$.
2) *If* $m = 1$ *and* $m_2 < +\infty$, *then for all* $s \in [0, 1[$, *the sequence* $1 - g_n(s)$ *is equivalent to* $\frac{2}{n(m_2-1)}$ *when* $n \to \infty$.

Proof The inequality $1 - g(s) \le m(1 - s)$ is obvious, since g' is increasing on $[0, 1]$ and $g'(1) = m$. By iterating this inequality, we immediately obtain assertion 1).

Let $a = \frac{m_2-1}{2} > 0$ (since $m = 1$). An expansion of g in the neighborhood of $s = 1$ shows that for all $s \in [0, 1[$,

$$1 - g(s) = (1 - s)(1 - (1 - s)(a + \gamma(s))), \quad \text{where } \lim_{s\to 1} \gamma(s) = 0,$$

since $g'(1) = 1$ and $g''(1) = m_2 - 1 = 2a$. Therefore,

$$\frac{1}{1 - g(s)} = \frac{1}{1 - s} + a + \epsilon(s), \quad \text{where } \lim_{s\to 1} \epsilon(s) = 0.$$

We deduce that

$$\frac{1}{1 - g_{j+1}(s)} = \frac{1}{1 - g_j(s)} + a + \epsilon(g_j(s))$$

and by repeating these equalities for $j = n - 1, \ldots, 2, 1$, we finally obtain

$$\frac{1}{1 - g_n(s)} = \frac{1}{1 - s} + na + \Delta_n(s), \quad \text{where } \Delta_n(s) = \sum_{j=0}^{n-1} \epsilon(g_j(s)).$$

For a fixed $s \in [0, 1[$, and when $j \to \infty$, we know that $g_j(s) \to 1$, therefore $\epsilon(g_j(s)) \to 0$. This implies that $\frac{\Delta_n(s)}{n}$ tends to 0 as n tends to infinity, as the Cesàro limit of the sequence $(\epsilon(g_n(s)))_n$. Thus, as n tends to infinity,

$$\begin{aligned} 1 - g_n(s) &= \frac{1}{na} \frac{1}{1 + \frac{1}{na(1-s)} + \frac{\Delta_n(s)}{na}} \underset{n\to+\infty}{\sim} \frac{1}{na} \\ &= \frac{2}{n(m_2 - 1)}. \end{aligned}$$

Assertion 2) has been proved. □

Proposition 3.2.6 *Let G_n be the generating function of X_n. Then, $\forall\, n \geq 0$,*

$$G_{n+1}(s) = G_n\,(g(s))\,.$$

In particular, when $X_0 = i$, we deduce that

$$G_n(s) = (g_n(s))^i \tag{3.2.4}$$

$$\mathbb{E}_i(X_n) = \mathbb{E}(X_n|X_0 = i) = m^n\, i. \tag{3.2.5}$$

Proof The independence of X_n and $(Y_{n,k}, n \geq 1, k \geq 1)$ implies that

$$G_{n+1}(s) = \sum_k \mathbb{E}(s^{\sum_{i=1}^k Y_{n,i}} \mathbf{1}_{\{X_n=k\}}) = \sum_k (g(s))^k \mathbb{P}(X_n = k) = G_n(g(s)).$$

Equation (3.2.4) follows easily from this computation. Then by differentiating this expression and letting s tend to 1, we obtain (3.2.5). □

This result will allow us to study the asymptotic behavior of the sequence $(X_n)_{n\geq 0}$ and to compute in particular the extinction probability.

Remark 3.2.7 Formula (3.2.5) implies that the population goes extinct when $m = \mathbb{E}(Y) < 1$. Indeed, it is easy to see that for all $i \geq 1$,

$$\mathbb{P}_i(X_n \neq 0) \leq \mathbb{E}_i(X_n) = m^n\, i.$$

We will now state a more complete result. The following study will lead us to distinguish three cases:

- The subcritical case, when $m < 1$.
- The critical case, when $m = 1$.
- The supercritical case, when $m > 1$.

Theorem 3.2.8

1) *In the subcritical and critical cases (when $m \leq 1$), the extinction probability is $p_i = \mathbb{P}_i(T_0 < +\infty) = 1$. The population goes extinct almost surely.*
2) *In the subcritical case, the average time of extinction $\mathbb{E}_i(T_0)$ is finite, while in the critical case, $\mathbb{E}_i(T_0) = +\infty$.*
3) *In the supercritical case ($m > 1$), the sequence X_n converges almost surely under $\mathbb{P}_i$ towards a limit X_∞ which only takes the values 0 and $+\infty$ with probability*

$$p_i = \mathbb{P}_i(T_0 < +\infty) = \mathbb{P}_i(X_\infty = 0) = (s_0)^i, \qquad \mathbb{P}_i(X_\infty = +\infty) = 1 - (s_0)^i,$$

where $s_0 \in]0,1[$ was defined in (3.2.3).

Proof Note that $X_{n+1} = 0$ as soon as $X_n = 0$; thus the events $\{X_n = 0\}$ increase with n. The event A = "extinction of the population" is therefore naturally defined as an increasing union of these sets,

$$A = \{T_0 < +\infty\} = \cup_n \{X_n = 0\}$$

and its probability is given by

$$\mathbb{P}(A) = \lim_n G_n(0).$$

The proof of the theorem therefore relies on the study of the sequence $(G_n(0))_n$, with $G_n(0) = (g_n(0))^i$. We use Lemma 3.2.4.

1) Since $\{T_0 \leq n\} = \{X_n = 0\}$, the probability

$$\mathbb{P}_i(T_0 \leq n) = \mathbb{P}_1(T_0 \leq n)^i = (g_n(0))^i \tag{3.2.6}$$

converges to s_0^i as n tends to infinity. This implies that $\mathbb{P}_i(T_0 < +\infty) = 1$ if $m \leq 1$, hence the first part of the theorem. Moreover, $\mathbb{P}_i(T_0 < +\infty) = (s_0)^i$ when $m > 1$.

2) Let us recall that

$$\mathbb{E}_i(T_0) = \sum_{n\geq 0} \mathbb{P}_i(T_0 > n) = \sum_{n\geq 0} (1 - g_n(0)^i).$$

Lemma 3.2.5 immediately implies that if $m < 1$, then $\mathbb{E}_1(T_0) < +\infty$, since in this case, the series with general term m^n is convergent. Moreover, we have seen that the extinction time of X_n given that $X_0 = i$ is the maximum of the extinction times of the sub-populations resulting from each individual of the initial population. Thus, for all $i \in \mathbb{N}^*$, $\mathbb{E}_i(T_0) < +\infty$.

If $m = 1$ and $m_2 < +\infty$, then by the second part of Lemma 3.2.5, we know that $\mathbb{E}_1(T_0) = +\infty$ (the harmonic series diverges), which then implies that $\mathbb{E}_i(T_0) = +\infty$ for all $i \geq 1$. Assertion (2) of Theorem 3.2.8 is therefore proven. If $m_2 = +\infty$, the proof is more delicate but the following intuitive idea can be justified mathematically. Indeed in this case, the branching law gives more weight to large integer values. The expected extinction time will therefore be greater than that of the case where m_2 is finite and we will still have $\mathbb{E}_i(T_0) = +\infty$.

3) It remains to study the asymptotic behavior of X_n when $m > 1$.

When $m > 1$, extinction is not certain since $\mathbb{P}(A) = s_0 < 1$. Let us recall that state 0 is absorbing and that the other states are transient. Therefore, only two cases are possible: either $X_n = 0$ from a certain rank (this is extinction), or $X_n \to +\infty$ (this is explosion). These two complementary events have probabilities respectively $(s_0)^i$ and $1 - (s_0)^i$. This concludes the proof of the theorem. □

Formula (3.2.6) actually gives the distribution function of the extinction time T_0, and therefore its law, under $\mathbb{P}_1$ and under $\mathbb{P}_i$. We have:

$$\mathbb{P}_1(T_0 = n) = \begin{cases} 0 & \text{if} \quad n = 0 \\ g_n(0) - g_{n-1}(0) & \text{if} \quad n \in \mathbb{N}^* \\ 1 - s_0 & \text{if} \quad n = +\infty. \end{cases}$$

Note that in the supercritical case, T_0 is a random variable with values in $\mathbb{N} \cup \{+\infty\}$.

Remark 3.2.9 We have shown that, in the case where the initial population is composed of i individuals, we find ourselves in one of the following three cases.

- When $m < 1$, the probability of "non-extinction at time n" tends to 0 at a geometric rate since, as we have seen,

$$\mathbb{P}_i(X_n \neq 0) \leq m^n i.$$

- When $m = 1$, this same probability tends much more slowly to 0. More precisely, we saw that if $m_2 < +\infty$,

$$\mathbb{P}_i(X_n \neq 0) \sim \frac{2i}{n(m_2 - 1)} \qquad \text{when } n \to \infty.$$

- When $m > 1$, X_n tends to 0 or to $+\infty$.

In the supercritical case, it is interesting to specify at what speed X_n tends to infinity, on the set $\{T_0 = +\infty\}$, which therefore has a strictly positive probability $1 - (s_0)^i$. We limit ourselves to the case where $m_2 < +\infty$. In the following, we will denote by σ^2 the variance of Y.

Theorem 3.2.10 *If $m > 1$ and $m_2 < +\infty$, the sequence $\frac{X_n}{m^n}$ converges $\mathbb{P}_i$-almost surely to a random variable U which is almost surely strictly positive on the set $\{T_0 = +\infty\}$ and such that $\mathbb{E}_i(U) = i$.*

This result tells us that on the set where there is no extinction, the growth of X_n to $+\infty$ is exponential (as m^n). This is the probabilistic expression of the Malthusian law.

Proof Let us set $U_n = \frac{X_n}{m^n}$. Consider the filtration $(\mathcal{F}_n)_n$ composed of the increasing σ-fields $\mathcal{F}_n = \sigma(X_k, k \leq n)$. As m and m_2 are the first two moments of the variables $Y_{n,k}$, Formula (3.2.1) implies that

$$\mathbb{E}_i(X_{n+1} \,|\, \mathcal{F}_n) = \mathbb{E}_i\left(\sum_{k=1}^{X_n} Y_{n,k} \,|\, \mathcal{F}_n\right) = m\, X_n.$$

Indeed, conditioning by $\mathcal{F}_n$ here amounts to fixing X_n and we then use the fact that the random variables $Y_{n,k}$ are independent of X_n. By similar reasoning, we also obtain that

$$\begin{aligned} E_i(X_{n+1}^2 | \mathcal{F}_n) &= \mathbb{E}_i\left(\left(\sum_{k=1}^{X_n} Y_{n,k}\right)^2 | \mathcal{F}_n\right) \\ &= \mathbb{E}_i\left(\sum_{k=1}^{X_n} (Y_{n,k})^2 \,|\, \mathcal{F}_n\right) + \mathbb{E}_i\left(\sum_{k \neq k'=1}^{X_n} Y_{n,k} Y_{n,k'} \,|\, \mathcal{F}_n\right) \\ &= m_2 X_n + m^2 X_n (X_n - 1) = \sigma^2\, X_n + m^2\, X_n^2, \end{aligned} \tag{3.2.7}$$

and so

$$\mathbb{E}_i(U_{n+1}|\mathcal{F}_n) = U_n \; ; \; \mathbb{E}_i(U_{n+1}^2|\mathcal{F}_n) = U_n^2 + \frac{\sigma^2}{m^{n+2}} U_n.$$

As a result, $\mathbb{E}_i(U_n) = i$ and, as $m > 1$, by iterating in n, we obtain that

$$\mathbb{E}_i(U_{n+1}^2) = \mathbb{E}_i(U_n^2) + \frac{i\sigma^2}{m^{n+2}} = i^2 + \frac{i\sigma^2}{m^2} \sum_{j=0}^{n} m^{-j} \le i^2 + \frac{i\sigma^2}{m(m-1)}.$$

We deduce from this that (U_n) is a martingale whose second-order moments are uniformly bounded. It therefore converges $\mathbb{P}_i$ almost surely and in expectation to U as n tends to infinity (see Theorem 2.5.5). In particular, $\mathbb{E}_i(U) = i$. It remains to show that $U > 0$, $\mathbb{P}_i$ almost surely on $\{T_0 = +\infty\}$. Let us set $r_i = \mathbb{P}_i(U = 0)$. By virtue of the branching property, the law of U under $\mathbb{P}_i$ is the same as that of $V^1 + \cdots + V^i$, where the V^j are independent random variables with the same law as U under $\mathbb{P}_1$. We deduce from this that $r_i = (r_1)^i$. The Markov property yields

$$\mathbb{P}_1(U = 0|\mathcal{F}_1) = \mathbb{P}_{X_1}(U = 0) = (r_1)^{X_1}.$$

Taking expectation gives $r_1 = \mathbb{E}_1(r_1^{X_1}) = g(r_1)$. Thus, r_1 is a solution of the equation $g(s) = s$, which admits in $[0,1]$ the two solutions s_0 and 1. However, $\mathbb{E}_1(U) = 1$ implies that $r_1 = \mathbb{P}_1(U = 0) < 1$, therefore necessarily $r_1 = s_0$ and $r_i = (s_0)^i$, that is to say, $\mathbb{P}_i(U = 0) = \mathbb{P}_i(T_0 < +\infty)$. But obviously, $\{T_0 < +\infty\} \subset \{U = 0\}$, and therefore necessarily the random variable U is strictly positive $\mathbb{P}_i$ almost surely on $\{T_0 = +\infty\}$.□

3.2.4 Subcritical Case: Detailed Analysis of Extinction

We consider the case $m < 1$ and we wish to study in detail the properties of the extinction time as a function of the initial condition. Suppose that $m_2 < +\infty$ and denote as before $\sigma^2 = \text{Var}(Y)$.

Proposition 3.2.11

1) *For all* $i, n \in \mathbb{N}^*$,

$$\frac{i(1-m)\,m^{n+1}}{\sigma^2(1-m^n) + m^{n+1}(1-m)} \left(1 - \frac{(i-1)m^n}{2}\right) \le \mathbb{P}_i(X_n > 0) \le i m^n, \quad (3.2.8)$$

which gives a precise estimate of $\mathbb{P}_i(X_n > 0)/m^n$.

2) *If* n *is large enough, we have*

$$i\frac{1-m}{\sigma^2} m^{n+1} \le \mathbb{P}_i(X_n > 0) \le i m^n.$$

Proof 1) We have already seen previously that

$$\mathbb{P}_i(X_n > 0) \le i\,\mathbb{P}_1(X_n > 0) \le i\,\mathbb{E}_1(X_n) = i\,m^n,$$

which proves the right inequality.

For the left inequality, let us first study the case where there is only one ancestor. Note that $\sigma^2 = m_2 - m^2$. We have

$$\left(\mathbb{E}_1(X_n \mathbb{1}_{X_n>0})\right)^2 \le \mathbb{E}_1(X_n^2)\,\mathbb{P}_1(X_n > 0),$$

hence

$$\mathbb{P}_1(X_n > 0) \ge \frac{\mathbb{E}_1(X_n)^2}{\mathbb{E}_1(X_n^2)} = \frac{m^{2n}}{\mathbb{E}_1(X_n^2)}.$$

However, as we have previously shown in (3.2.7),

$$\mathbb{E}_1(X_n^2|\mathcal{F}_{n-1}) = m_2 X_{n-1} + m^2 X_{n-1}(X_{n-1} - 1).$$

Hence,

$$\begin{aligned}\mathbb{E}_1(X_n^2) &= (m_2 - m^2)\mathbb{E}_1(X_{n-1}) + m^2\,\mathbb{E}_1((X_{n-1})^2)\\ &= \sigma^2 m^{n-1} + m^2\,\mathbb{E}_1((X_{n-1})^2)\\ &= \sigma^2\left(m^{n-1} + m^n + \ldots + m^{2(n-1)}\right) + m^{2n}\mathbb{E}_1((X_0)^2)\\ &= \sigma^2\, m^{n-1}\,\frac{1-m^n}{1-m} + m^{2n}.\end{aligned}$$

Let us now prove the general case of an initial condition $i \in \mathbb{N}^*$.
Note that

$$\begin{aligned}\mathbb{P}_i(X_n > 0) &= 1 - \mathbb{P}_i(X_n = 0) = 1 - (\mathbb{P}_1(X_n = 0))^i\\ &= 1 - (1 - \mathbb{P}_1(X_n > 0))^i.\end{aligned}$$

We then use the fact that $\forall\ k \in \mathbb{N}, x \in [0,1]$,

$$1 - (1-x)^k \ge kx - \frac{k(k-1)}{2}x^2,$$

due to the fact that $kx - (1-(1-x)^k) = k(k-1)\int_0^x \mathrm{d}u \int_0^u (1-y)^{k-2}\,\mathrm{d}y$. We then have

$$\mathbb{P}_i(X_n > 0) \ge i\,\mathbb{P}_1(X_n > 0)\left(1 - \frac{i-1}{2}\,\mathbb{P}_1(X_n > 0)\right),$$

from which the first assertion of the proposition follows.

2) When n is large, the dominant term in the left term of inequality (3.2.8) is $\frac{1-m}{\sigma^2}\,m^{n+1}$, which gives the second assertion. □

From an ecological point of view, it is interesting to study the average extinction time of a subcritical population as a function of the initial population size. The following result will answer this question for a very large initially subcritical population.

Proposition 3.2.12 *Consider a subcritical BGW process* $(X_n)_n$ *($m < 1$) such that* $X_0 = i$*. Also assume that* $m_2 < +\infty$ *and let* $c_1 = \frac{(1-m)m}{\sigma^2}$.

Then, for sufficiently large i,

$$\left(\frac{\ln i - \ln\ln i}{|\ln m|}\right)\left(1 - \frac{1}{i^{c_1}}\right) \leq \mathbb{E}_i(T_0) \leq \frac{\ln i}{|\ln m|} + \frac{2-m}{1-m}.$$

Corollary 3.2.13 *Under the same conditions as above, we have*

$$\mathbb{E}_i(T_0) \sim_{i\to\infty} \frac{\ln i}{|\ln m|}.$$

Remark 3.2.14 This corollary has significant ecological implications: even if at time 0, the population is extremely large, it will quickly go extinct if the average number of children per individual is less than 1. Indeed, $\ln i$ is negligible compared to i when $i \to +\infty$.

Proof Let us introduce $\phi(i) = \frac{\ln i}{|\ln m|}$ and $\psi(i) = \frac{\ln\ln i}{|\ln m|}$. We have

$$\begin{aligned}
\mathbb{E}_i(T_0) &= \sum_n \mathbb{P}_i(T_0 > n) \\
&= \sum_{n=0}^{[\phi(i)]} \mathbb{P}_i(T_0 > n) + \sum_{n=[\phi(i)]+1}^{\infty} \mathbb{P}_i(T_0 > n) \\
&\leq \phi(i) + 1 + i \sum_{n=[\phi(i)]+1}^{\infty} m^n \\
&\leq \phi(i) + 1 + i\,\frac{m^{\phi(i)}}{1-m}, \qquad \text{since } m < 1, \\
&\leq \frac{\ln i}{|\ln m|} + \frac{2-m}{1-m},
\end{aligned}$$

since

$$i m^{[\phi(i)]} \leq i \mathrm{e}^{-\frac{\ln i}{\ln m}\ln m} = 1.$$

This gives us the right inequality of Proposition 3.2.12.

For the left inequality, first note that

$$\mathrm{e}^{-c_1 i m^{\phi(i)-\psi(i)}} = \mathrm{e}^{-c_1 m^{-\psi(i)}} = \mathrm{e}^{-c_1 \mathrm{e}^{-\psi(i)\ln m}} = \mathrm{e}^{-c_1 \ln i} = \mathrm{e}^{c_1 \ln\frac{1}{i}} = \frac{1}{i^{c_1}}.$$

We deduce that for $\phi(i)-\psi(i)$ large enough,

$$\begin{aligned}\mathbb{P}_i(T_0 \le \phi(i)-\psi(i)) &= (1-\mathbb{P}_1(T_0 > \phi(i)-\psi(i)))^i \\ &\le \mathrm{e}^{-i\mathbb{P}_1(T_0>\phi(i)-\psi(i))} \le \mathrm{e}^{-ic_1 m^{\phi(i)-\psi(i)}} = \frac{1}{i^{c_1}},\end{aligned}$$

since $(1-x)^k \le \mathrm{e}^{-kx}$ and by Proposition 3.2.11 2).

Thus, if $n \le \phi(i)-\psi(i)$, we have

$$\begin{aligned}\mathbb{P}_i(T_0 > n) &\ge \mathbb{P}_i(T_0 > \phi(i)-\psi(i)) \\ &\ge 1-\mathbb{P}_i(T_0 \le \phi(i)-\psi(i)) \\ &\ge 1-\frac{1}{i^{c_1}},\end{aligned}$$

hence

$$\begin{aligned}\mathbb{E}_i(T_0) \ge \sum_{n=0}^{[\phi(i)-\psi(i)]} \mathbb{P}_i(T_0 > n) &\ge \left(1-\frac{1}{i^{c_1}}\right)([\phi(i)-\psi(i)]+1) \\ &\ge \left(\frac{\ln i - \ln\ln i}{|\ln m|}\right)\left(1-\frac{1}{i^{c_1}}\right).\end{aligned}$$

□

Example (Application to the extinction of North Atlantic right whales) (See Caswell et al. [15] and Haccou–Jagers–Vatutin [39].) We are interested in the reproduction law of female whales. The unit of time is one year. The probability of dying is $\beta \in [0,1]$ and the probability of having a female offspring (in one year) is $\alpha \in [0,1]$. The reproduction law is therefore given by $q_0 = \beta$, $q_1 = (1-\beta)(1-\alpha)$, $q_2 = \alpha(1-\beta)$. In 1994, estimates gave $q_0 = 0.06$ and $\alpha = 0.038$. Then $m = q_1 + 2q_2 = (1-\beta)(1+\alpha) = 0.976 < 1$, and the BGW process is subcritical, with $\sigma^2 = q_1 + 4q_2 = 0.095$. On the other hand, statistical studies estimate the number of whales in 1994 to be 150.

Let us find out how many years we can be sure, with a 99% chance, that extinction will not occur. We do the calculation by applying Proposition 3.2.11.

$$\begin{aligned}\mathbb{P}(X_n > 0 | X_0 = 150) \ge 0.99 \ \text{ as soon as } \ & i(1-m)\frac{m^n+1}{\sigma^2} \ge 0.99 \\ \Longleftrightarrow \quad & n \le 150.\end{aligned}$$

Thus, with a 99% chance and if there is no environmental change, the whales will survive until 2144.

Let us now find out in how many years the whales will have a 99% chance of having disappeared.

$$\begin{aligned}\mathbb{P}(X_n = 0 | X_0 = 150) \ge 0.99 \ \text{ as soon as } \ & i m^n \le 0.01 \\ \Longleftrightarrow \quad & n \ge 395.\end{aligned}$$

Thus, we can assure with a 99% chance that if there is no environmental change, the whales will have disappeared by 2389.

Corollary 3.2.13 applied to $m = 0.976$ and $i = 150$ gives that $\mathbb{E}_i(T_0) \approx 206$.

Remark 3.2.15 Of course, the branching model is very approximate for describing the behavior of the whale population and these computations give an order of magnitude of the extinction time. Moreover, when the population becomes small, the reproduction processes more easily lead to the accumulation of deleterious mutations that increase the whales' poor adaptation to the environment and could lead to an acceleration of their extinction. This acceleration is called an "extinction vortex". Research is currently being developed to model this extinction spiral. See for example [29].

3.3 Relationship Between the Bienaymé–Galton–Watson Process and the Genealogy Model

Let us study the behavior of a BGW process conditioned to remain of constant size N. We will see that in this case and when the reproduction law of the BGW is a Poisson distribution, the conditional law is related to the Wright–Fisher model, which we will study in detail in Chapter 6.

Theorem 3.3.1 *Let N be a non-zero integer. Consider a BGW process X whose reproduction law is a Poisson distribution with parameter m and suppose that $X_0 = N$. So, conditionally on the fact that X keeps the same value N ($X_n = N$, for all n), the distribution of the descendants $(Y_1, \ldots, Y_N)$ of each individual from the previous generation follows a multinomial distribution with parameters $(N, \frac{1}{N}, \ldots, \frac{1}{N})$.*

Proof Let D_i be the number of descendants of the individual i in the BGW model conditioned to remain of constant size N. Let $(k_1, \ldots, k_N)$ be an N-tuple of integers summing to N. We then have

$$\begin{aligned}\mathbb{P}_N(D_1 = k_1, \ldots, D_N = k_N) &= \mathbb{P}_N(Y_1 = k_1, \ldots, Y_N = k_N | X_1 = N)\\ &= \frac{\prod_{i=1}^N \mathbb{P}(Y_i = k_i)}{\mathbb{P}_N(X_1 = N)},\end{aligned}$$

where the random variables Y_i, denoting the number of descendants of the individual i in the BGW model, are independent. Let us call g the generating function of the reproduction law, which is a Poisson distribution with parameter m. For all $s \in [0, 1]$

$$g(s) = \exp(-m(1-s)),$$

and the generating function of X_1 under $\mathbb{P}_N$ is $g(s)^N = \exp(-Nm(1-s))$. The law of X_1 under $\mathbb{P}_N$ is therefore a Poisson distribution with parameter mN. We deduce that

$$\mathbb{P}(D_1 = k_1, \ldots, D_N = k_N) = \frac{N!}{e^{-mN}(Nm)^N} \prod_{i=1}^{N} e^{-m} \frac{m^{k_i}}{k_i!}$$
$$= \frac{N!}{k_1! \cdots k_N!} \left(\frac{1}{N}\right)^N.$$

We recognize here the multinomial distribution with parameters $(N, \frac{1}{N}, \ldots, \frac{1}{N})$. □

3.4 Quasi-Stationary Behavior

Let us return here to the case of a monotype branching process. We have seen that in the subcritical or critical cases, the BGW process $(X_n)_n$ almost surely goes extinct. But the time required for extinction can be very long (for example, in the critical case). If the extinction time is much larger than the demographic time scale, the population process can present an apparent stationarity before extinction. For example, a study done on rattlesnakes in Arizona (cf. Renault, Ferrière, Porter, unpublished), shows that the average number of offspring for the rattlesnake is $m = 0.91$. The population process is therefore clearly subcritical. The experimental study, however, shows that after a phase of violent population decline, the population size seems to stabilize.

Our goal in this section is to give a precise mathematical meaning to this notion of stability before extinction. By analogy with the search for an invariant (or stationary) probability measure, the question can be posed in one of the following forms:

1. Is there a probability measure on $\mathbb{N}^*$ that leaves invariant the conditional law of X_n given that the population is not extinct at time n? Such a probability measure, if it exists, will be called a quasi-stationary distribution.
2. Does the limit of the conditional law of X_n given that the population is not extinct at time n,
$$\lim_{n\to\infty} \mathrm{Law}(X_n | X_n > 0),$$
exist?

We will introduce the notions of quasi-stationary distribution and Yaglom limit (named after the mathematician who introduced this notion for branching processes [79]) for an absorbed Markov chain at 0 and give some of their fundamental properties. We will then study more precisely the case of BGW processes.

The notion of quasi-stationarity is very rich. For a more in-depth study of its properties, we refer to Collet–Martinez–San Martin [24] or to Méléard–Villemonais [61].

3.4.1 Quasi-Stationary Distributions and the Yaglom Limit

Let $(X_n)_n$ be a Markov chain on $\mathbb{N}$ absorbed at 0. We further assume that it almost surely reaches 0. As before, let T_0 denote the hitting time of 0. Thus,

$$\mathbb{P}(T_0 < +\infty) = 1. \tag{3.4.1}$$

For any probability measure $\nu = (\nu_j)_{j \geq 1}$ on $\mathbb{N}^*$, we will denote $\sum_{j \geq 1} \nu_j \mathbb{P}_j$ by $\mathbb{P}_\nu$.

Definition 3.4.1 A probability measure ν on $\mathbb{N}^*$ is a *quasi-stationary distribution* for the process X if it satisfies that for all $j \geq 1$ and for all n,

$$\mathbb{P}_\nu(X_n = j \,|\, T_0 > n) = \nu_j.$$

Remark 3.4.2 Note that the conditioning event $\{T_0 > n\} = \{X_n \neq 0\}$ has a probability that tends to 0 as n tends to infinity since T_0 is almost surely finite. We are therefore conditioning by an event that becomes increasingly rare as n increases.

Definition 3.4.3 The Markov process $(X_n)_n$ admits the *Yaglom limit* μ if μ is a probability measure on $\mathbb{N}^*$ such that for all $j \geq 1$ and for all $i \geq 1$,

$$\lim_{n\to\infty} \mathbb{P}_i(X_n = j \,|\, T_0 > n) = \mu_j.$$

Proposition 3.4.4 *A Yaglom limit for the process is a quasi-stationary distribution.*

Remark 3.4.5 Note that, when it exists, the Yaglom limit is unique. This is not the case for quasi-stationary distributions, as we will see in the following section. Thus, some quasi-stationary distributions are not Yaglom limits.

Proof Let μ be the Yaglom limit of an integer-valued Markov chain $(X_n)_n$. It is therefore a probability measure on $\mathbb{N}^*$. Let us show that it is a quasi-stationary distribution. Consider a function f, defined and bounded on $\mathbb{N}^*$. By definition and for all $i \geq 1$, we have

$$\mu(f) = \lim_{n\to\infty} \mathbb{P}_i(f(X_n) \,|\, T_0 > n) = \lim_{n\to\infty} \frac{\mathbb{P}_i(f(X_n)\,;\, T_0 > n)}{\mathbb{P}_i(T_0 > n)}.$$

Applying this property to the function $f(l) = \mathbb{P}_l(T_0 > k)$, where $k \geq 1$, we deduce from the Markov property that

$$\mathbb{P}_\mu(T_0 > k) = \lim_{n\to\infty} \frac{\mathbb{P}_i(T_0 > k+n)}{\mathbb{P}_i(T_0 > n)}.$$

If we now consider the function $f(l) = \mathbb{P}_l(X_k \in A; T_0 > k)$, where A is a subset of $\mathbb{N}^*$ and applying Markov property again, we obtain

$$\begin{aligned}\mathbb{P}_\mu(X_k \in A; T_0 > k) &= \lim_{n\to\infty} \frac{\mathbb{P}_i(X_{k+n} \in A; T_0 > k+n)}{\mathbb{P}_i(T_0 > n)} \\ &= \lim_{n\to\infty} \frac{\mathbb{P}_i(X_{k+n} \in A; T_0 > k+n)}{\mathbb{P}_i(T_0 > k+n)} \frac{\mathbb{P}_i(T_0 > k+n)}{\mathbb{P}_i(T_0 > n)}.\end{aligned}$$

By Definition 3.4.3, we know that $\frac{\mathbb{P}_i(X_{k+n} \in A; T_0 > k+n)}{\mathbb{P}_i(T_0 > k+n)}$ converges to $\mu(A)$ as n tends to infinity and that $\frac{\mathbb{P}_i(T_0 > k+n)}{\mathbb{P}_i(T_0 > n)}$ converges to $\mathbb{P}_\mu(T_0 > k)$. We deduce that for any set A of $\mathbb{N}^*$ and for all $k \geq 1$,

$$\mu(A) = \mathbb{P}_\mu(X_k \in A \,|\, T_0 > k).$$

Thus, the probability measure μ is a quasi-stationary distribution for the process. □

Proposition 3.4.6 *Let ν be a quasi-stationary distribution on $\mathbb{N}^*$. Then there exists a real number $\rho(\nu) \in (0,1)$ such that*

$$\mathbb{P}_\nu(T_0 > n) = \rho(\nu)^n.$$

Thus, if the process originates from a quasi-stationary distribution, its extinction time follows a geometric distribution of parameter $\rho(\nu)$.

Proof The Markov property allows us to write

$$\begin{aligned}\mathbb{P}_\nu(T_0 > n+k) &= \mathbb{E}_\nu\big(\mathbb{P}_{X_n}(T_0 > k)\mathbb{1}_{T_0 > n}\big) \\ &= \mathbb{P}_\nu(T_0 > n)\,\mathbb{E}_\nu(\mathbb{P}_{X_n}(T_0 > k) \,|\, T_0 > n).\end{aligned}$$

By Definition 3.4.1,

$$\mathbb{E}_\nu(\mathbb{P}_{X_n}(T_0 > k) \,|\, T_0 > n) = \mathbb{P}_\nu(T_0 > k).$$

Thus, for all n, k,

$$\mathbb{P}_\nu(T_0 > n+k) = \mathbb{P}_\nu(T_0 > n)\,\mathbb{P}_\nu(T_0 > k).$$

A basic argument invoking (3.4.1) yields the result. □

Proposition 3.4.7 *Let ν be a quasi-stationary distribution on $\mathbb{N}^*$. Then for all $0 < \gamma < -\log \rho(\nu)$,*

$$\mathbb{E}_\nu(\mathrm{e}^{\gamma T_0}) < +\infty.$$

In particular, there exists $i \in \mathbb{N}^*$ such that $\mathbb{E}_i(\mathrm{e}^{\gamma T_0}) < +\infty$.

This proposition suggests that if, with positive probability, the population can go extinct after an extremely long time, then the process will not have a quasi-stationary distribution. This will be the case for a critical BGW process (see Theorem 3.4.8 below).

Proof Since under ν, T_0 follows a geometric distribution of parameter $\rho(\nu)$, the exponential moment $\mathbb{E}_\nu(e^{\gamma T_0})$ is finite if $\gamma < -\log\rho(\nu)$, as we see in the computation below:

$$\begin{aligned}\mathbb{E}_\nu(e^{\gamma T_0}) &= \sum_{n\geq 1} e^{\gamma n}(\rho(\nu))^n(1-\rho(\nu))\\ &= \frac{e^{\gamma}(1-\rho(\nu))}{1-e^{\gamma}\rho(\nu)}.\end{aligned}$$

The last assertion follows from $\mathbb{E}_\nu(e^{\gamma T_0}) = \sum_{i\geq 1}\nu_i\,\mathbb{E}_i(e^{\gamma T_0})$. □

3.4.2 Quasi-Stationary Distributions for a BGW Chain

Let us study the quasi-stationary distributions of a BGW process and the existence of a Yaglom limit.

Theorem 3.4.8 *Let $(X_n)_n$ be a BGW process and m the mean of its reproduction law.*

(i) *There is no quasi-stationary distribution if the chain is critical or supercritical.*
(ii) *If the chain is subcritical, there exists a Yaglom limit $\mu = (\mu_j)_{j\geq 1}$ on $\mathbb{N}^*$ such that $\forall i, j \geq 1$,*

$$\lim_{n\to\infty} \mathbb{P}_i(X_n = j \mid X_n > 0) = \mu_j. \tag{3.4.2}$$

Furthermore, the generating function $\hat{g}$ of μ is the unique solution of the equation: for all $s \in [0, 1]$,

$$1 - \hat{g}(g(s)) = m(1 - \hat{g}(s)). \tag{3.4.3}$$

(iii) *There are an infinite number of quasi-stationary distributions for a subcritical BGW process.*

Proof The proof is adapted from Athreya and Ney [5, pp. 13–14] and from Seneta and Vere-Jones [74]. It mainly uses generating functions, which we have seen are a particularly suitable tool, as soon as we have the branching property.

(i) In the critical and supercritical cases, we have $\mathbb{E}_1(T_0) = +\infty$, which implies that $\mathbb{E}_\nu(T_0) = +\infty$ for any probability measure ν on $\mathbb{N}^*$. This contradicts the existence of a quasi-stationary distribution, as seen in Proposition 3.4.7.

(ii) As before, let us denote by g_n the generating function of X_n. For each $n \in \mathbb{N}^*$, we introduce the conditional generating function defined for $s \in [0, 1]$ by

$$\begin{aligned}\hat{g}_n(s) = \mathbb{E}_1(s^{X_n} \mid X_n > 0) &= \frac{\mathbb{E}_1(s^{X_n} \mathbb{1}_{X_n>0})}{\mathbb{P}_1(X_n > 0)} \\ &= \frac{\mathbb{E}_1(s^{X_n}) - \mathbb{P}_1(X_n = 0)}{1 - \mathbb{P}_1(X_n = 0)} \\ &= \frac{g_n(s) - g_n(0)}{1 - g_n(0)} = 1 - \frac{1 - g_n(s)}{1 - g_n(0)} \in [0,1].\end{aligned}$$

Note that by definition, this conditional law gives zero probability to point 0.

Let us show that for fixed s, the sequence $n \longrightarrow \frac{1-g_n(s)}{1-g_n(0)}$ is increasing. We write

$$\frac{1 - g_{n+1}(s)}{1 - g_{n+1}(0)} = \frac{1 - g_{n+1}(s)}{1 - g_n(s)} \frac{1 - g_n(s)}{1 - g_n(0)} \frac{1 - g_n(0)}{1 - g_{n+1}(0)}.$$

Generating functions are convex and therefore their rate of increase is increasing. Thus, the function $u \longrightarrow \frac{1-g(u)}{1-u}$ is increasing, and we deduce that

$$\frac{1 - g_{n+1}(s)}{1 - g_n(s)} \geq \frac{1 - g_{n+1}(0)}{1 - g_n(0)} \quad \text{and} \quad \frac{1 - g_{n+1}(s)}{1 - g_{n+1}(0)} \geq \frac{1 - g_n(s)}{1 - g_n(0)}.$$

Thus, the sequence $\left(\frac{1-g_n(s)}{1-g_n(0)}\right)_n$ is increasing and bounded by 1, because $g_n(s) \geq g_n(0)$. The sequence $(\hat{g}_n(s))_n$ is therefore decreasing and bounded below by 0 and converges to $\hat{g}(s) \in [0,1]$. The function $\hat{g}$ is then the generating function of a finite measure of mass less than or equal to 1. If we can show that $\hat{g}$ is continuous at 1, then we can conclude by the continuity theorem of Paul Lévy that $\hat{g}$ is the generating function of a probability measure μ that assigns zero probability to 0 (see [10, p. 303]).

For this, we first show that $\hat{g}$ is the solution of a functional equation. We have

$$\begin{aligned}1 - \hat{g}(s) &= \lim_n \frac{1 - g_{n+1}(s)}{1 - g_{n+1}(0)} \\ &= \lim_n \frac{1 - g_{n+1}(s)}{1 - g_n(0)} \frac{1 - g_n(0)}{1 - g_{n+1}(0)}.\end{aligned}$$

But

$$\lim_{n\to\infty} \frac{1 - g_n(0)}{1 - g_{n+1}(0)} = \lim_{u\to 1} \frac{1 - u}{1 - g(u)} = \frac{1}{m},$$

because $m = g'(1)$. Moreover,

$$\lim_n \frac{1 - g_n(g(s))}{1 - g_n(0)} = 1 - \hat{g}(g(s)),$$

by definition of $\hat{g}$. We then obtain that $1 - \hat{g}(g(s)) = m(1 - \hat{g}(s))$ and $\hat{g}$ is the solution of (3.4.3). When s tends to 1, the limit in the equation (3.4.3) implies the continuity of $\hat{g}$ at 1.

A priori, the function $\hat{g}$ depends on the starting point of the process. Let us show that this is not the case. Suppose there are initially i individuals. Then, as before, we have

$$\hat{g}_{n,i}(s) = \mathbb{E}_i(s^{X_n}\,|\,X_n > 0) = \frac{\mathbb{E}_i(s^{X_n}) - \mathbb{P}_i(X_n = 0)}{1 - \mathbb{P}_i(X_n = 0)} = \frac{(\mathbb{E}_1(s^{X_n}))^i - (\mathbb{P}_1(X_n = 0))^i}{1 - (\mathbb{P}_1(X_n = 0))^i}.$$

As the process $(X_n)_n$ is subcritical, the quantities $\mathbb{E}_1(s^{X_n})$ and $\mathbb{P}_1(X_n = 0)$ tend to 1 as n tends to infinity. We deduce that

$$\begin{aligned}\mathbb{E}_i(s^{X_n}) &= 1 + i\,(\mathbb{E}_1(s^{X_n}) - 1) + o(\mathbb{E}_1(s^{X_n}) - 1),\\ \mathbb{P}_i(X_n = 0) &= 1 - i\,\mathbb{P}_1(X_n > 0) + o(\mathbb{P}_1(X_n > 0)).\end{aligned}$$

Therefore,

$$\mathbb{E}_i(s^{X_n}\,|\,X_n > 0) \sim_{n\to\infty} \frac{\mathbb{E}_1(s^{X_n}) - 1 + \mathbb{P}_1(X_n > 0)}{1 - 1 + \mathbb{P}_1(X_n > 0)} = \frac{\mathbb{E}_1(s^{X_n}) - \mathbb{P}_1(X_n = 0)}{1 - \mathbb{P}_1(X_n = 0)},$$

hence $\hat{g}_{n,i}(s)$ converges to $\hat{g}(s)$ as n tends to infinity.

Thus, μ is the Yaglom limit of the BGW process and satisfies (3.4.2).

Let us now prove that $\hat{g}$ is the unique solution of (3.4.3). Suppose there exist two generating functions $\hat{g}$ and $\hat{h}$ satisfying (3.4.3). By induction, we can show that for all $n \geq 1$ and $s \in [0,1]$,

$$\begin{aligned}\hat{g}(g_n(s)) &= m^n \hat{g}(s) - (m^{n-1} + \cdots + m + 1)(m-1),\\ \hat{h}(g_n(s)) &= m^n \hat{h}(s) - (m^{n-1} + \cdots + m + 1)(m-1).\end{aligned}$$

We deduce that

$$\hat{g}'(g_n(s))\,g_n'(s) = m^n \hat{g}'(s)\,;\, \hat{h}'(g_n(s))\,g_n'(s) = m^n \hat{h}'(s).$$

Since in the subcritical case, the sequence $(g_n(0))$ tends to 1 in an increasing manner, there exists, for all $s \in [0,1]$, an integer k such that

$$g_k(0) \leq s \leq g_{k+1}(0).$$

Thus we have

$$\frac{\hat{g}'(s)}{\hat{h}'(s)} = \frac{\hat{g}'(g_n(s))}{\hat{h}'(g_n(s))} \leq \frac{\hat{g}'(g_{n+k+1}(0))}{\hat{g}'(g_{n+k}(0))} = \frac{\hat{g}'(0)}{\hat{h}'(0)}\,\frac{m\,g_{n+k}'(0)}{g_{n+k+1}'(0)} = \frac{\hat{g}'(0)}{\hat{h}'(0)}\,\frac{m}{g'(g_{n+k}(0))}.$$

When n tends to infinity, we deduce that for all $s \in [0,1]$, $\frac{\hat{g}'(s)}{\hat{h}'(s)} \leq \frac{\hat{g}'(0)}{\hat{h}'(0)}$. We can establish the inverse inequality in a similar way and deduce the equality of the two expressions. Moreover, since $\hat{g}$ and $\hat{h}$ are two generating functions on $\mathbb{N}^*$, they satisfy $\hat{g}(0) = \hat{h}(0) = 0$ and $\hat{g}(1) = \hat{h}(1) = 1$. This finally implies the equality of the two functions.

(iii) We will now conclude the proof of the theorem by showing that there are infinitely many quasi-stationary distributions for the subcritical BGW process.

Let $a \in]0,1[$ be a constant and define μ_a to be the probability measure on $\mathbb{N}^*$ with generating function $g^{\mu_a}(s) = 1-(1-s)^a$. We have

$$\mathbb{E}_{\mu_a}\left(s^{X_n} \mid X_n > 0\right) = 1 - \frac{1-g^{\mu_a} \circ g_n(s)}{1-g^{\mu_a} \circ g_n(0)} = 1 - \left(\frac{1-g_n(s)}{1-g_n(0)}\right)^a.$$

If we denote by ν_a the probability measure with generating function $g^{\nu_a}(s) = 1-(1-\hat{g}(s))^a$, we then deduce that

$$\lim_{n\to\infty} \mathbb{E}_{\mu_a}(f(X_n) \mid X_n > 0) = \nu_a(f),$$

for any bounded function f on $\mathbb{N}^*$. By a proof similar to that of Proposition 3.4.4, we deduce that ν_a is a quasi-stationary distribution of the process. Note that $\nu = \nu_1$. Thus the family $(\nu_a)_{a\in]0,1]}$ is an infinite collection of quasi-stationary distributions for the process $(X_n)_n$. □

3.5 Extension 1: Density-Dependent Chains

We will now consider, in a somewhat superficial way, BGW processes that are density-dependent and without immigration. Of course, these are much more credible models from a biological point of view. We will develop them in more detail, in the context of continuous time, in the following chapter. The state 0 is still an absorbing state, and the extinction time is still $T_0 = \inf\{n : X_n = 0\}$. However, the "branching property" is no longer valid.

Typically, such a process models the evolution of a BGW type population taking into account limited resources that decrease the possibilities of reproduction as the effective i of the population increases. It is then natural to suppose that the averages $m(i)$ of the reproduction laws $Q(i) = (q_\ell(i))_{\ell\in\mathbb{N}}$ and the probabilities of death without offspring $q_0(i)$ satisfy:

$$m(i+1) \le m(i), \qquad q_0(i) \le q_0(i+1). \tag{3.5.1}$$

For example, consider the "logistic" interaction framework:

$$q_0(i) = \frac{q_0 + Ci}{1+Ci}; \quad q_\ell(i) = \frac{q_\ell}{1+Ci},$$

where $(q_\ell)_{\ell\in\mathbb{N}}$ is a probability measure on $\mathbb{N}$ and C a strictly positive constant. This model satisfies Properties (3.5.1).

We will content ourselves with showing a partial extinction result. We will see in the proof that density-dependence leads to major technical complications. Recall that $m(i) < 1$ implies $q_0(i) > 0$, since $m(i) = \sum_{\ell\ge1} \ell\, q_\ell(i) \ge \sum_{\ell\ge1} q_\ell(i) = 1 - q_0(i)$.

Theorem 3.5.1 *Suppose that the reproduction law satisfies* $q_0(i) > 0$ *and* $m(i) < +\infty$ *for all* $i \geq 1$*. Suppose further that there exists* $a \in]0,1[$ *such that* $m(i) \leq a$ *for all* $i \geq 1$ *except (possibly) for a finite number. Then we have almost sure extinction of the process:* $\mathbb{P}_i(T_0 < +\infty) = 1$ *for all* i*.*

Note that under (3.5.1) the hypotheses of this theorem are reduced to $q_0(1) > 0$ and $\lim_{i\to\infty} m(i) < 1$.

Proof As $q_0(i) \geq 1 - m(i) \geq 1 - a$ when $m(i) \leq a$, it turns out that

$$b = \inf_{\{i;m(i)\leq a\}} q_0(i) \in]0,1[.$$

A similar computation as in the density-independent case shows that $\mathbb{E}_i(X_{n+1}|\mathcal{F}_n) = X_n\, m(X_n)$ (with $m(0) = 0$). Thus, the sequence

$$Z_n = \begin{cases} X_0, & \text{if } n = 0 \\ X_n / \prod_{l=0}^{n-1} m(X_l) & \text{if } n \geq 1 \end{cases}$$

is a positive martingale, with expectation $\mathbb{E}_i(Z_n) = i$. It then converges $\mathbb{P}_i$–almost surely to a finite limit. Consequently, the random variable $M = \sup_n Z_n$ is finite $\mathbb{P}_i$–almost surely and we have

$$X_n \leq M U_n, \qquad \text{where } U_n = \prod_{l=0}^{n-1} m(X_l). \tag{3.5.2}$$

For $N \in \mathbb{N}^*$, let $\alpha_N(i,j)$ be the number of times X_n passes in $j \geq 1$ between times 0 and N, starting from i. As 0 is an absorbing state, this number is equal to the number of times $X_n = j$ and $X_{n+1} > 0$ for n between 0 and $N-1$, plus possibly 1. Therefore for $j \geq 1$,

$$\alpha_N(i,j) = \mathbb{E}_i\left(\sum_{n=0}^{N} 1_{\{X_n=j\}}\right) \leq 1 + \mathbb{E}_i\left(\sum_{n=0}^{N-1} 1_{\{X_n=j, X_{n+1}>0\}}\right).$$

As $P_{j,0} = q_0(j)^j \geq b^j$ if $m(j) \leq a$, we can deduce by the Markov property that

$$\mathbb{P}_i(X_n = j, X_{n+1} > 0) = (1 - q_0(j)^j)\, \mathbb{P}_i(X_n = j) \leq (1 - b^j)\, \mathbb{P}_i(X_n = j),$$

and then

$$\alpha_N(i,j) \leq 1 + (1 - b^j)\alpha_{N-1}(i,j).$$

By iterating this relation and as $\alpha_0(i,j) \leq 1$, we obtain

$$\alpha_N(i,j) \leq 1 + (1 - b^j) + \cdots + (1 - b^j)^N \leq 1/b^j.$$

By letting N tend to infinity, we deduce that $\mathbb{E}_i(\sum_n 1_{\{X_n=j\}}) < +\infty$. The number of visits of the chain in j starting from i is integrable.

Let C be the (finite) set of states j such that $m(j) > a$. We have, if $X_0 = i$,

$$U_n \leq a^{\sum_{k=0}^{n-1} 1_{\{X_k \in C^c\}}} \prod_{j \in C} m(j)^{\alpha_{n-1}(i,j)}.$$

But as each random variable $\sum_n 1_{\{X_n=j\}}$ is almost surely finite for $j \in C$, the number of passages through C is finite almost surely. Thus, $\sum_{k=0}^{n-1} 1_{\{X_k \in C^c\}}$ tends to infinity with n. Consequently, U_n converges to 0 almost surely as n tends to infinity and (3.5.2) implies that the same is true for the sequence (X_n). Since X_n takes integer values and 0 is absorbing, this gives the result. □

3.6 Extension 2: BGW Process With Immigration

This section is essentially derived from Lambert [56].

We have seen that the BGW chain has a "degenerate" limit behavior in a certain sense and there is no invariant probability measure other than the Dirac mass at 0. The situation is different with immigration, as we will see below.

Let us go back to the model introduced in Section 3.1, assuming that for all $i \in \mathbb{N}$,

$$Q^i = Q; \quad \eta^i = \eta.$$

Thus, the birth and death process is density-independent with immigration and is described by

$$X_{n+1} = \sum_{k=1}^{i} Y_{n,k} + Z_{n+1} \ \text{ if } X_n = i, \quad \text{for } i \in \mathbb{N}.$$

The random variables $(Z_n)_{n\geq 1}$ have the same law η (the distribution of a random variable Z). The main difference with the simple BGW process (without immigration) is that 0 is no longer an absorbing state, since $P_{0,j} = \eta_j$ is strictly positive for at least one $j \geq 1$.

In the following, we assume that

$$0 < q_0 < 1, \ \ q_1 < 1, \ \ \eta_0 < 1.$$

As in the previous section, let us denote by m the expectation of the reproduction law. As soon as it appears, each migrant behaves like the individuals already present in the population and therefore generates a BGW tree composed of its descendants, with the same reproduction law.

The following theorems describe the long-term behavior of the BGW process with immigration. Their proofs are more subtle than in the case without immigration and we will only give partial proofs.

Theorem 3.6.1 *Suppose* $m < 1$. *Let* $\log^+ Z = \sup(\log Z, 0)$. *We then have the following dichotomy.*

$$\mathbb{E}(\log^+ Z) < +\infty \implies (X_n) \text{ converges in law.}$$
$$\mathbb{E}(\log^+ Z) = +\infty \implies (X_n) \text{ converges in probability to} + \infty.$$

Note that Theorem 3.6.1 implies in particular the convergence in law of $(X_n)_n$ for $m < 1$ and $\mathbb{E}(Z) < +\infty$. Thus, in a subcritical case where the population without immigration would almost surely go extinct, immigration will lead to either convergence to a non-trivial limit state, or explosion of the population.

Theorem 3.6.2 *Suppose* $m > 1$. *We then have the following dichotomy.*

$$\mathbb{E}(\log^+ Z) < +\infty \implies \lim_n m^{-n} X_n \text{ exists and is finite a.s.}$$
$$\mathbb{E}(\log^+ Z) = +\infty \implies \limsup_n c^{-n} X_n = +\infty \text{ for all constant } c > 1.$$

Theorem 3.6.2 explores the behavior of (X_n) in the case of supercritical reproduction. In the first case, X_n behaves in an exponential manner as n tends to infinity, as in the case of the BGW process without immigration. On the other hand, when $\mathbb{E}(\ln^+ Z) = +\infty$, the population size explodes faster than an exponential function.

To prove these theorems, we will use the following lemma:

Lemma 3.6.3 *Let* $\zeta_1, \zeta_2, \ldots$ *be positive independent random variables with the same distribution as* ζ. *Then for all* $c > 1$,

$$\mathbb{E}(\log^+ \zeta) < +\infty \implies \sum_{n \geq 1} c^{-n} \zeta_n < +\infty \ a.s.$$
$$\mathbb{E}(\log^+ \zeta) = +\infty \implies \limsup_n c^{-n} \zeta_n = +\infty \ a.s.$$

Proof of Lemma 3.6.3 (i) Let us first show that for any sequence of positive independent random variables $W_1, W_2, \ldots$ with the same distribution (same law as W), $\limsup_n \frac{W_n}{n}$ is almost surely 0 or infinity, depending on whether $\mathbb{E}(W)$ is finite or not. Indeed, let $\varepsilon > 0$. We then have

$$\sum_n \mathbb{P}\left(\frac{W_n}{n} > \varepsilon\right) = \sum_n \sum_{i > n\varepsilon} \mathbb{P}(W_n = i) = \sum_i \sum_{n < \frac{i}{\varepsilon}} \mathbb{P}(W_n = i)$$
$$= \sum_i \frac{i}{\varepsilon} \mathbb{P}(W = i) = \frac{1}{\varepsilon} \mathbb{E}(W).$$

If $\mathbb{E}(W) < +\infty$, the Borel–Cantelli Lemma implies that $\mathbb{P}(\limsup_n \{\frac{W_n}{n} > \varepsilon\}) = 0$ and thus, the sequence $(\frac{W_n}{n})_n$ converges almost surely to 0. If on the contrary $\mathbb{E}(W) = +\infty$, and since the W_i are independent, then the probability above is 1, and therefore $\frac{W_n}{n}$ tends to infinity almost surely.

(ii) Let us now return to the proof of Lemma 3.6.3. Consider a sequence $(\zeta_n)_n$ of independent random variables with the same law. Note that if the asymptotic results stated in Lemma 3.6.3 are true for the subsequence of ζ_n which are strictly larger than 1, they will still be true for all ζ_n. Let us restrict ourselves to this subsequence and let $W_n = \log(\zeta_n)$. Also let $c > 1$ and $a = \log c > 0$. Then for all n such that $\zeta_n > 1$, we have $c^{-n}\zeta_n = \exp -n(a - \frac{W_n}{n})$. The application of (i) to the sequence $(W_n)_n$ leads to the expected result. □

Proof of Theorem 3.6.1 Note that the BGW process with immigration $(X_n)_n$ originating from 0 and evaluated at generation n has the same law as

$$\sum_{k=0}^{n} \xi_{n,k},$$

where the integer-valued random variables $\xi_{n,k}$ are independent and for each k, $\xi_{n,k}$ describes the contribution to generation n of immigrants arriving at the generation $n-k$. Thus $\xi_{n,k}$ is the value at generation k of a BGW process without immigration linked to the reproduction law Y and therefore does not depend on n.

We seek to determine whether $X_\infty = \sum_{k=0}^{\infty} \xi_{n,k}$ is a finite or infinite random variable. It is possible to show (thanks to a result called the zero-one law, cf. [44, Theorem 10.6] or [10, p. 50]), that X_∞ is almost surely finite or infinite. Let us recall that Z_k denotes the number of migrants arriving at generation k. Note that $\xi_{n,0} = Z_n$, $\xi_{n,1}$ is equal to the number of children of the Z_{n-1} migrants from generation $n-1$, etc. Furthermore, the random variables $(Z_k)_{k\geq 1}$ are independent and have the same distribution as Z, so $\mathbb{E}(\log^+ Z) = \mathbb{E}(\log^+ Z_k)$, for all $k \geq 1$. Let $\mathcal{G}$ be the σ-field generated by the random variables Z_k, $k \in \mathbb{N}^*$. Suppose first that $\mathbb{E}(\log^+ Z) < +\infty$. We then have

$$\mathbb{E}(X_\infty | \mathcal{G}) = \sum_{k=0}^{\infty} Z_k m^k.$$

Indeed, $\xi_{n,0}$ has the distribution Z, the conditional expectation of $\xi_{n,1}$ given $\mathcal{G}$ is equal to the expectation of the number of individuals from the first generation starting from Z_1 individuals, that is mZ_1, etc. We can then apply Lemma 3.6.3 (with $c = \frac{1}{m}$ since $m < 1$), and deduce that the conditional expectation is almost surely finite, which implies that X_∞ is almost surely finite.

By a slightly more complicated argument that we will not develop here, it is possible to show the converse, namely that if X_∞ is almost surely finite, $\mathbb{E}(\log^+ Z) < +\infty$. Note that if $\mathbb{E}(X_\infty) < +\infty$ almost surely, this converse is immediate thanks to Lemma 3.6.3. □

Proof of Theorem 3.6.2 Let us first consider the case where $\mathbb{E}(\log^+ Z) = +\infty$. Thanks to Lemma 3.6.3, we know that $\limsup_n c^{-n} Z_n = +\infty$, for all $c > 1$. As $X_n \geq Z_n$, the result follows.

Let us now assume that $\mathbb{E}(\log^+ Z) < +\infty$.

Let $\mathcal{H}_n^{\mathcal{Z}}$ be the σ-field generated by $X_0, X_1, \ldots, X_n$ as well as by all the random variables Z_k, $k \in \mathbb{N}^*$. We then have

$$\mathbb{E}\left(\frac{X_{n+1}}{m^{n+1}}\middle|\mathcal{H}_n^{\mathcal{Z}}\right) = \frac{1}{m^{n+1}}\mathbb{E}\left(\sum_{i=1}^{X_n} Y_i + Z_{n+1}\middle|\mathcal{H}_n^{\mathcal{Z}}\right) = \frac{X_n}{m^n} + \frac{Z_{n+1}}{m^{n+1}} \geq \frac{X_n}{m^n},$$

since $Z_{n+1} \geq 0$. We deduce that $\left(\frac{X_n}{m^n}\right)_n$ is an $\mathcal{H}_n^{\mathcal{Z}}$-submartingale and, by immediate iteration, that

$$\mathbb{E}\left(\frac{X_n}{m^n}\middle|\mathcal{H}_0^{\mathcal{Z}}\right) = X_0 + \sum_{k=1}^{n} \frac{Z_k}{m^k}, \qquad \text{for all } n \geq 1.$$

In the case where $\mathbb{E}(Z) < +\infty$, we deduce that

$$\sup_n \mathbb{E}\left(\frac{X_n}{m^n}\right) \leq \mathbb{E}(X_0) + \sum_{k=1}^{\infty} \mathbb{E}\left(\frac{Z_k}{m^k}\right) \leq \mathbb{E}(X_0) + \frac{\mathbb{E}(Z)}{m-1}.$$

The sequence $(X_n/m^n)_n$ is thus a submartingale with uniformly bounded expectation. It then converges almost surely to a random variable that is almost surely finite. (For this convergence result for submartingales, which generalizes Theorem 2.5.5, see for example [49, Theorem 5.1] or [64]).

In the case where we only have the assumption $\mathbb{E}(\log^+ Z) < +\infty$, the proof is significantly more subtle. Lemma 3.6.3, applied with $\zeta = Z$ and $c = m$ $(m > 1)$, implies that

$$\mathbb{E}\left(\frac{X_n}{m^n}\middle|\mathcal{H}_0^{\mathcal{Z}}\right) \leq X_0 + \sum_{k=1}^{\infty} \frac{Z_k}{m^k} < +\infty \quad a.s.$$

We then need to apply a theorem of conditional convergence, because conditionally on X_0 and Z_k, $k \geq 1$, the sequence $(X_n/m^n)_n$ has uniformly bounded expectation.□

3.7 Extension 3: The Multi-Type GW process

As suggested in the examples of Section 3.1, it is of practical interest to consider the dynamics of a population composed of sub-populations of individuals of different types. The type of an individual is defined as an attribute (or a set of attributes) that remains fixed during the individual's life. Classic examples of such types are birth size, sex, genotype, stage of maturity (juvenile, breeder)... The type can affect the distribution of the number of offspring. For example, juvenile individuals can become breeding individuals (able to reproduce) or remain juveniles (unable to reproduce), but breeding individuals will always produce juvenile individuals. The offspring distributions will therefore be different (they do not have the same support). Another example is that of genotypes: a parent's genotype obviously affects the genotype of its descendants and may also affect their number. In a model with clonal (asexual) reproduction and mutations, an individual will have the same genotype as its parent with a probability close to 1, but it can also mutate with a small probability. In that case, the mutant type individual will in turn produce individuals carrying this new type, with a reproduction law that may be different from the initial reproduction law. These differences between the reproduction laws of the different genotypes can explain some evolutionary scenarii.

We assume that individuals only take a finite number K of distinct types. The multi-type BGW model consists in describing a population in which for each type $j \in \{1, \ldots, K\}$, an individual of type j generates at the next generation, and independently of all others, a K-tuple describing the distribution of its descendants in the K types.

Example The population is composed of two types $\heartsuit$ and $\spadesuit$. We have 5 individuals for a certain generation and in the next generation we get:

$$\begin{array}{rccccc} \text{first generation:} & \heartsuit & \heartsuit & \spadesuit & \spadesuit & \spadesuit \\ \text{second generation:} & (\heartsuit,\heartsuit) & (\heartsuit,\spadesuit) & (\heartsuit,\spadesuit,\spadesuit) & (\heartsuit,\spadesuit,\spadesuit) & (\spadesuit). \end{array}$$

We will further assume that for each fixed type j, the random K-tuples from each individual of type j are independent and identically distributed, and that the offspring of all individuals are independent of each other.

More precisely, we will introduce the vector-valued process (columns)

$$X_n = \left(X_n^{(1)}, \cdots, X_n^{(K)}\right)^t$$

(V^t denotes the column vector that is the transpose of the row vector V) describing the number of individuals of each type in generation n, which takes values in $\mathbb{N}^K$. Suppose that the initial distribution is defined by the vector $\left(X_0^{(1)}, \ldots, X_0^{(K)}\right)^t$. The first generation will then be defined by the vector $\left(X_1^{(1)}, \ldots, X_1^{(K)}\right)^t$, where for $k \in \{1, \ldots, K\}$,

$$X_1^{(k)} = \sum_{j=1}^{K} \sum_{i=1}^{X_0^{(j)}} Y_j^{(k)}(i).$$

The random variable $Y_j^{(k)}(i)$ represents the number of children of type k from the i-th individual of type j. We have independence of the random descendants corresponding to different types j. For a fixed j, and conditionally on $X_0^{(j)}$, the random vectors $Y_j^{(1)}, \ldots, Y_j^{(k)}, \ldots Y_j^{(K)}$ are independent. Moreover, for each j, k, each vector $\left(Y_j^{(k)}(i), 1 \le i \le X_0^{(j)}\right)$ is composed of independent random variables with the same distribution.

In the "heart and spade" example, we have $X_0 = \begin{pmatrix} 2 \\ 3 \end{pmatrix}$ and $X_1 = \begin{pmatrix} 5 \\ 6 \end{pmatrix}$.

Remark 3.7.1 Let us underline that the descendants of each individual of a certain type are independent and have the same distribution, and that the descendants of distinct individuals at the n-th generation are independent from each other.

Definition 3.7.2 These processes are called *multi-type Galton–Watson processes* and are part of the class of multi-type branching processes.

For these multi-type Galton–Watson processes, the generating functions will play the same fundamental role as in the monotype case.

For all $n \in \mathbb{N}$, any type $j \in \{1, \ldots, K\}$ and $\mathbf{s} = (s_1, \ldots, s_K) \in [0,1]^K$, we define the j-th generating function $f_n^j(\mathbf{s})$ which determines the distribution of the number of offspring of each type at generation n, produced by a particle of type j:

$$\begin{aligned} f_n^j(\mathbf{s}) &= \mathbb{E}\left(s_1^{X_n^{(1)}} \cdots s_K^{X_n^{(K)}} \,\middle|\, X_0^{(j)} = 1, X_0^{(m)} = 0, \forall m \neq j \right) \\ &= \sum_{i_1, \ldots, i_K \geq 0} p_n^j(i_1, \cdots, i_K)\, s_1^{i_1} \cdots s_K^{i_K}, \end{aligned}$$

where $p_n^j(i_1, \ldots, i_K)$ is the probability that, at generation n, a parent of type j produces i_1 descendants of type 1, i_2 descendants of type 2, ..., i_K descendants of type K.

Example 3.7.3 Consider a population of lynxes. These animals are first juveniles. As they grow, most lynx become capable of reproduction but are unable to reproduce if they are not in a pack. They are then said to be floating. Once they are in a pack, they can reproduce. However, it happens that an individual cannot integrate into the pack and remains floating. The lynx population is therefore composed of 3 types: juvenile (J), floating (F), reproducer (R). Given σ, σ', ρ in $(0,1)$, an individual passes

- from state J to state F with probability σ,
- from state J to state J with probability $1-\sigma$,
- from state F to state R with probability σ',
- from state F to state F with probability $1-\sigma'$,
- from state R to state F with probability $1-\rho$ (the lynx leaves the pack),
- finally, an individual of type R reproduces with probability ρ and gives a random number of juveniles following a Poisson distribution with parameter m.

At the first generation, we will therefore have

$$\begin{aligned} f_1^1(s_1, s_2, s_3) &= \sigma s_2 + (1-\sigma) s_1 \\ f_1^2(s_1, s_2, s_3) &= \sigma' s_3 + (1-\sigma') s_2 \\ f_1^3(s_1, s_2, s_3) &= \mathrm{e}^{m(s_1-1)} \rho s_3 + (1-\rho) s_2. \end{aligned}$$

By reasoning similar to that of the monotype case, we can show the following recursive formula.

Proposition 3.7.4 *For all $j \in \{1, \ldots, K\}$, for all $n \in \mathbb{N}^*$ and $\mathbf{s} = (s_1, \ldots, s_K) \in [0,1]^K$,*

$$\begin{aligned} f_n^j(\mathbf{s}) &= f_{n-1}^j\left(f_1^1(\mathbf{s}), f_1^2(\mathbf{s}), \ldots, f_1^K(\mathbf{s})\right) \\ &= f_1^j\left(f_{n-1}^1(\mathbf{s}), f_{n-1}^2(\mathbf{s}), \ldots, f_{n-1}^K(\mathbf{s})\right). \end{aligned}$$

The population will go extinct at a certain time n if each type goes extinct and therefore if for all $j \in \{1, \ldots, K\}$, $X_n^{(j)} = 0$. Reasoning as in the monotype case, it is possible to compute these extinction probabilities in terms of the solutions of the system given in the following theorem. (We refer to Athreya–Ney [5] for the proof).

Theorem 3.7.5 *Let $(X_n)_n = \left(X_n^{(1)}, \ldots, X_n^{(K)}\right)_n$ be a multi-type process. Then*

$$\mathbb{P}\left(X_n^{(j)} = 0, \forall j \,\middle|\, X_0^{(j)} = k_j, \forall j\right) = \prod_{j=1}^{K} \left(U_n^{(j)}\right)^{k_j},$$

where $U_n^{(j)}$, $j \in \{1, \ldots, K\}$, is the unique solution of the system: for all $j \in \{1, \ldots, K\}$,

$$\begin{cases} U_0^{(j)} = 0 \\ U_n^{(j)} = f_1^j\left(U_{n-1}^{(1)}, \ldots, U_{n-1}^{(K)}\right). \end{cases}$$

It is then possible to deduce results describing the long-term behavior of the process, following the same reasoning as in the monotype case, see Athreya–Ney [5] for more details.

We will only be interested here in the asymptotic behavior of the process's expectation matrix and deduce a generalization of the subcritical, critical, and supercritical cases developed in Section 3.2.

The average reproduction is described by a matrix

$$M = \left(m_{kj}\right)_{k,j \in \{1,\ldots,K\}},$$

where

$$m_{kj} = \mathbb{E}(Y_j^{(k)})$$

is the average number of k-type individuals produced by a j-type individual. The sum of the coefficients of the k-th row therefore represents the average number of k-type individuals at each reproduction and the sum of the coefficients of the j-th column represents the average number of individuals produced by a j-type individual.

Proposition 3.7.6 *The expectation of the number of k-type individuals at generation n satisfies*

$$\mathbb{E}(X_n^{(k)}) = \mathbb{E}(\mathbb{E}(X_n^{(k)}|X_{n-1})) = \sum_{j=1}^{K} m_{kj}\mathbb{E}(X_{n-1}^{(j)}) = \Big(M\,\mathbb{E}(X_{n-1})\Big)_k.$$

Thus, we have the vector equality

$$\mathbb{E}(X_n) = M\,\mathbb{E}(X_{n-1}) = M^n\,\mathbb{E}(X_0).$$

Proof It suffices to write

$$\mathbb{E}(X_n^{(k)}\,|\,X_{n-1}) = \mathbb{E}\left(\sum_{j=1}^{K}\sum_{i=1}^{X_{n-1}^{(j)}} Y_j^{(k)}(i)\,\middle|\,X_{n-1}\right) = \sum_{j=1}^{K}\mathbb{E}(Y_j^{(k)})\,X_{n-1}^{(j)}.$$ □

There will be subcriticality and extinction if the matrix $M^n \to 0$ and supercriticality if $M^n \to +\infty$ (in a sense to be defined). A question that arises is whether we can characterize criticality by a single scalar parameter.

Obviously, the matrix M is positive, in the sense that all its terms are positive.

In the following, we will make the additional assumption (see Athreya–Ney [5], Haccou–Jagers–Vatutin [39]):

There exists an integer n_0 such that all the elements of the matrix M^{n_0} are strictly positive. (3.7.1)

Such matrices have interesting properties (cf. Serre [75]).

From a biological point of view, this means that any initial configuration can lead in finite time to any other composition of the population in the different types.

Let us state the following fundamental theorem.

Theorem 3.7.7 (Perron–Frobenius Theorem) *(cf. [75], [39]). Let M be a square matrix $K \times K$ with positive coefficients and satisfying* (3.7.1). *Then*

1) *there exists a unique real number $\lambda_0 > 0$ such that*

- *λ_0 is a simple eigenvalue of M,*
- *every eigenvalue λ of M (real or complex) is such that $|\lambda| \leq \lambda_0$.*

The eigenvalue λ_0 is called the dominant eigenvalue of M.

2) *the left eigenvector $u = (u_1, \ldots, u_K)$ and the right eigenvector $v = (v_1, \ldots, v_K)$ corresponding to the eigenvalue λ_0 can be chosen such that*

$$u_k > 0;\, v_k > 0;\, \sum_{k=1}^{K} u_k = 1; \qquad \sum_{k=1}^{K} u_k v_k = 1.$$

Under these conditions, the eigenvectors are unique. Moreover,

$$M^n = \lambda_0^n A + B^n,$$

where $A = (v_k u_j)_{k,j\in\{1,\dots,K\}}$ *and* B *are matrices such that:*

- $AB = BA = 0$,
- *there exist constants* $\rho \in]0, \lambda_0[$ *and* $C > 0$ *such that none of the elements of the matrix* B^n *exceeds* $C\rho^n$.

Note that it is easy to check that $A = (v_k u_j)_{k,j\in\{1,\dots,K\}}$ satisfies: $Av = v$, $A^t u = u$ and that $A^n = A$ for all n. (A^t denotes the transpose matrix of A.)

For easier reading of the results, we will actually apply Theorem 3.7.7 to the transpose matrix M^t of M. Let λ_0 denote the dominant eigenvalue common to M and M^t. We then immediately obtain the following proposition.

Proposition 3.7.8 *The multi-type branching process with reproduction matrix* M *(satisfying* (3.7.1)*) is*

- *subcritical if* $\lambda_0 < 1$,
- *critical if* $\lambda_0 = 1$,
- *supercritical if* $\lambda_0 > 1$.

In particular, if $\lambda_0 < 1$, *the multi-type branching process almost surely goes extinct. More precisely,*

$$\mathbb{E}(X_n) \sim_{n\to\infty} (\lambda_0)^n A^t\, \mathbb{E}(X_0).$$

If the initial population is reduced to an individual of type j, *we have*

$$\mathbb{E}_j(X_n^{(k)}) \sim_{n\to\infty} (\lambda_0)^n\, u_k v_j.$$

For an initial population of a single individual of a given type, the limit of the average number of individuals of type k divided by the average size of the population, as time tends to infinity, is given by

$$\lim_{n\to\infty} \frac{\mathbb{E}(X_n^{(k)})}{\mathbb{E}(|X_n|)} = \frac{u_k}{u_1 + \dots + u_K} = u_k,$$

where $|X_n|$ *denotes the total number of individuals in the population at generation* n.

Proof The proof is an immediate application of the Perron–Frobenius Theorem,

$$\mathbb{E}_j(X_n^{(k)}) = \left(\lambda_0^n A^t_{kj} + (B^n)^t_{kj}\right)\mathbb{E}_j(X_0) \sim_{n\to\infty} \lambda_0^n\, u_k v_j.$$

The computation of the limit proportions is then immediate. □

Thus, u_k represents the proportion of individuals of type k and v_j represents the fertility of an individual of type j. From a biological point of view, we deduce that in a population satisfying the assumption (3.7.1), all types grow (on average) at the same rate λ_0 and the proportions of individuals of each type will stabilize over time towards a strictly positive value. However, the average numbers of individuals can differ depending on the type of the initial individual: the more fertile the type of the ancestor, the greater is its offspring.

Let us go back to Example 3.7.3. In this case, the average reproduction matrix has the form

$$M = \begin{pmatrix} 1-\sigma & 0 & \rho m \\ \sigma & 1-\sigma' & 1-\rho \\ 0 & \sigma' & 0 \end{pmatrix}.$$

It is easy to check that M^2 has strictly positive coefficients and the matrix M satisfies the assumptions (3.7.1) of the Perron–Frobenius Theorem. It is difficult to find the dominant eigenvalue by elementary computation except in special cases. For example, if $m = 1$, we can show that $\lambda = 1$ is an eigenvalue and that the characteristic polynomial for $\lambda \in \mathbb{C}$ is

$$\det(M - \lambda I) = -(\lambda - 1)\Big(\lambda^2 + (\sigma + \sigma' - 1)\lambda + \sigma'(\sigma + \rho - 1)\Big).$$

Thus, if in particular $(\sigma + \sigma' - 1)^2 - 4\sigma'(\sigma + \rho - 1) < 0$ (which will be the most common case because σ, σ', ρ are numbers close to 1), then 1 is the simple root of the matrix. This will therefore imply that the process is critical.

We simplify the problem by forgetting the floating individuals. We therefore have two types, juveniles and breeders, and the average reproduction matrix is then

$$M = \begin{pmatrix} s_1 & m \\ s_2 & 0 \end{pmatrix}.$$

It still satisfies the assumptions (3.7.1). Let $\lambda \in \mathbb{C}$. Then

$$\det(M - \lambda I) = -\lambda(s_1 - \lambda) - s_2 m\sigma.$$

The discriminant is $\Delta = (s_1)^2 + 4\rho m s_2 > 0$ and the positive root is $\lambda_0 = \frac{s_1 + \sqrt{\Delta}}{2}$.

An immediate analysis shows that

$$\begin{aligned} \lambda_0 > 1 &\iff 1 - s_1 - m s_2 < 0 \\ \lambda_0 < 1 &\iff 1 - s_1 - m s_2 > 0 \\ \lambda_0 = 1 &\iff 1 - s_1 - m s_2 = 0. \end{aligned}$$

Thus, the lynx population will go extinct if $1 - s_1 - m s_2 > 0$ and will develop with positive probability if $1 - s_1 - m s_2 < 0$. The critical value for the average of the reproduction law of the breeders is therefore $\frac{1-s_1}{s_2}$.

Proposition 3.7.8 gives an average behavior, but it is also possible to study the almost-sure behavior of the stochastic process, as was done in the context of the Galton–Watson process (cf. Athreya [5, chap. 5]). We can summarize the results in the following theorem.

Theorem 3.7.9 *If $\lambda_0 \le 1$, the population goes extinct almost surely.*

If $\lambda_0 > 1$, the population survives with positive probability and grows geometrically, with growth rate λ_0 on its persistence set. The distribution in each type is given by the vector v.

3.8 Exercises

Exercise 3.8.1 (An epidemiological model)

At time n, a population consists of S_n healthy individuals and I_n infected individuals. At each time step, independently,

- each healthy individual is infected with probability τ, and with probability $1-\tau$ is replaced by a random number of healthy individuals, with generating function denoted by g;
- each infected individual dies with probability a, and with probability $1-a$ is replaced by a random number of infected individuals, with generating function g.

The goal of the problem is to study the asymptotic behavior of this population and in particular its extinction. We assume that $a, \tau \in]0,1[$ to avoid degenerate cases.

We denote by $\mathbb{P}_{(i,j)}$ the law of the random walk (S_n, I_n) when $(S_0, I_0) = (i, j)$.

1. How do we interpret the fact that the function g is the same for both reproduction laws?

   Set $u_n = \mathbb{P}_{(1,0)}(S_n + I_n = 0)$, $v_n = \mathbb{P}_{(0,1)}(S_n + I_n = 0)$.

2. Prove that for all $i, j, n \in \mathbb{N}$,

$$\mathbb{P}_{(i,j)}(S_n = 0, I_n = 0) = u_n^i v_n^j.$$

3. Show the following:

$$\begin{cases} u_{n+1} &= \tau v_n + (1-\tau)\, g(u_n) \\ v_{n+1} &= a + (1-a)\, g(v_n). \end{cases}$$

   Deduce the equations satisfied by q_1 and q_2, where q_1 (respectively q_2) denotes the extinction probability of a population originating from a healthy individual (respectively infected).

   Discuss the values of q_1 and q_2 depending on τ, in the case where $g(s) = s^2$ and $a = \frac{1}{2}$.

4. Assume that g is differentiable at 1.

   Compute the average number $(\mathbb{E}(S_n), \mathbb{E}(I_n))$ of healthy and infected individuals at time n in terms of $(\mathbb{E}(S_{n-1}), \mathbb{E}(I_{n-1}))$.

Give the maximum eigenvalue of the average reproduction matrix. Can we directly deduce the global extinction criterion of the population from this?
In which case does the average size of the healthy population tend to infinity as $n \longrightarrow \infty$?
In which case does the average size of the infected population tend to infinity as $n \longrightarrow \infty$?

Exercise 3.8.2
We study the impact of vaccination on the transmission of an infectious disease in a population. The propagation of this disease is modeled by a BGW process. Per unit of time, each sick individual can, independently of others, infect k ($k \geq 0$) individuals (previously healthy) and transmit the disease to them with probability p_k. We denote by $\mathbb{P}_i$ the law of this process starting from i individuals.

We now assume that in this population, each healthy individual is vaccinated with probability α, $0 \leq \alpha \leq 1$, and therefore cannot be contaminated. (We assume that the size of the population is large enough that only this probability matters.)

1. What is the probability $p_{k,\alpha}$ (in terms of $p_j, j \in \mathbb{N}$ and α) that an infected individual actually transmits the disease to k individuals.
2. Fix $\alpha \in [0, 1]$. Let g be the generating function associated with the distribution (p_k), with mean m. Give the generating function g_α associated with the distribution $(p_{k,\alpha})$ as a function of g.
   Show that for a fixed s, the function $\alpha \longrightarrow g_\alpha(s)$ is increasing.
3. Deduce the average number of individuals infected per patient, as a function of m.
   What is the minimum number α_{min} that must be chosen to ensure that the disease is completely eradicated?
   Comment on this result.
4. We now assume that $\alpha \geq \alpha_{min}$. Let T_α be the time of extinction of the disease.
   Show that for $n \in \mathbb{N}$, the function $\alpha \longrightarrow \mathbb{P}_i(T_\alpha \leq n)$ is an increasing function.
   How do you analyze this result?
5. Suppose that the initial contamination process is a BGW process, and that each individual can infect $0, 1, 2$ individuals with probabilities p_0, p_1, p_2 respectively.
   Provide a condition on p_0 and p_2 for $\alpha_{min} > 0$.
   Provide the value of $\alpha_{min} > 0$ in this case.
   What is the value of
   $$\mathbb{P}_1(T_{\alpha_{min}} < \infty)?$$

Exercise 3.8.3 (An island model with immigration)
We are interested in a population located on an island. The population size at time n is described by a BGW process $(X_n)_{n\geq 0}$:

$$X_0 = 1, \qquad X_{n+1} = \sum_{i=1}^{X_n} Y_{n,i},$$

where the random variables $Y_{n,i}$ are independent and identically distributed like a random variable Y with finite variance σ^2. We also assume that $\mathbb{P}(Y = 1) < 1$.

1. What modeling assumptions are made here? When does the population survive with strictly positive probability?

We assume for the rest of the exercise that $\mathbb{E}(Y) = 1$ and $\mathrm{Var}(Y) = \sigma^2 < +\infty$. Recall that $\lim_{n\to\infty} n\,\mathbb{P}(X_n > 0) = \frac{2}{\sigma^2}$ (see Lemma 3.2.5).

2. We want the population to be able to survive until a large time n with a probability of at least $\varepsilon \in (0,1)$. What should the minimum initial size x_0 of the population be?

We now want to take into account random immigrations from a continent to the island. To do this, we assume that the population size is described by the process $(Z_n)_{n\geq 0}$ defined by:

$$Z_0 = 1, \qquad Z_{n+1} = \sum_{i=1}^{Z_n} Y_{n,i} + I_{n+1},$$

where the variables I_n are independent and have the same law, distributed like a random variable $I \geq 0$ and independent of all the $Y_{k,j}$, $k, j \geq 0$. We also assume that $\mathbb{E}(I) < +\infty$.

3. What modeling assumptions are made here?
4. Show that $(Z_n - n\mathbb{E}(I))_n$ is a martingale. What could we expect for the evolution of the population size on the island?

We study in detail the asymptotic behavior of Z_n. For this, let g be the generating function of Y and f that of I.

5. Let h_n be the generating function of Z_n (starting from $Z_0 = 1$).

   a) Show that for all $n \geq 0$ and $s \in [0,1]$:

   $$h_{n+1}(s) = h_n(g(s))f(s).$$

   Conclude that, if we denote by $g_k(s) = g \circ g \circ \cdots \circ g$ the k-th composition of g, then

   $$h_n(s) = g_n(s) \prod_{k=0}^{n-1} f(g_k(s)),$$

   where, by convention, $g_0(s) = s = h_0(s)$.

   b) Show that for all $\lambda > 0$, there exists a sequence of integers $(m_n)_{n\geq 0}$ such that

   $$\mathbb{P}(X_{m_{n+1}} > 0) \leq 1 - \exp(-\lambda/n) \leq \mathbb{P}(X_{m_n} > 0).$$

   Then show that

   $$\lim_{n\to\infty} \frac{m_n}{n} = \frac{2}{\sigma^2 \lambda}.$$

c) Deduce that

$$\mathbb{P}(X_{n+m_n}=0)\prod_{k=0}^{n-1} f\big(\mathbb{P}(X_{k+m_n}=0)\big) \le \mathbb{E}\big(e^{-\lambda Z_n/n}\big)$$
$$\le \mathbb{P}(X_{n+m_{n+1}}=0)\prod_{k=0}^{n-1} f\big(\mathbb{P}(X_{k+m_{n+1}}=0)\big).$$

6. We assume that

$$\lim_{n\to\infty}\log\Big(\mathbb{E}\big(e^{-\lambda Z_n/n}\big)\Big)=\lim_{n\to\infty}-\frac{2f'(1)}{\sigma^2}\sum_{k=0}^{n-1}\frac{1}{k+m_n}=-\frac{2f'(1)}{\sigma^2}\ln\Big(1+\frac{\sigma^2\lambda}{2}\Big).$$

a) Specify the convergence of Z_n/n when n tends to infinity. One can use the fact that a Gamma law G with parameters (α,θ), whose density is equal to $\frac{x^{\alpha-1}e^{-x/\theta}}{\Gamma(\alpha)\theta^\alpha}\mathbf{1}_{\{x>0\}}$, is characterized by its Laplace transform

$$\mathbb{E}\big(e^{-\lambda G}\big)=(1+\theta\lambda)^{-\alpha}=\exp\big\{-\alpha\ln(1+\theta\lambda)\big\},\qquad \forall\lambda\ge 0.$$

b) What can you say about the asymptotic behavior of $Z_n-n\mathbb{E}(I)$? Comment on this result in connection with Question 4.

Exercise 3.8.4 (BGW process conditioned to extinction)
Let $(Z_n;n\ge 0)$ be a BGW process with reproduction law $\mu=(\mu(k),k\in\mathbb{N})$ satisfying $\mu(0)>0$ and $\sum_{k\ge 0}k\mu(k)\in\,]1,+\infty[$.

We denote by g the generating function of μ, by E the event $E=\{\lim_{n\to\infty}Z_n=0\}$ and by $(Z_n^e;n\ge 0)$ the process conditioned to extinction: the law of $(Z_n^e;n\ge 0)$ is defined by

$$\mathbb{P}(Z_0^e=z_0,\dots,Z_n^e=z_n)=\mathbb{P}(Z_0=z_0,\dots,Z_n=z_n|E)\quad\text{for all } n,z_0,\dots,z_n\in\mathbb{N}.$$

Part I

1. What can be said about the extinction probability $q=\mathbb{P}(E|Z_0=1)$?
   For $i\in\mathbb{N}$, give $\mathbb{P}(E|Z_0=i)$ in terms of q.
2. For all $n,i,j,z_0,\dots,z_{n-1}\in\mathbb{N}$ such that $\mathbb{P}(Z_0=z_0,\dots,Z_{n-1}=z_{n-1},Z_n=i)>0$, show that

$$\mathbb{P}(Z_{n+1}=j|Z_0=z_0,\dots,Z_{n-1}=z_{n-1},Z_n=i,E)=\frac{\mathbb{P}(E|Z_0=j)\mathbb{P}(Z_1=j|Z_0=i)}{\mathbb{P}(E|Z_0=i)}.$$

3. Deduce an expression for $\mathbb{P}(Z_{n+1}=j|Z_0=z_0,\dots,Z_{n-1}=z_{n-1},Z_n=i,E)$ in terms of i, j, q and $\mathbb{P}(Z_1=j|Z_0=i)$.
4. Show that $(Z_n^e;n\ge 0)$ is a Markov chain and identify its transition kernel.
5. Deduce that $(Z_n^e;n\ge 0)$ is a BGW process whose generating function g^e of the reproduction law μ^e is defined by $g^e(s)=q^{-1}g(qs)$ for all $s\in[0,1]$.

Part II

We now assume that μ is a Poisson distribution with parameter $\theta > 1$: $\mu(k) = \mathrm{e}^{-\theta}\theta^k/k!$ for all $k \in \mathbb{N}$.

1. Give the generating function g of the distribution μ. What is the equation satisfied by $q = \mathbb{P}(E|Z_0 = 1)$?
2. Show that $\mathrm{e}^{\theta-1} > \theta$. Deduce that $q < 1/\theta$.
3. What is the reproduction law μ^e of the chain $(Z_n^e; n \geq 0)$?

Chapter 4
Brownian Motion and Diffusion Processes

Randomness: comes from the old French hunting term "randon", referring to "the impetuous and rapid course of a wild animal around its territory". This term also gave rise to the French word "randonnée".

Littré Dictionary

The examples of random walks developed in Chapter 2 are very simple, but they show that the computations giving the probabilities of reaching barriers, the average times before being trapped, and the invariant probabilities quickly become complicated. One can imagine that more complex dynamics (higher dimension, complicated shape of the spatial domain,...) can lead to great technical difficulties. We will overcome this difficulty by considering approximations of these random walks in well-chosen spatial and time scales. These approximations are random processes defined for all time in $\mathbb{R}_+$ and almost surely with continuous trajectories. Their prototype is Brownian motion, fundamental in probabilistic modeling, obtained as the limit of rescaled symmetric simple random walks. This process is almost surely continuous and nowhere differentiable. Brownian motion is a fundamental process in probability, with countless properties. However, we only give its most important features, related to the martingale property and the Markov property. Thanks to Brownian motion, we will be able to model more sophisticated dynamics and define randomly perturbed differential equations, called stochastic differential equations. For this, it is necessary to give a meaning to an integral with respect to Brownian motion. This is the object of stochastic calculus, of which we will give the essential elements. Thanks to stochastic differential equations, we will be able to propose new models of population size dynamics, allele proportion and spatial displacements.

There is a very rich literature on Brownian motion, stochastic calculus and stochastic differential equations, and our list of references is not exhaustive. We refer the reader to the books by Comets–Meyre [26], Ikeda–Watanabe [42], Karatzas–Shreve [48], Le Gall [57], Protter [69], and Revuz–Yor [71].

S. Méléard, *Random Models in Biology, Ecology and Evolution*, Texts in Applied Mathematics 87, https://doi.org/10.1007/978-3-662-73483-4_4

4.1 Finite-Dimensional Convergence of Renormalized Random Walks

We will provide an intuitive approach to the definition and existence of Brownian motion, by outlining its construction as a limit of renormalized random walks in time and space.

A simple symmetric random walk on $\mathbb{Z}$, null at 0, jumps by one unit of space during each unit of time. This type of model seems inadequate when considering very erratic movements that seem to vary almost continuously, such as the movements of microorganisms, insects, or pollen particles on the surface of water. In such situations, the size scales are very small and the positions change very fast. Nevertheless, the displacements can be seen as those of an accelerated random walk viewed from afar. A change in space and time scales will allow us, from a simple symmetric random walk starting at 0, to define the limit object describing this type of behavior, namely Brownian motion. The correct scale will be given by the central limit theorem. As the space step and the time step will tend towards 0, the limit object will be a random function defined on $\mathbb{R}_+$ and with real values.

Let us fix a time interval $[0,t]$. We will perform a scale change on a simple symmetric random walk to obtain a non-degenerate limiting object.

Let $(Z_n)_n$ be a sequence of independent and equidistributed random variables, such that

$$\mathbb{P}(Z_1 = 1) = \mathbb{P}(Z_1 = -1) = \frac{1}{2}.$$

We change the time so that the jumps occur at all times $\frac{1}{n}$. Therefore, there will be $[nt]$ jumps on the interval $[0,t]$, where $[nt]$ denotes the integer part of nt. We then study the random walk X, defined at times $[nt]$:

$$X_{[nt]} = \sum_{k=1}^{[nt]} Z_k; \quad X_0 = 0.$$

Our goal is to describe its asymptotic behavior, as n tends to infinity. Note that if this random walk models displacements, the jumps at every time $\frac{1}{n}$ made in the limit positions change very fast.

As the expectation of $X_{[nt]}$ is zero and its variance is $[nt]$, the natural scaling as n tends to infinity is a spatial scale of order $\frac{1}{\sqrt{n}}$. This intuitive approach is confirmed by the central limit theorem. Indeed, the random variables Z_k are independent equidistributed with mean 0 and variance 1. So when t is fixed,

$$\frac{X_{[nt]}}{\sqrt{[nt]}} \xrightarrow[n\to\infty]{\text{in law}} W,$$

where W denotes a random variable of normal distribution $\mathcal{N}(0,1)$. We can then write

$$\frac{X_{[nt]}}{\sqrt{n}} = \frac{X_{[nt]}}{\sqrt{[nt]}} \frac{\sqrt{[nt]}}{\sqrt{n}},$$

and by using that $\frac{\sqrt{[nt]}}{\sqrt{n}}$ converges to $\sqrt{t}$ as n tends to infinity, we deduce the following

Proposition 4.1.1 *For any fixed t, the sequence of random variables $(X_t^n)_n$ defined by*

$$X_t^n = \frac{1}{\sqrt{n}} \sum_{k=1}^{[nt]} Z_k = \frac{X_{[nt]}}{\sqrt{n}}$$

converges in law, as n tends to infinity, to a random variable with a Gaussian distribution, centered and with variance t.

Remark 4.1.2 1) Note that if the random variables Z_k were of mean $m \neq 0$, the asymptotic behavior would be different. Indeed, the law of large numbers would lead to the almost-sure convergence of the sequence of random variables $\frac{X_{[nt]}}{n}$ to mt. Thus, the random variable $X_{[nt]}$ would be of order n and the limit of $\frac{X_{[nt]}}{n}$ would be deterministic.

2) Let b and σ be two real numbers. For any fixed t, the sequence of random variables defined by

$$W_t^n = \sum_{k=1}^{[nt]} \left(\frac{\sigma}{\sqrt{n}} Z_k + \frac{b}{n} \right)$$

converges in law to $\sigma W_t + bt$ as n tends to infinity, where W_t follows the normal distribution $\mathcal{N}(0,t)$, centered and with variance t. The limit is then the sum of a deterministic term that represents the average behavior of the process and a random normal variable with variance $\sigma^2 t$ that models the fluctuations around this average trend.

We can also consider the linear approximation

$$Y_t^n = \frac{X_{[nt]}}{\sqrt{n}} + \frac{nt - [nt]}{\sqrt{n}} Z_{[nt]+1}.$$

Proposition 4.1.3 *For each fixed t, the sequence of random variables $(Y_t^n)_n$ converges in law, as n tends to infinity, to a random variable with a normal distribution, centered and with variance t.*

Proof Since $\frac{nt-[nt]}{\sqrt{n}} \leq \frac{1}{\sqrt{n}}$, the quantity $\mathbb{E}\left(\left(\frac{nt-[nt]}{\sqrt{n}} Z_{[nt]+1}\right)^2\right)$ tends to 0 as n tends to infinity. Thus, the sequence of random variables $\left(\frac{nt-[nt]}{\sqrt{n}} Z_{[nt]+1}\right)_n$ converges in mean square and therefore in law to 0. Moreover, for each n, the random variables $\frac{X_{[nt]}}{\sqrt{n}}$ and $\frac{nt-[nt]}{\sqrt{n}} Z_{[nt]+1}$ are independent. We can then use Lévy's Theorem (see [59, Theorem 6.3.9]) which characterizes the convergence in law of the sequence $(Y_t^n)_n$ by the convergence of the characteristic functions. Let $u \in \mathbb{R}$. We have

$$\mathbb{E}\left(e^{iuY_t^n}\right) = \mathbb{E}\left(e^{iu\frac{X_{[nt]}}{\sqrt{n}}}\right)\mathbb{E}\left(e^{iu\frac{nt-[nt]}{\sqrt{n}}Z_{[nt]+1}}\right),$$

and this quantity converges to $e^{-\frac{tu^2}{2}}$ as n tends to infinity. This implies that for each fixed t, the sequence of random variables $(Y_t^n)_n$ converges in law to a random variable with a normal distribution $\mathcal{N}(0,t)$, centered and with variance t. □

These convergences in law at fixed t are not sufficient to characterize the convergence of a random function of time. We will see in Section 4.2 that it is necessary to have at least the convergence of any p-tuple of coordinates, which we now prove in this case.

Proposition 4.1.4

(i) *For any p-tuple of times $0 \le t_1 < t_2 < \cdots < t_p$, the sequence of random vectors $(X_{t_1}^n, X_{t_2}^n, \ldots, X_{t_p}^n)$ converges in law, as n tends to infinity, to the p-tuple $(B_{t_1}, B_{t_2}, \ldots, B_{t_p})$. The law of the vector $(B_{t_1}, B_{t_2}, \ldots, B_{t_p})$ is entirely characterized by the fact that the increments $(B_{t_1}, B_{t_2} - B_{t_1}, \ldots, B_{t_p} - B_{t_{p-1}})$ are independent and that for each t_i, the random variable B_{t_i} follows the centered normal distribution with variance t_i.*
(ii) *The sequence of random vectors $(Y_{t_1}^n, Y_{t_2}^n, \ldots, Y_{t_p}^n)$ also converges in law to $(B_{t_1}, B_{t_2}, \ldots, B_{t_p})$.*

Proof Let us consider a finite sequence of times $0 \le t_1 < t_2 < \cdots < t_p$. The variables $X_{t_1}^n, X_{t_2}^n - X_{t_1}^n, \ldots, X_{t_p}^n - X_{t_{p-1}}^n$ are independent, and moreover,

$$\begin{aligned} X_{t_j}^n - X_{t_{j-1}}^n &= \frac{1}{\sqrt{n}}\left(Z_{[nt_{j-1}]+1} + \cdots + Z_{[nt_j]}\right) \\ &= \frac{\sqrt{[nt_j]-[nt_{j-1}]}}{\sqrt{n}} \times \frac{Z_{[nt_{j-1}]+1} + \cdots + Z_{[nt_j]}}{\sqrt{[nt_j]-[nt_{j-1}]}}. \end{aligned}$$

The first term on the right-hand side tends to $\sqrt{t_j - t_{j-1}}$ and the second converges in law to a random variable with law $\mathcal{N}(0,1)$. Since the coordinates are independent, we deduce (by a new use of Lévy's Theorem), that

$$(X_{t_1}^n, X_{t_2}^n - X_{t_1}^n, \ldots, X_{t_p}^n - X_{t_{p-1}}^n) \underset{n\to\infty}{\overset{\text{in law}}{\longrightarrow}} (W_{t_1}, W_{t_2-t_1}, \ldots, W_{t_p-t_{p-1}}),$$

where the limit coordinates are independent and follow respectively the distributions $\mathcal{N}(0,t_1)$, $\mathcal{N}(0,t_2-t_1)$, ..., $\mathcal{N}(0,t_p-t_{p-1})$. But then,

$$X_{t_2}^n = X_{t_1}^n + (X_{t_2}^n - X_{t_1}^n);\ \ldots;\ X_{t_p}^n = X_{t_1}^n + \cdots + (X_{t_p}^n - X_{t_{p-1}}^n)$$

and consequently,

$$(X_{t_1}^n, X_{t_2}^n, \ldots, X_{t_p}^n) \underset{n\to\infty}{\overset{\text{in law}}{\longrightarrow}} (B_{t_1}, B_{t_2}, \ldots, B_{t_p}).$$

The vector $(B_{t_1}, B_{t_2}, \ldots, B_{t_p})$ is such that the increments

$$(B_{t_1}, B_{t_2} - B_{t_1}, \ldots, B_{t_p} - B_{t_{p-1}})$$

are independent. Moreover, for each t_i, the random variable B_{t_i} follows the normal distribution, centered with variance t_i.

The proof of (ii) follows from (i) and the reasoning of Proposition 4.1.3. □

For each finite family $0 \leq t_1 < t_2 < \cdots < t_p$, we have obtained a random vector $(B_{t_1}, B_{t_2}, \ldots, B_{t_p})$, unique in law. Is it enough to define a random function $t \to B_t$? Can we generalize this convergence to an infinite number of coordinates? More generally, we can ask if the collection of random variables $(X_t^n, t \in \mathbb{R}_+)$ converges (and in what sense?) towards a uniquely defined collection of random variables $(B_t, t \in \mathbb{R}_+)$. This question gave rise to a host of mathematical problems and was finally resolved in 1933 by Kolmogorov.

4.2 Random Processes and Brownian Motion

Definition 4.2.1 A *random process* is a family $(X_t)_{t \geq 0}$ of random variables indexed by $\mathbb{R}_+$, all these variables being defined on the same probability space $(\Omega, \mathcal{A}, \mathbb{P})$. We assume here that each X_t is real-valued.

We can also view this process as a random variable X defined on $(\Omega, \mathcal{A}, \mathbb{P})$ and with values in the set of functions $x : t \mapsto x_t$ from $\mathbb{R}_+$ to $\mathbb{R}$, which associates with each ω the function $t \mapsto X_t(\omega)$.

The modern theory of probability measures is based on the fundamental results of Kolmogorov (1933), which allow us to construct a σ-field on the set of functions from $\mathbb{R}_+$ to $\mathbb{R}$ that makes the mapping X measurable from $(\Omega, \mathcal{A})$ to $\mathbb{R}^{\mathbb{R}_+}$ (and thus allows us to speak of a random function). This σ-field is the smallest σ-field that makes the functions $x \mapsto x_t$ measurable from $\mathbb{R}^{\mathbb{R}_+}$ to $\mathbb{R}$ for all $t > 0$ and is called the *product σ-field*. Its construction is described in [26] and detailed in [10] and [63]. In particular, it will allow us to characterize the law of the process (i.e., a probability on a space of trajectories) in terms of the laws of random vectors (probabilities on $\mathbb{R}^k$).

Theorem 4.2.2 (Kolmogorov, 1933) *The law of a process $(X_t, t \geq 0)$, which is a probability measure on $\mathbb{R}^{\mathbb{R}_+}$, is characterized by its finite-dimensional marginal laws, defined as the laws of the k-tuples $(X_{t_1}, ..., X_{t_k})$, for all times $t_1, ..., t_k$.*

This theorem, which we will not prove, is extremely useful for proving uniqueness in law for processes. It will allow us in particular to characterize the law of a process $(B_t, t \in \mathbb{R}_+)$, called Brownian motion, whose finite-dimensional coordinates are characterized (in law) by Proposition 4.1.4. But first, we will generalize the notion of a process with stationary independent increments (PSII), already seen for random walks (see Definition 2.1.2), to the framework of continuous time processes.

Definition 4.2.3 A process $(X_t, t \in \mathbb{R}_+)$ is a *process with stationary independent increments* (PSII) if for all positive real numbers $0 \leq t_1 < t_2 < \cdots < t_k$, the random variables $(X_{t_1}, X_{t_2} - X_{t_1}, \ldots, X_{t_k} - X_{t_{k-1}})$ are independent and if $X_{t_2} - X_{t_1}$ has the same law as $X_{t_2 - t_1} - X_0$.

In this case, the law of the process is characterized by the laws of its increments.

Proposition 4.2.4 *Let $(X_t, t \geq 0)$ be a process with stationary independent increments. The law of the process $(X_t, t \geq 0)$ is characterized by the laws of the random variables $X_t - X_0$, for $t \geq 0$, and by that of X_0.*

Proof Consider an n-tuple $0 \leq t_1 \leq \cdots \leq t_n$. As the random variables X_{t_1}, $X_{t_2} - X_{t_1}, \ldots, X_{t_n} - X_{t_{n-1}}$ are independent, the law of the vector

$$(X_{t_1}, X_{t_2} - X_{t_1}, \ldots, X_{t_n} - X_{t_{n-1}})$$

is known as soon as we know the laws of the coordinates: it is the tensor product of these laws. Moreover, the law of the coordinate $X_{t_k} - X_{t_{k-1}}$ is equal to the law of $X_{t_k - t_{k-1}} - X_0$, which is assumed to be known. It is then enough to use the fact that the law of the random vector $(X_{t_1}, X_{t_2}, \ldots, X_{t_n})$ is characterized by its characteristic function. Indeed, we see that for $(u_1, \ldots, u_n) \in \mathbb{C}$ and by setting $X_{t_0} = 0$,

$$\mathbb{E}\left(e^{i\sum_{k=1}^{n} u_k X_{t_k}}\right) = \mathbb{E}\left(e^{i\sum_{k=1}^{n}(u_k + \cdots + u_n)(X_{t_k} - X_{t_{k-1}})}\right) = \Pi_{k=1}^{n} \mathbb{E}\left(e^{i(u_k + \cdots + u_n)(X_{t_k} - X_{t_{k-1}})}\right).$$

Theorem 4.2.2 completes the proof. □

We are now able to define a Brownian motion.

Definition 4.2.5 A *Brownian motion* is a process $(B_t, t \in \mathbb{R}_+)$, with stationary independent increments, such that for any fixed time $t > 0$, the random variable B_t has the centered normal distribution $\mathcal{N}(0, t)$ with variance t.

More precisely, for all $s, t \geq 0$, the variable $B_{t+s} - B_t$ is independent of the variables $(B_r : r \leq t)$, $B_0 = 0$, and moreover, the law of the increment $B_{t+s} - B_t$ is the normal law $\mathcal{N}(0, s)$. It therefore depends only on s.

Theorem 4.2.6 *The law of a process $(B_t, t \geq 0)$ with stationary independent increments, starting at* 0 *and such that B_t has the normal distribution $\mathcal{N}(0, t)$ for all $t > 0$, is unique. It is the law of a Brownian motion.*

Proof We adapt to this particular case the proof of Proposition 4.2.4. As we saw in Theorem 4.2.2, it is enough to uniquely characterize the law of the vector $(B_{t_1}, \ldots, B_{t_k})$, for any integer k and for all times $t_1, \ldots, t_k$. We can limit ourselves to assuming that $0 \leq t_1 < t_2 < \cdots < t_k$, up to a change of notation. The random variables $B_{t_1}, B_{t_2} - B_{t_1}, \ldots, B_{t_k} - B_{t_{k-1}}$ are independent and have respective laws $\mathcal{N}(0, t_1), \mathcal{N}(0, t_2 - t_1), \ldots, \mathcal{N}(0, t_k - t_{k-1})$. The distribution of the k-tuple $B_{t_1}, B_{t_2} - B_{t_1}, \ldots, B_{t_k} - B_{t_{k-1}}$ is therefore the tensor product of the normal distributions. It is now enough to observe that the distribution of $(B_{t_1}, \ldots, B_{t_k})$ is the image measure of the previous tensor product by the map $(x_1, x_2, \ldots, x_k) \longrightarrow (x_1, x_1 + x_2, \ldots, x_1 + x_2 + \ldots + x_k)$. It is uniquely characterized. □

Historically, Brownian motion was associated with the analysis of movements whose dynamics over time are so disordered that they are hard to predict, even for a very short time. It was Robert Brown, an English botanist, who described in 1827 the erratic movement of fine organic particles (pollen particles) suspended in a gas or a fluid. The trajectories of this very irregular movement seemed not to admit a tangent in any point; one could therefore not speak of its speed, nor *a fortiori* apply the laws of mechanics to it. Brownian motion was introduced by Bachelier in 1900 to model the dynamics of stock prices on the Stock Exchange. It was rediscovered by Einstein in 1905 and has become one of the major tools in statistical physics and probabilistic modeling.

Remark 4.2.7 There are several proofs of existence of the Brownian motion. Our choice was to construct Brownian motion as the limit of renormalized random walks.

We will formalize the results of Section 3.1 a bit more.

Definition 4.2.8 We say that the sequence of processes $(Z_t^n, t \geq 0)_n$ *converges in law in the sense of finite-dimensional marginals* towards the process $(Z_t, t \geq 0)$ if for any p-tuple of times $0 \leq t_1 < t_2 < \cdots < t_p$, the p-tuple of random variables $(Z_{t_1}^n, Z_{t_2}^n, \ldots, Z_{t_p}^n)$ converges in law, as n tends to infinity, towards the p-tuple $(Z_{t_1}, Z_{t_2}, \ldots, Z_{t_p})$.

Remark 4.2.9 As recalled above (Theorem 4.2.2), the law of a stochastic process Z is characterized by the law of all the p-tuples $(Z_{t_1}, Z_{t_2}, \ldots, Z_{t_p})$. This therefore implies that the convergence in law of a sequence of processes in the sense of finite-dimensional marginals allows us to characterize its limit (in law).

Theorem 4.2.10 *The two sequences of processes $(X_t^n, t \geq 0)_n$ and $(Y_t^n, t \geq 0)_n$ introduced in Section 4.1 converge in law, in the sense of finite-dimensional marginals, towards a Brownian motion $(B_t, t \geq 0)$.*

Proof It suffices to apply Proposition 4.1.4 and Theorem 4.2.6. □

We would like to know more about the convergence of the random functions $(t \longrightarrow X_t^n)_n$, which jump by an amplitude of $\frac{1}{\sqrt{n}}$ every $\frac{1}{n}$ units of time, and their continuous linear interpolations $(t \longrightarrow Y_t^n)_n$. We have seen that both sequences of processes converge in the sense of finite-dimensional marginals to the same Brownian motion limit. But this convergence doesn't give any information concerning, for example, the convergence of $\sup_{t \leq T} X_t^n$ or of $\int_0^t f(X_s^n)\,ds$. To deduce such information, we need to define a stronger notion of convergence in law for a sequence of stochastic processes. This convergence is related to the functional metric that can be defined for continuous processes. The proof of such functional convergence exceeds the mathematical scope of this book, so we recommend the following works to the reader: [11], [34] or [48]. We have only shown here the convergence in law in the sense of finite-dimensional marginals. However, if the reader accepts that there is functional convergence, he can observe that the limit must be a process with continuous time trajectories. Indeed, the amplitude of the jumps of X_t^n is $\frac{1}{\sqrt{n}}$ and therefore tends to 0 as n tends to infinity. Moreover, the intervals between jump times are equal to $\frac{1}{n}$ and also tend towards 0.

The convergence is illustrated by the simulations given in Figures 4.1 and 4.2.

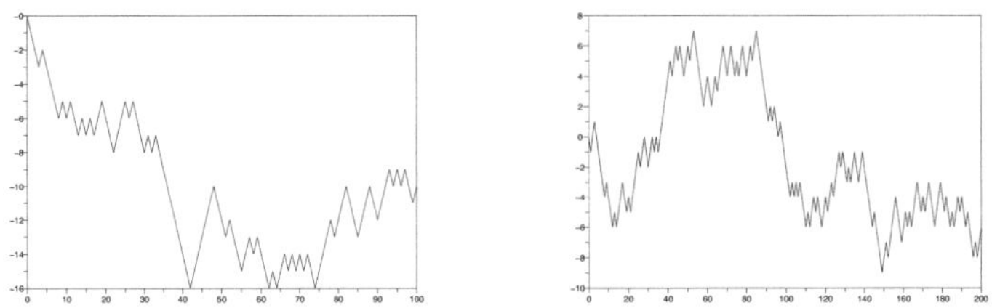

Fig. 4.1 Random walk in dimension 1, $n = 100$ and $n = 200$.

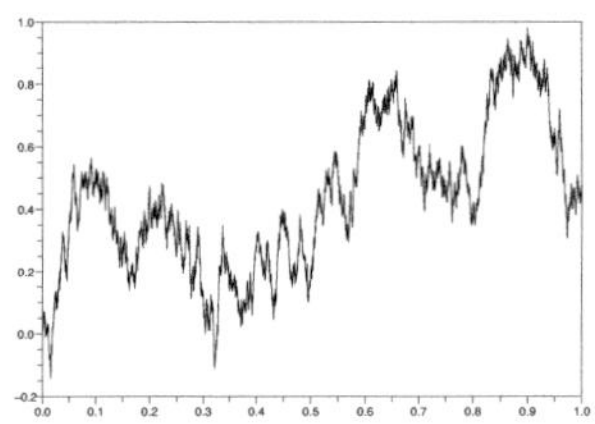

Fig. 4.2 Brownian motion in dimension 1.

4.3 Some Properties of Brownian Motion

Using the definition of Brownian motion (Definition 4.2.5), we can show the following two fundamental properties, the proofs of which are left as an exercise (Exercise 4.8.1).

Proposition 4.3.1 *Let* $(B_t, t \geq 0)$ *be a Brownian motion.*

(i) *Symmetry property: The process* B'*, defined for all* $t > 0$ *by* $B'_t = -B_t$*, is a Brownian motion.*
(ii) *Self-similarity property: For all* $c > 0$*, the process* $B^{(c)}$*, defined for all* $t > 0$ *by* $B_t^{(c)} = \frac{1}{c} B_{c^2 t}$*, is a Brownian motion.*

This self-similarity property justifies the fractal aspect of the trajectories of a Brownian motion.

Note that we can easily compute the covariance of B_t and B_s for all $t, s > 0$:

$$\mathbb{E}(B_t B_s) = t \wedge s.$$

Indeed, if $s < t$ and since the increments are independent,

$$\mathbb{E}(B_t B_s) = \mathbb{E}((B_t - B_s)B_s + B_s^2) = \mathbb{E}(B_t - B_s)\mathbb{E}(B_s) + \mathbb{E}(B_s^2) = s.$$

It is also possible to describe the long-term behavior of a Brownian motion thanks to the following “law of the iterated logarithm”.

Proposition 4.3.2

$$\mathbb{P}\left(\limsup_{t\to+\infty} \frac{B_t}{\sqrt{2t\ln\ln t}} = 1 \text{ and } \liminf_{t\to+\infty} \frac{\mathrm{B_t}}{\sqrt{2\mathrm{t}\ln\ln\mathrm{t}}} = -1\right) = 1.$$

Thus, $\limsup_{t\to+\infty} |B_t|$ behaves almost surely like $\sqrt{2t\ln\ln t}$ as t tends to infinity. The proof of Proposition 4.3.2 will be developed in Exercise 4.8.4.

Let us now study the regularity properties of the trajectories of a Brownian motion. First of all, note that it is not necessary to include a trajectory regularity property in the definition of a Brownian motion. Indeed, if a process satisfies all the properties of Definition 4.2.5, it admits a modification with continuous trajectories a.s., in the sense that there exists a random function W with values in the set of continuous functions from $\mathbb{R}_+$ to $\mathbb{R}$ such that for all $t \geq 0$, $\mathbb{P}(B_t = W_t) = 1$. The proof of this result is based on the following criterion, called Kolmogorov’s continuity criterion.

Theorem 4.3.3 (Kolmogorov’s continuity criterion) *Let $X : \Omega \times [0, T] \to \mathbb{R}$ be a stochastic process and suppose that there are three strictly positive constants γ, β, C with, for all $0 \leq s, t \leq T$,*

$$\mathbb{E}(|X_t - X_s|^\gamma) \leq C\,|t - s|^{1+\beta}.$$

Then X admits a modification almost surely with continuous trajectories.

The great strength of this theorem is that it characterizes a trajectory regularity property (thus ω by ω) by a property of the moments of the process. It requires a little more than simple Lipschitz continuity of these moments, but not much more.

Proof We take $T = 1$ for simplicity. We show that X admits a modification $\tilde{X}$ which has a Hölderian trajectory for any order $\alpha < \frac{\beta}{\gamma}$. We introduce the set D of dyadic points of $[0, 1]$: $D = \cup_m D_m$, where $D_m = \{2^{-m}i; i \text{ integer } \in [0, 2^m]\}$. Let $\Delta_m = \{(s,t) \in D_m^2; |s-t| = 2^{-m}\}$. There are less than 2^{m+1} such pairs. We also define $K_m = \sup_{(s,t)\in\Delta_m} |X_s - X_t|$.

By hypothesis, we have

$$\mathbb{E}(K_m^\gamma) \leq \sum_{(s,t)\in\Delta_m} \mathbb{E}(|X_s - X_t|^\gamma) \leq C\,2^{-m(1+\beta)}\,2^{m+1} = 2C\,2^{-m\beta}.$$

Let us fix m. For every point s (resp. t) of D, there exists an increasing sequence (s_n) of D, (resp. (t_n)), such that for all $n \geq m$, $s_n \leq s$, $t_n \leq t$ with $s_n, t_n \in D_n$ and $s_n = s$ (resp. $t_n = t$) for some n. If $(s,t) \in D^2$ with $|s-t| \leq 2^{-m}$, we set $s_m = t_m$ or $(s_m, t_m) \in \Delta_m$. In all cases, we have

$$X_s - X_t = \sum_{i=m}^{\infty}(X_{s_{i+1}} - X_{s_i}) + X_{s_m} - X_{t_m} + \sum_{i=m}^{\infty}(X_{t_{i+1}} - X_{t_i}),$$

where the series are in fact finite sums. We deduce that

$$|X_s - X_t| \leq K_m + 2\sum_{i=m+1}^{\infty} K_i \leq 2\sum_{i=m}^{\infty} K_i.$$

Let us define $M_\alpha = \sup\{|X_s - X_t|/|s-t|^\alpha;\, s,t \in D, s \neq t\}$. We have

$$\begin{aligned} M_\alpha &\leq \sup_{m\in\mathbb{N}}\left\{2^{(m+1)\alpha} \sup_{2^{-(m+1)}\leq|s-t|\leq 2^{-m}} |X_s - X_t|;\, s,t \in D, s\neq t\right\} \\ &\leq \sup_{m\in\mathbb{N}}\left\{2.2^{(m+1)\alpha}\left(\sum_{i=m}^{\infty} K_i\right)\right\} \\ &\leq 2^{(\alpha+1)}\sum_{i=0}^{\infty} 2^{i\,\alpha} K_i. \end{aligned}$$

For $\gamma \geq 1$ and $\alpha < \beta/\gamma$, we have

$$\mathbb{E}((M_\alpha)^\gamma)^{1/\gamma} \leq 2^{(\alpha+1)}\sum_{i=0}^{\infty} 2^{i\,\alpha}\mathbb{E}((K_i)^\gamma)^{1/\gamma} \leq 2^{(\alpha+1)}(2C)^{1/\gamma}\sum_{i=0}^{\infty} 2^{i\,(\alpha-(\beta/\gamma))} < +\infty.$$

For $\gamma < 1$, the same reasoning applies to $\mathbb{E}((M_\alpha)^\gamma)$.

Thus, for almost every ω, X is uniformly continuous on D. We will then set, for such ω and $t \in [0,1]$,

$$\tilde{X}_t(\omega) = \lim_{s\to t, s\in D} X_s(\omega).$$

Indeed, uniform continuity and the Cauchy criterion imply that $X_{s_n}(\omega)$ has a limit, independent of the choice of sequence $s_n \in D$ converging to t. Note that trivially, $\tilde{X}_t = X_t$ for $t \in D$. Moreover, if $t \notin D$ and if $s_n \in D$ is a sequence converging to t, then by definition, X_{s_n} converges almost surely to $\tilde{X}_t$. Furthermore, by hypothesis, $\mathbb{P}(|X_t - X_{s_n}| > \varepsilon) \leq C\varepsilon^{-\gamma}|t - s_n|^{1+\beta}$, and X_{s_n} converges in probability to X_t. We deduce that $\tilde{X}$ is clearly the desired modification. □

In the case of Brownian motion, the random variable $B_t - B_s$ is a centered normal random variable with variance $|t-s|$, therefore

$$\mathbb{E}(|B_t - B_s|^4) = 3\,(t-s)^2.$$

From this, we immediately deduce the following corollary.

Corollary 4.3.4 *There exists a process B almost surely with continuous trajectories, with independent increments and such that $B_0 = 0$ and for all $t > 0$, the random variable has a normal distribution, centered and with variance t. Moreover, the trajectories of the process are Hölderian of order α for all $\alpha < 1/2$ in the sense that, for almost all ω, there exists a constant $C(\omega)$ such that for all $s,t \geq 0$, $|B_t(\omega) - B_s(\omega)| \leq C(\omega)\,|t-s|^\alpha$.*

Such a process is called a standard Brownian motion. *In the following, we will only consider such processes, which we will simply call* Brownian motions.

Apart from these continuity results, the regularity properties of Brownian motion are very poor. Let us first note that B is unlikely to be of finite variation. Indeed, since $E(|B_{i/n} - B_{(i-1)/n}|^2) = \frac{1}{n}$, $|B_{i/n} - B_{(i-1)/n}|$ is of the order of magnitude $\frac{1}{\sqrt{n}}$. Thus $\sum_{i=1}^{[nt]} |B_{i/n} - B_{(i-1)/n}|$ is of the order of magnitude $\sqrt{n}$ and cannot converge as n tends to infinity. However, we will show that the situation is different if we consider the "approximate quadratic variation of B at level n", defined as the process

$$V(B,n)_t = \sum_{i=1}^{[nt]} (B_{i/n} - B_{(i-1)/n})^2. \tag{4.3.1}$$

The independence of the increments of the Brownian motion will allow us to prove the following proposition.

Proposition 4.3.5 *For any fixed $t > 0$, the sequence of random variables $(V(B,n)_t)_n$ defined by* (4.3.1) *converges to t in $\mathbb{L}^2$, as n tends to infinity.*

Note that this limit is deterministic. This gives an intrinsic character to Brownian motion which will play a main role in the construction of the stochastic integral.

Proof Each increment $B_{i/n} - B_{(i-1)/n}$ is of normal distribution $\mathcal{N}(0, 1/n)$ and therefore $\mathbb{E}(V(B,n)_t) = \frac{[nt]}{n}$. We then write

$$\mathbb{E}\left((V(B,n)_t - t)^2\right) \leq 2\left(\mathbb{E}\left(V(B,n)_t - \frac{[nt]}{n}\right)^2 + \left(\frac{[nt]}{n} - t\right)^2\right).$$

As the increments $(B_{i/n} - B_{(i-1)/n})^2 - \frac{1}{n}$ are independent and $\mathbb{E}\left((B_{i/n} - B_{(i-1)/n})^2\right) = \frac{1}{n}$ and $\mathbb{E}\left((B_{i/n} - B_{(i-1)/n})^4\right) = \frac{3}{n^2}$, we have

$$\begin{aligned}
\mathbb{E}\left(\left(\sum_{i=1}^{[nt]} (B_{i/n} - B_{(i-1)/n})^2 - \frac{[nt]}{n}\right)^2\right) &= \sum_{i=1}^{[nt]} \mathbb{E}\left(\left((B_{i/n} - B_{(i-1)/n})^2 - \frac{1}{n}\right)^2\right) \\
&= \sum_{i=1}^{[nt]} \left(\mathbb{E}\left((B_{i/n} - B_{(i-1)/n})^4\right) - \frac{1}{n^2}\right) \\
&= \frac{2}{n^2} \times [nt].
\end{aligned}$$

Thus, $\mathbb{E}\left((V(B,n)_t - t)^2\right)$ is equivalent to $2t/n$ and converges to 0 as n tends to infinity. We deduce that $(V(B,n)_t)_n$ converges to t in $\mathbb{L}^2$. □

A fundamental consequence, which shows the great irregularity of the trajectories of Brownian motion, is the following non-differentiability property.

Proposition 4.3.6 *The trajectories $t \mapsto B_t$ are almost surely nowhere differentiable.*

Remark 4.3.7 For this reason, one cannot simply define the integral $\int_0^t f(s)\,\mathrm{d}B_s(\omega)$, for a given ω, as we could do for finite variation functions. This will justify the subsequent construction of the stochastic integral.

Proof The convergence in $\mathbb{L}^2$ of a sequence of random variables implies the almost sure convergence of a subsequence. Thus, we deduce from Proposition 4.3.5 that outside a null probability set, and for any pair of rational numbers p, q, there exists a subdivision $(t_i)_i$ of step $\frac{1}{n}$ of $[p,q]$ such that

$$\lim_{n \to +\infty} \sum_{i=[np]}^{[nq]} \left(B_{t_i}(\omega) - B_{t_{i-1}}(\omega)\right)^2 = q - p.$$

If we had $|B_t(\omega) - B_s(\omega)| \leq k|t-s|$ for $p \leq s < t \leq q$, we would deduce that

$$\sum_{i=[np]}^{[nq]} \left(B_{t_i}(\omega) - B_{t_{i-1}}(\omega)\right)^2 \leq k^2\ (q-p)\ \sup_i |t_i - t_{i-1}| = \frac{k^2(q-p)}{n}.$$

When $n \to \infty$, the fact that the left-hand term tends to a finite non-null limit would lead to a contradiction. The trajectories of the Brownian motion therefore cannot be Lipschitz continuous on any subinterval $[p,q]$ of $\mathbb{R}_+$, with p and q rational numbers. By density of $\mathbb{Q}$ in $\mathbb{R}$, we deduce that the trajectories of the Brownian motion are almost surely nowhere differentiable. □

4.4 The Markov Property and Brownian Motion

In Section 2.1 we defined the filtration associated with a Markov chain, or more generally associated with a process indexed by discrete time. It is an increasing sequence (with respect to inclusion) of σ-fields that models the evolution of the information given by the process over time. We will generalize this notion for a process indexed by continuous time.

Remark 4.4.1 In the following, we will consider processes with almost sure left-limited and right-continuous trajectories. This property is fundamental for the Markov and martingale properties. It will be implicit in the rest of this book.

Definition 4.4.2 Let $X = (X_t)_{t\geq 0}$ be a random process defined on a probability space $(\Omega, \mathcal{A}, P)$. The *filtration generated by* X is the increasing family $(\mathcal{F}_t^X)_{t\geq 0}$, where $\mathcal{F}_t^X$ is the σ-field generated by the random variables $\{X_s, s \leq t\}$. The σ-field $\mathcal{F}_t^X$ is the smallest σ-field making the variables X_s, for $s \leq t$, measurable. It describes the events realized by the process before time t.

We can generalize the notion of Markov property to continuous time.

Definition 4.4.3 A process $(X_t, t \geq 0)$ satisfies the *Markov property* if for all $s, t \geq 0$, and for any bounded measurable function f on $\mathbb{R}$,

$$\mathbb{E}(f(X_{t+s})|\mathcal{F}_t^X) = \mathbb{E}(f(X_{t+s})|X_t). \tag{4.4.1}$$

The process X is then called a *Markov process*.

The formula (4.4.1) implies in particular that the conditional expectation of $f(X_{t+s})$ given $\mathcal{F}_t^X$ is a measurable function of X_t. This function depends *a priori* on s and t. An important special case is when it only depends on time s, that is, on the increment between t and $t+s$.

Definition 4.4.4 If for all $s \geq 0$, there exists a bounded measurable function h_s on $\mathbb{R}$ such that, for all $t \geq 0$,

$$\mathbb{E}(f(X_{t+s})|\mathcal{F}_t^X) = h_s(X_t),$$

the process X is called a *time-homogeneous Markov process*.

We now limit ourselves to the study of time-homogeneous Markov processes. Let us introduce the family of operators $(P_t)_{t\geq 0}$, defined for any bounded measurable function f and any $x \in \mathbb{R}$ by

$$P_t f(x) = \mathbb{E}(f(X_t)|X_0 = x). \tag{4.4.2}$$

We then have that $h_s(x) = P_s f(x)$ and therefore

$$\mathbb{E}(f(X_{t+s})|\mathcal{F}_t^X) = P_s f(X_t). \tag{4.4.3}$$

Proposition 4.4.5 *The family of operators $(P_t)_{t\geq 0}$ is sufficient to characterize the law of the process X.*

Proof Indeed, if $0 < s < t$ and if f and g are two bounded measurable functions and $x \in \mathbb{R}$, we have, thanks to the Markov property and to (4.4.3),

$$\begin{aligned}
\mathbb{E}(f(X_s)\, g(X_t)|X_0 = x) &= \mathbb{E}(\mathbb{E}(f(X_s)\, g(X_t)\, |\mathcal{F}_s^X)\, |X_0 = x) \\
&= \mathbb{E}(f(X_s)\, \mathbb{E}(g(X_t)\, |\mathcal{F}_s^X)\, |X_0 = x) \\
&= \mathbb{E}(f(X_s)\, P_{t-s} g(X_s)\, |X_0 = x) \\
&= P_s(f\, P_{t-s} g)(x).
\end{aligned}$$

The bi-dimensional marginals of X are therefore characterized by the semi-group $(P_t)_t$. It is easy to show that the same is true for the d-dimensional marginals, which characterize the law of the process (see Theorem 4.2.2). □

We can also show that the family $(P_t)_t$ satisfies the so-called semi-group property:

$$P_{t+s}f(x) = P_t(P_s f)(x) = P_s(P_t f)(x); \quad P_0 f(x) = f(x).$$

Indeed, in the same way as above, we have

$$\begin{aligned} P_{t+s}f(x) &= \mathbb{E}(f(X_{t+s})\,|X_0 = x) \\ &= \mathbb{E}(\,\mathbb{E}(f(X_{t+s})\,|\mathcal{F}_t^X)\,|X_0 = x) \\ &= \mathbb{E}(P_s f(X_t)\,|X_0 = x) = P_t(P_s f)(x). \end{aligned}$$

Definition 4.4.6 The family of operators $(P_t)_{t\geq 0}$ defined by (4.4.2) is called the *transition semi-group* of the time-homogeneous Markov process X.

Theorem 4.4.7 *Brownian motion is a homogeneous Markov process, with a transition semi-group defined for all measurable and bounded function f and for all $x \in \mathbb{R}$, $t > 0$ by*

$$P_t f(x) = \mathbb{E}(f(B_t)\,|B_0 = x) = \int_{\mathbb{R}} f(y)\,\frac{1}{\sqrt{2\pi t}}\exp\left(-\frac{(y-x)^2}{2t}\right)\mathrm{d}y. \tag{4.4.4}$$

Proof Consider a Brownian motion B on $(\Omega, \mathcal{A}, P)$, and $(\mathcal{F}_t^B)_{t\geq 0}$ the associated filtration. Due to the independence of the increments, the variable $Y = B_{t+s} - B_s$ is independent of the σ-field $\mathcal{F}_s^B$ and has for distribution the normal law $\mathcal{N}(0,t)$. Thus for each measurable and bounded function f on $\mathbb{R}$, for all $t, s > 0$,

$$\mathbb{E}(f(B_{t+s})|\mathcal{F}_s^B) = \mathbb{E}(f(B_s + Y)|\mathcal{F}_s^B) = \int_{\mathbb{R}} \frac{1}{\sqrt{2\pi t}}\mathrm{e}^{-x^2/2t} f(B_s + x)\,\mathrm{d}x.$$

This formula shows that conditionally to $\mathcal{F}_s^B$, the law of B_{t+s} depends on the past (i.e. of all the variables B_r for $r \leq s$) by the law of the "present" B_s of the process, through a function of the time increment t. The Brownian motion is therefore a homogeneous Markov process, with the transition semi-group given by (4.4.4). □

Note that for Brownian motion, $P_t f(x) = \int_{\mathbb{R}} f(y)\,p(t,x,y)\,\mathrm{d}y$, with

$$p(t,x,y) = \frac{1}{\sqrt{2\pi t}}\exp\left(-\frac{(y-x)^2}{2t}\right). \tag{4.4.5}$$

The function $p(t,x,\cdot)$ is the density at time t of the Brownian motion, starting from x at time 0.

Using (4.4.5), it is easy to show by simple computation that if f is a bounded continuous function, the function $u(t,x) = P_t f(x) = \int p(t,x,y)\,f(y)\,\mathrm{d}y$ is the unique regular solution of the partial differential equation

$$\frac{\partial u}{\partial t}(t,x) = \frac{1}{2}\frac{\partial^2 u}{\partial x^2}(t,x); \qquad u(0,x) = f(x),\; x \in \mathbb{R},\; t > 0.$$

This partial differential equation is called the *heat equation*. We have therefore exhibited a probabilistic solution to this equation, namely $u(t,x) = \mathbb{E}(f(B_t)|B_0 = x)$.

The definition of Brownian motion generalizes to the multi-dimensional case.

Definition 4.4.8 A *d-dimensional Brownian motion* is a d-tuple $B = (B^i)_{1\le i\le d}$ of real-valued Brownian motions $B^i = (B^i_t)_{t\ge 0}$, independent of each other.

This process is still a homogeneous Markov process with independent and stationary increments. A simulation of its trajectory is given in Figure 4.4 in the bi-dimensional case. The process can also be obtained as the limit of renormalized random walks, as suggested by Figure 4.3.

The semi-group of a d-dimensional Brownian motion is defined for ϕ measurable and bounded on $\mathbb{R}^d$, $x \in \mathbb{R}^d$ and $t > 0$, by

$$P_t\phi(x) = \int_{\mathbb{R}^d} \phi(y) \frac{1}{(2\pi t)^{d/2}} \exp\left(-\frac{\|y-x\|^2}{2t}\right) \mathrm{d}y,$$

where $\|\cdot\|$ denotes the Euclidean norm and $\mathrm{d}y$ the Lebesgue measure on $\mathbb{R}^d$. As in dimension 1, the Brownian motion admits a density at each time $t > 0$, which allows us to obtain the explicit solution of the heat equation in dimension d

$$\frac{\partial u}{\partial t}(t,x) = \frac{1}{2}\Delta u(t,x); \qquad u(0,x) = \phi(x),\ x \in \mathbb{R}^d,\ t > 0.$$

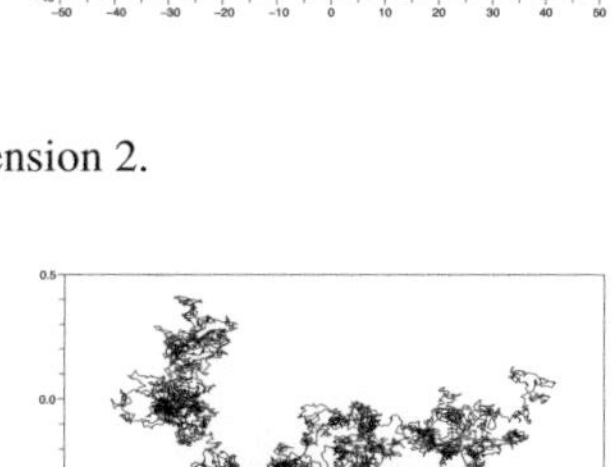

Fig. 4.3 Random walk in dimension 2.

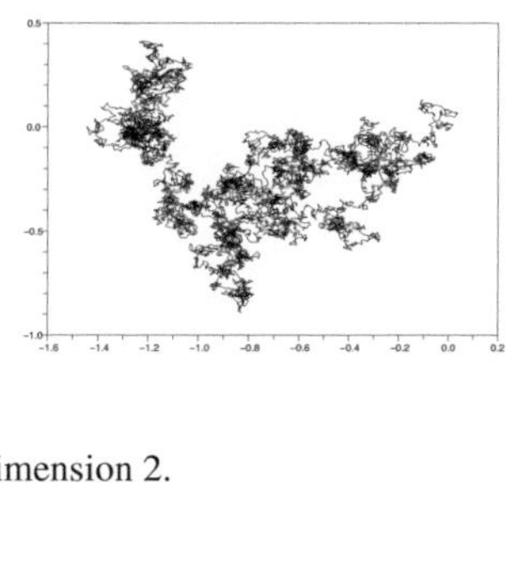

Fig. 4.4 Brownian motion in dimension 2.

4.5 Continuous Time Martingales and Stopping Times

4.5.1 Continuous Time Martingales

We will highlight a fundamental property of Brownian motion concerning the behavior of its fluctuations, namely the martingale property.

Let $(B_t, t \geq 0)$ be a Brownian motion. We introduced in the previous section the filtration $(\mathcal{F}_t^B)_t$ generated by this Brownian motion: $\mathcal{F}_t^B = \sigma(B_s, s \leq t)$ (see Definition 4.4.2). In fact, this filtration is not big enough to develop the stochastic calculus in the next chapter. We will introduce the augmented filtration of B, namely a filtration $(\mathcal{F}_t)_t$ for which B is still a Brownian motion, which contains negligible sets and which is right-continuous, in the sense that for all $t \geq 0$,

$$\mathcal{F}_t = \cap_{\varepsilon > 0} \mathcal{F}_{t+\varepsilon}.$$

The σ-field $\mathcal{F}_t$ will therefore allow us to take into account events related to the behavior of the process immediately after t. The construction of the filtration $(\mathcal{F}_t)_t$ is developed, for example, in [48, Chap. 2.7].

Since for all $t \geq 0$, B_t is $\mathcal{F}_t$-measurable and B has independent and stationary increments, we easily see that for all $s \leq t$,

$$\mathbb{E}(B_t \,|\mathcal{F}_s) = B_s, \tag{4.5.1}$$

$$\mathbb{E}(B_t^2 - t \,|\mathcal{F}_s) = B_s^2 - s. \tag{4.5.2}$$

Indeed,

$$\mathbb{E}(B_t \,|\mathcal{F}_s) = \mathbb{E}(B_s + B_t - B_s) = B_s + \mathbb{E}(B_t - B_s) = B_s;$$

$$\mathbb{E}(B_t^2 - t \,|\mathcal{F}_s) = \mathbb{E}((B_t - B_s)^2 + 2B_s(B_t - B_s) + B_s^2 - t) = t - s + B_s^2 - t = B_s^2 - s.$$

The property (4.5.1) means that the future increments of the Brownian motion conditionally on the information given by the process up to the present time are of zero mean. This property generalizes to continuous time the definition of martingale given for discrete time (Definition 2.5.2).

Definition 4.5.1 A process $(X_t, t \geq 0)$ is said to be *adapted to a filtration* $(\mathcal{F}_t)$ if for all $t \geq 0$, X_t is $\mathcal{F}_t$-measurable, that is, for every Borel set A of $\mathbb{R}$, the event $\{X_t \in A\}$ belongs to $\mathcal{F}_t$. (The information given by $(X_s, s \leq t)$ can be described by an event of $\mathcal{F}_t$.)

In this section, every process X is assumed to be adapted to the filtration $(\mathcal{F}_t)_t$ of the Brownian motion B. Intuitively, this means that the realization (or not) of events of the type $\{X_t \in A\}$ can be decided from the knowledge of the Brownian trajectories up to time t.

Definition 4.5.2 A real-valued process $M = (M_t)_{t\geq 0}$, adapted to the filtration $(\mathcal{F}_t)$, is a *martingale* if for any $t \geq 0$, the variable M_t is integrable and if for all $s, t \geq 0$,

$$s \leq t \quad \Longrightarrow \quad \mathbb{E}(M_t|\mathcal{F}_s) = M_s. \tag{4.5.3}$$

In particular, the expectation $\mathbb{E}(M_t)$ of a martingale does not depend on t: for all t,

$$\mathbb{E}(M_t) = \mathbb{E}(M_0). \tag{4.5.4}$$

For more details, we refer to the following books (but this is not exhaustive): Shreve–Karatzas [48] and Revuz–Yor [71].

Let us give some fundamental examples of martingales adapted to the Brownian filtration.

Proposition 4.5.3 *The following three $\mathcal{F}_t$-adapted processes are martingales.*

1. *The Brownian motion $(B_t, t \geq 0)$.*
2. *The process $(M_t, t \geq 0)$ defined for all t by $M_t = B_t^2 - t$.*
3. *The process $(M_t^\lambda, t \geq 0)$ defined by $M_t^\lambda = \exp(\lambda B_t - \frac{\lambda^2}{2}t)$, for all $\lambda \in \mathbb{R}$.*

Proof The properties (4.5.1) and (4.5.2) immediately imply that the processes B and M are martingales. Moreover, it is clear that, for all $\lambda > 0$, M^λ is $\mathcal{F}_t$-adapted and that M_t^λ is integrable, since $\mathbb{E}(|M_t^\lambda|) = \mathbb{E}(M_t^\lambda) = 1$. For $s \leq t$, we can write $\mathbb{E}\left(M_t^\lambda\, |\mathcal{F}_s\right) = M_s^\lambda\ \mathbb{E}\left(\exp(\lambda(B_t - B_s) - \frac{\lambda^2}{2}(t-s))\right)$, since $B_t - B_s$ is independent of $\mathcal{F}_s$. Using that for a random variable U with normal distribution $\mathcal{N}(0, \sigma^2)$, $\mathbb{E}(\exp(\lambda U)) = \exp\left(\frac{\lambda^2\sigma^2}{2}\right)$, we obtain $\mathbb{E}\left(\exp(\lambda(B_t - B_s) - \frac{\lambda^2}{2}(t-s))\right) = 1$, from which we deduce that $\mathbb{E}\left(M_t^\lambda|\mathcal{F}_s\right) = M_s^\lambda$. □

This proposition has a converse, which is a characterization of Brownian motion due to Paul Lévy.

Proposition 4.5.4 *Let $(M_t)_{t\geq 0}$ be a real-valued stochastic process with almost surely continuous trajectories and starting at 0. The following two properties are equivalent:*

(i) *$(M_t)_{t\geq 0}$ and $(M_t^2 - t)_{t\geq 0}$ are martingales for the filtration generated by M.*
(ii) *$(M_t)_{t\geq 0}$ is a Brownian motion.*

We only need to prove (i)⟹(ii). To do this, we show that for all $0 \leq s \leq t$, $E(\mathrm{e}^{\lambda(M_t - M_s)}|\mathcal{F}_s) = \mathrm{e}^{\frac{\lambda^2}{2}(t-s)}$, which is enough to characterize a Brownian motion law. The intuitive idea of the proof is simple: expand $\mathrm{e}^{\lambda x}$ to order 2 of and use (i). However, as the technical details are numerous, we refer, for example, to [42, Theorem 6.1] or to [26] for a detailed proof.

4.5.2 Fundamental Inequalities and Behavior at Infinity

One of the reasons why martingales are central to probability theory is that they allow us to obtain very useful inequalities in practice. We have already proved these inequalities in discrete time in Proposition 2.5.6. They are therefore true for a continuous time process restricted to rational times. Since all processes are right-continuous and left-limited, the results generalize to all times by passage to a limit, the details of which we will not give. The inequalities below are known as the Maximal Lemma and Doob's inequality.

Proposition 4.5.5 *Consider a martingale $(M_t, t \geq 0)$. We have for all $\alpha > 0$ and all $p > 1$,*

$$\mathbb{P}(\sup_{s \leq t} |M_s| \geq \alpha) \leq \frac{\mathbb{E}(|M_t|)}{\alpha}$$

and

$$\mathbb{E}((\sup_{s \leq t} |M_s|)^p) \leq \left(\frac{p}{p-1}\right)^p \mathbb{E}(|M_t|^p). \tag{4.5.5}$$

The great strength of these inequalities is that it enables us, in a sense, to exchange the expectation and the supremum, which is impossible in a general context.

Note that by Jensen's inequality, for $s < t$, $\mathbb{E}(|M_t| | \mathcal{F}_s) \geq |M_s|$ and therefore $t \to \mathbb{E}(|M_t|)$ is increasing.

One of the most powerful properties of martingales is the following theorem, which describes the long-term behavior of a martingale whose first moments $E(|M_t|)$ are uniformly bounded (in t). This theorem has many applications. The proof generalizes that of the analogous theorem for discrete-time martingales and uses some variations of the fundamental martingale inequalities. We will not present it in this book, but refer the reader to the classic references Shreve–Karatzas [48] and Revuz–Yor [71].

Theorem 4.5.6 *If $(M_t, t \geq 0)$ is a martingale satisfying*

$$\sup_{t \in \mathbb{R}_+} \mathbb{E}(|M_t|) < +\infty, \tag{4.5.6}$$

then there exists an integrable random variable $M_\infty \in]-\infty, +\infty[$ such that M_t converges almost surely to M_∞ as t tends to infinity.

Remark 4.5.7 This theorem is one of the rare results of almost sure convergence in the long term. That's why many proofs that demonstrate the convergence of a process as time tends to infinity consist of highlighting a martingale and then applying Theorem 4.5.6 to it.

Note that the integrability property (4.5.6) is not true for the Brownian motion B, for which $\mathbb{E}(|B_t|) = \sqrt{2t/\pi}$, nor for the process M of Proposition 4.5.3, for which $\mathbb{E}(|M_t|) = ct$ for a constant $c > 0$, while it is true for the third example. Indeed, for all λ, $\mathbb{E}(|M_t^\lambda|) = \mathbb{E}(M_t^\lambda) = 1$. In this case, Theorem 4.5.6 implies that for each real number λ, there exists $M_\infty^\lambda \in]-\infty,+\infty[$ such that M_t^λ almost surely converges to M_∞^λ as t tends to infinity. Furthermore, in this last case, using the law of the iterated logarithm (stated in Proposition 4.3.2), we can show that $M_\infty^\lambda = 0$ if $\lambda \neq 0$, and that otherwise, $M_\infty^\lambda = 1$ if $\lambda = 0$, since $M_t^0 = 1$ for all t. Thus, the equality (4.5.4) is clearly false when $t = +\infty$, except in the case where $\lambda = 0$.

An important question is when the equality (4.5.3) remains true for $t = +\infty$. As we just noted from Example 3 of Proposition 4.5.3, the assumptions must be strictly stronger than the integrability assumption (4.5.6). The right assumption is a uniform integrability assumption that we now define.

Definition 4.5.8 We say that a process $(M_t, t \in \mathbb{R}_+)$ is *uniformly integrable* if

$$\lim_{A\to+\infty} \sup_{t\in\mathbb{R}_+} \mathbb{E}(|M_t|\, \mathbb{1}_{|M_t|>A}) = 0. \tag{4.5.7}$$

Let us give some simple cases where a martingale is uniformly integrable.

Corollary 4.5.9 *The uniform integrability property* (4.5.7) *holds as soon as the martingale M satisfies one of the following three properties:*

1) *The martingale M is bounded, in the sense that there exists $K > 0$ such that*

$$\forall\, \omega,\ \forall\, t,\ |M_t(\omega)| \leq K. \tag{4.5.8}$$

2) *The martingale M is uniformly square-integrable, in the sense that*

$$\sup_{t\in\mathbb{R}_+} \mathbb{E}(M_t^2) < +\infty.$$

3) *The martingale M is only defined on a bounded time interval* $[0,T]$.

Proof It suffices to verify that each of the three cases leads to the uniform integrability of the martingale M. This is obvious for a bounded martingale. The second case follows from Markov's inequality: $A\,\mathbb{E}(|M_t|\, \mathbb{1}_{|M_t|>A}) \leq \mathbb{E}(M_t^2)$. In the third case, we first note that for all $t \in [0,T]$, $\mathbb{E}(|M_t|) \leq \mathbb{E}(|M_T|)$ and therefore $\mathbb{P}(|M_t| > A)$ converges to 0 uniformly on $[0,T]$, as A tends to infinity. As M_T is integrable, this also implies by dominated convergence that the same is true for $\mathbb{E}(|M_T|\, \mathbb{1}_{|M_t|>A})$. The conclusion comes from the following inequality:

$$\begin{aligned}\mathbb{E}(|M_t|\, \mathbb{1}_{|M_t|>A}) &= \mathbb{E}(|\mathbb{E}(M_T|\mathcal{F}_t)|\, \mathbb{1}_{|M_t|>A}) \\ &\leq \mathbb{E}(\mathbb{E}(|M_T|\,|\mathcal{F}_t)\, \mathbb{1}_{|M_t|>A}) \\ &= \mathbb{E}(|M_T|\, \mathbb{1}_{|M_t|>A}).\end{aligned}$$

□

Theorem 4.5.10 *If the martingale M is uniformly integrable, then M_t almost surely converges to M_∞ as t tends to infinity and for all $t > 0$,*

$$M_t = \mathbb{E}(M_\infty|\mathcal{F}_t). \tag{4.5.9}$$

In particular,

$$\mathbb{E}(M_\infty) = \mathbb{E}(M_0).$$

Proof Since M is uniformly integrable, we can write

$$\mathbb{E}(|M_t|) = \mathbb{E}(|M_t|\,\mathbb{1}_{|M_t|\le A}) + \mathbb{E}(|M_t|\,\mathbb{1}_{|M_t|>A}) \le A + \mathbb{E}(|M_t|\,\mathbb{1}_{|M_t|>A}),$$

from which we deduce that (4.5.6) is satisfied and that M_∞ exists and has finite expectation. Let us show (4.5.9). Let $F \in \mathcal{F}_s$ be a random event. We want to show that

$$\mathbb{E}(M_\infty\,\mathbb{1}_F) = \mathbb{E}(M_s\,\mathbb{1}_F).$$

We know that for all $t > s$,

$$\mathbb{E}(M_t\,\mathbb{1}_F) = \mathbb{E}(M_s\,\mathbb{1}_F).$$

So, we need to let t tend to infinity. Note that this is immediate if M is bounded, by applying the dominated convergence theorem. Under the assumption of uniform integrability, the intuitive idea is as follows: on the event $|M_t| \le A$, we can revert to the previous argument and the complementary event $|M_t| > A$ is "uniformly" negligible for sufficiently large A.

Let us develop this idea more precisely: let $\varepsilon, A > 0$,

$$\begin{aligned}\mathbb{E}(|M_t - M_\infty|\,\mathbb{1}_F) &= \mathbb{E}(|M_t - M_\infty|\,\mathbb{1}_F\,\mathbb{1}_{|M_t-M_\infty|\le\varepsilon}) + \mathbb{E}(|M_t - M_\infty|\,\mathbb{1}_F\,\mathbb{1}_{|M_t-M_\infty|>\varepsilon})\\ &\le \varepsilon + \mathbb{E}(|M_\infty|\,\mathbb{1}_F\,\mathbb{1}_{|M_t-M_\infty|>\varepsilon}) + \mathbb{E}(|M_t|\,\mathbb{1}_F\,\mathbb{1}_{|M_t-M_\infty|>\varepsilon}).\end{aligned}$$

The almost-sure convergence of M_t to M_∞ implies its convergence in probability and $\mathbb{P}(|M_t - M_\infty| > \varepsilon)$ tends to 0 as t tends to infinity (for any fixed $\varepsilon > 0$). Since M_∞ is integrable, the second term of the expression above therefore tends to 0 (by the dominated convergence theorem). Let us study the third term. We have

$$\begin{aligned}&\mathbb{E}(|M_t|\,\mathbb{1}_F\,\mathbb{1}_{|M_t-M_\infty|>\varepsilon})\\ &\quad = \mathbb{E}(|M_t|\,\mathbb{1}_F\,\mathbb{1}_{|M_t-M_\infty|>\varepsilon}\,\mathbb{1}_{|M_t|>A}) + \mathbb{E}(|M_t|\,\mathbb{1}_F\,\mathbb{1}_{|M_t-M_\infty|>\varepsilon}\,\mathbb{1}_{|M_t|\le A}).\end{aligned}$$

The first term is less than $\mathbb{E}(|M_t|\,\mathbb{1}_{|M_t|>A})$ and tends to 0 uniformly in t as A tends to infinity and the second term is less than $A\,\mathbb{P}(|M_t - M_\infty| > \varepsilon)$. Gathering all this information, with ε sufficiently small, we deduce the result. □

Remark 4.5.11 For Y an integrable random variable, the process M defined by $M_t = \mathbb{E}(Y|\mathcal{F}_t)$ is a martingale satisfying the property (4.5.6). Indeed, for all $t > 0$,

$$\mathbb{E}(|M_t|) = \mathbb{E}(|\mathbb{E}(Y|\mathcal{F}_t)|) \le \mathbb{E}(|Y|) < +\infty.$$

Moreover, $\mathbb{E}(Y^2) < +\infty$ implies that (4.5.9) is satisfied and M is therefore a martingale up to infinity. This applies in particular if Y is the indicator function of a random event.

Application: assume that the size of a population at any time t is modeled by a Markov process (X_t) and that Y is the indicator function of the event "Population Extinction", that is, the event $\{\lim_{t\to\infty} X_t = 0\}$. Then, if $u(x)$ denotes the extinction probability when the initial population is of size x, we deduce from the Markov property that $M_t = u(X_t)$ defines a bounded martingale.

4.5.3 Stopping Theorems

As in the discrete case, stopping times play a very important role in the continuous time context. We will show in this section that the martingale property extends under certain assumptions to stopping times, and we will see important applications to hitting times of domains. The definition of a stopping time in the continuous time context generalizes that of the discrete case.

Definition 4.5.12 Given a filtration $(\mathcal{F}_t)$, a random variable $T \in [0, +\infty]$ is an $(\mathcal{F}_t)$*-stopping time* if the event $\{T \le t\}$ belongs to $\mathcal{F}_t$ for all $t \ge 0$.

Apart from constant times, the fundamental example of a stopping time is the hitting time of a closed set A by a continuous adapted process X.

Proposition 4.5.13 *Let $(X_t, t \ge 0)$ be a continuous and adapted process. Let A be a closed set of $\mathbb{R}$. We define the entry time of X into A by*

$$T_A = \inf\{t \ge 0; X_t \in A\}$$

(with the convention $\inf \emptyset = +\infty$*). The random time T_A is a stopping time for the filtration generated by X.*

Proof For all t, we have

$$\{T_A \le t\} = \left\{\omega;\ \inf_{s\in\mathbb{Q}, s\le t} d(X_s(\omega), A) = 0\right\},$$

where $d(x, A)$ is the distance from a point $x \in \mathbb{R}$ to the set A. It is obvious that the term on the right belongs to $\mathcal{F}_t$. □

This theorem becomes very delicate as soon as A is only a Borel set of $\mathbb{R}$ or as soon as X is only a right-continuous process with left limits. In this case, the property remains true, but for the augmented filtration of the process.

Be careful that the last time before a fixed time s where X visits A, defined by $S = \sup\{t \le s, X_t \in A\}$ (where $\sup \emptyset = 0$), is not a stopping time. Indeed, for $t < s$, the event $\{\omega, S(\omega) \le t\}$ belongs to $\mathcal{F}_s$ but not to $\mathcal{F}_t$: it depends on the entire trajectory of $X(\omega)$ up to time s.

Proposition 4.5.14 *If S and T are two stopping times, then the random time $S \wedge T$ defined by $(S \wedge T)(\omega) = S(\omega) \wedge T(\omega) = \min(S(\omega), T(\omega))$ is a stopping time.*

The proof of this proposition is left to the reader.

If M is a martingale, we denote by M_T the random variable defined on Ω by $M_T(\omega) = M_{T(\omega)}(\omega)$. Note that as soon as M_∞ is well defined, the stopping times can take the value infinity, provided that we set $M_T = M_\infty$ on the set $\{T = \infty\}$.

We will generalize Theorem 2.5.4 to continuous time and give a converse.

Theorem 4.5.15

1) *Let (M_t) be a martingale. If T is a bounded stopping time, we have*

$$\mathbb{E}(M_T) = \mathbb{E}(M_0). \tag{4.5.10}$$

2) *Let X be an $\mathcal{F}_t$-adapted process such that for all t, $\mathbb{E}(|X_t|) < +\infty$. Then X is a martingale if and only if for every bounded stopping time T, $X_T \in \mathbb{L}^1$ and $\mathbb{E}(X_T) = \mathbb{E}(X_0)$.*

Proof 1) Consider the sequence of random times T_n defined by

$$T_n(\omega) = \frac{k}{2^n}, \quad \text{if} \quad \frac{k-1}{2^n} \leq T(\omega) < \frac{k}{2^n}.$$

It is easy to check that the times T_n are finite-valued stopping times, since T is bounded. Thus, there exists a K such that for all n,

$$\begin{aligned}
\mathbb{E}(M_{T_n}) &= \sum_{k=0}^{K2^n} \mathbb{E}(M_{\frac{k}{2^n}} \mathbb{1}_{\{T_n = \frac{k}{2^n}\}}) \\
&= \sum_{k=0}^{K2^n} \mathbb{E}(\mathbb{E}(M_K | \mathcal{F}_{\frac{k}{2^n}}) \mathbb{1}_{\{T_n = \frac{k}{2^n}\}}) \\
&= \mathbb{E}(M_K) = E(M_0).
\end{aligned} \tag{4.5.11}$$

We want to take the limit as $n \to +\infty$ in (4.5.11). By construction, the sequence (T_n) converges almost surely and in decreasing order towards T. As M is right-continuous, it follows that M_{T_n} converges almost surely towards M_T. As T is bounded by K, we can limit ourselves to considering the martingale M on $[0, K]$, where it is uniformly integrable. Thus, the sequence $(M_{T_n})_n$ is uniformly integrable and converges almost surely towards M_T, it then converges in mean towards M_T and the conclusion follows.

2) We only have to show that if $\mathbb{E}(X_T) = \mathbb{E}(X_0)$ for every bounded stopping time T, then X is a martingale. Let $s < t$ and $A \in \mathcal{F}_s$. Define the random time T by

$$T = t\,\mathbb{1}_{A^c} + s\,\mathbb{1}_A.$$

It is easy to check that T is a bounded stopping time. Thus,

$$\mathbb{E}(X_T) = \mathbb{E}(X_t \mathbb{1}_{A^c}) + \mathbb{E}(X_s \mathbb{1}_A) = \mathbb{E}(X_0).$$

Furthermore, as t is (obviously) a bounded stopping time, we also have

$$\mathbb{E}(X_t) = \mathbb{E}(X_t \mathbb{1}_{A^c}) + \mathbb{E}(X_t \mathbb{1}_A) = \mathbb{E}(X_0).$$

We deduce that $\mathbb{E}(X_s \mathbb{1}_A) = \mathbb{E}(X_t \mathbb{1}_A)$ and $\mathbb{E}(X_t | \mathcal{F}_s) = X_s$. Thus the process X is a martingale. □

We can easily deduce the following proposition, extremely useful in practice, as we will see in the next subsection.

Proposition 4.5.16 *Consider a martingale (M_t) and a stopping time T. Then the process (M_t^T) defined by $M_t^T = M_{t\wedge T}$ is a martingale, called the* martingale stopped at time *T.*

Proof For all t, $M_{T\wedge t}$ is $\mathcal{F}_t$-measurable and in $\mathbb{L}^1$. Let S be a bounded stopping time. Then $T \wedge S$ is a bounded stopping time and $\mathbb{E}(M_{T\wedge S}) = \mathbb{E}(M_0)$. We deduce that $\mathbb{E}(M_S^T) = \mathbb{E}(M_0^T)$, which ensures by Theorem 4.5.15 that M^T is a martingale. □

Remark 4.5.17 Note that (4.5.10) can be false if T is not bounded. For example, for the Brownian motion B and $T = \inf\{t \geq 0, B_t = 1\}$, we have $\mathbb{E}(B_0) = 0 < \mathbb{E}(B_T) = 1$. In this case, the stopping time T is almost surely finite, but is not bounded and even has an infinite expectation (see (4.5.14)).

However, in the case of a bounded martingale (satisfying (4.5.8)), we have the following

Theorem 4.5.18 *If M is a bounded martingale, then (4.5.10) occurs for any stopping time T.*

Proof The proof is immediate thanks to Proposition 4.5.16. Indeed, since for all t, we have $\mathbb{E}(M_{T\wedge t}) = \mathbb{E}(M_0)$, it suffices to let t tend to infinity and use the dominated convergence theorem, thanks to the boundedness of M. □

4.5.4 Applications to Brownian Motion

In this subsection, we will show how the martingale property and the use of stopping times allow us to perform computations and in particular to obtain the laws of certain functionals of Brownian motion.

We first generalize the Markov property of Brownian motion to stopping times. For this, we need to describe the information accumulated up to a stopping time T. Suppose that a random event A is part of this information. Its occurrence or non-occurrence will be decided at time T. In particular, $A \cap \{T \leq t\}$ must belong to $\mathcal{F}_t$. This leads us the following definition.

Definition 4.5.19 Let T be a stopping time for the filtration $(\mathcal{F}_t)_t$. The *σ-field $\mathcal{F}_T$ of events prior to T* is the set of random events such that for all $t \geq 0$, $A \cap \{T \leq t\} \in \mathcal{F}_t$.

A – The Strong Markov Property

First, note that a Brownian motion satisfies the strong Markov property, namely the Markov property generalizes to a stopping time on the set where the latter is finite. The following theorem gives much more information.

Theorem 4.5.20 *If T is a stopping time, then on the set $\{T < +\infty\}$, the process defined by $B_s^{(T)} = B_{T+s} - B_T$ is still a Brownian motion, independent of $\mathcal{F}_T$. The Brownian motion is a strong Markov process: For any Borel set A,*

$$\mathbb{E}_x(B_{T+t} \in A \,|\, \mathcal{F}_T) = \mathbb{E}_x(B_{T+t} \in A \,|\, B_T) \quad \textit{a.s. on } \{T < +\infty\}. \tag{4.5.12}$$

Proof Let us only prove the theorem for a bounded stopping time. By virtue of Paul Lévy's characterization from Proposition 4.5.4, we show that both processes $(B_s^{(T)}, s \geq 0)$ and $((B_s^{(T)})^2 - s, s \geq 0)$ are martingales. We use characterization 2) of Theorem 4.5.15. Consider another bounded stopping time S. It suffices to show that

$$\mathbb{E}(B_{T+S} - B_T) = 0; \;\; \mathbb{E}((B_{T+S} - B_T)^2 - S) = 0.$$

These equalities are immediate thanks to part 1) of Theorem 4.5.15. We immediately deduce (4.5.12). □

B – Law of the Hitting Time of a Point

1) Use of the martingale stopping theorem

Let B be a Brownian motion, and introduce for $a > 0$ the stopping time

$$T_a = \inf\{t > 0, B_t = a\},$$

where $T_a = +\infty$ if the set is empty. Note that, thanks to Proposition 4.3.2 (law of the iterated logarithm), the set $\{t > 0, B_t = a\}$ is almost surely unbounded.

Consider the martingale $M_t^\lambda = \exp(\lambda B_t - \frac{\lambda^2}{2} t)$ for an arbitrary $\lambda > 0$ and $N = (M^\lambda)^{T_a}$ the martingale M^λ stopped at time T_a. Then $N_t = M^\lambda_{T_a \wedge t}$, from which we deduce that $0 < N_t \leq e^{a\lambda}$, implying that N is bounded.

We can therefore apply Theorem 4.5.15 and (4.5.10) to the martingale N and the stopping time T_a, which gives $\mathbb{E}(N_{T_a}) = \mathbb{E}(N_0) = 1$. Since $N_{T_a} = \exp(\lambda a - \frac{\lambda^2}{2} T_a)$ and by setting $\theta = \lambda^2/2$, we deduce that

$$\mathbb{E}(e^{-\theta T_a}) = e^{-a\sqrt{2\theta}}. \tag{4.5.13}$$

By letting θ tend to 0 and by the dominated convergence theorem, it is easy to obtain

$$\mathbb{P}(T_a < +\infty) = 1. \tag{4.5.14}$$

Similarly, by differentiating with respect to θ (which can be justified by the monotone convergence theorem), then letting θ tend to 0, we obtain $\mathbb{E}(T_a \mathbb{1}_{T_a<+\infty}) = +\infty$ and therefore *a fortiori*

$$\mathbb{E}(T_a) = +\infty. \tag{4.5.15}$$

Since, as seen in Proposition 4.3.1, if B is a Brownian motion then $-B$ is a Brownian motion, (4.5.14) and (4.5.15) are also true for $a < 0$.

We can conclude from this result that the Brownian motion hits each point with probability 1, but it will take an infinite time to reach them. Intuitively, this property can be understood as related to the fractal irregularity of the trajectories.

Thanks to Theorem 4.5.20, it is easy to deduce from (4.5.14) the recurrence property of Brownian motions, by conditioning on the successive passage times at point a. The proof of the following proposition is left as an exercise for the reader.

Proposition 4.5.21 *With probability* 1, *for all* $t \geq 0$, *the process* B *visits every real number* a *infinitely often after time* t.

The expression (4.5.13) gives the Laplace transform of the variable T_a in θ, which can be inverted (see [26, p. 187]). This shows that T_a admits a density on $\mathbb{R}_+$ given by

$$f_a(u) = \frac{a}{\sqrt{2\pi u^3}} \mathrm{e}^{-a^2/2u}, \; u \geq 0. \tag{4.5.16}$$

The law of T_a is called a *stable law of index* $1/2$.

We will provide a second method, more probabilistic, to find the law of T_a. This one is based on a property of the Brownian motion, called the reflection principle.

2) The reflection principle

Let us introduce the continuous and increasing process $(S_t, t \geq 0)$ defined by $S_t = \sup_{s\leq t} B_s$. This process and the stopping times T_a are inverses of each other in the sense that $\{T_a \leq t\} = \{S_t \geq a\}$ and therefore

$$T_a = \inf\{t; S_t \geq a\} \quad \text{and} \quad S_t = \inf\{a, T_a \geq t\}.$$

Theorem 4.5.20 allows us to prove the following reflection principle.

Proposition 4.5.22 *For $a > 0$, for all $t > 0$,*

$$\mathbb{P}(S_t \geq a) = \mathbb{P}(T_a \leq t) = 2\mathbb{P}(B_t \geq a) = \mathbb{P}(|B_t| \geq a) = 2\Big(1 - \Phi\Big(\frac{a}{\sqrt{t}}\Big)\Big), \qquad (4.5.17)$$

where Φ denotes the distribution function of the law $\mathcal{N}(0,1)$, defined by $\Phi(t) = P(U \leq t)$ if U is a standard normal variable.

Remark 4.5.23 1) This result gives us the distribution function of T_a and by a classic differentiation argument, we can find (or deduce) that the law of T_a has the density given by (4.5.16).

2) The property (4.5.17) implies in particular that, for all t, the random variables S_t and $|B_t|$ have the same law since they have the same distribution function. However, the processes $(S_t, t \geq 0)$ and $(|B_t|, t \geq 0)$ do not have the same law, since the first one is increasing and the second one almost surely returns to 0, as we have seen above.

Proof The name of this result comes from the following heuristic argument. Among the trajectories that reach a before time t, "half" will be greater than a at time t. Indeed,

$$\mathbb{P}(T_a \leq t) = \mathbb{P}(T_a \leq t, B_t \geq a) + \mathbb{P}(T_a \leq t, B_t < a).$$

But $\mathbb{P}(T_a \leq t, B_t \geq a) = \mathbb{P}(B_t \geq a)$, because a Brownian motion, having continuous trajectories and starting from 0, must necessarily pass through a to become greater than a. Furthermore, by the strong Markov property (Theorem 4.5.20) and by symmetry of the Brownian motion, we have

$$\begin{aligned}\mathbb{P}(T_a \leq t, B_t < a) &= \mathbb{E}(\mathbb{1}_{T_a \leq t}\,\mathbb{E}(B_t < a|\mathcal{F}_{T_a}))\\ &= \mathbb{E}(\mathbb{1}_{T_a \leq t}\,\mathbb{E}(B_{t-T_a+T_a} < B_{T_a}|\mathcal{F}_{T_a}))\\ &= \int_0^t \mathbb{E}(B_{t-s} < 0)\,\mathbb{P}(T_a \in \mathrm{d}s) = \frac{1}{2}\mathbb{P}(T_a \leq t).\end{aligned}$$

We deduce (4.5.17). □

C – Time to reach the bounds of an interval

Now consider $a > 0$ and $b > 0$ and let $T = T_a \wedge T_{-b}$. According to (4.5.14), we have $\mathbb{P}(T < +\infty) = 1$. Let us study more precisely the exit time from the interval $]-b, a[$.

Proposition 4.5.24 *For a Brownian motion starting from 0, the probabilities of exiting the interval $]-b, a[$ through either bound are given by*

$$\mathbb{P}(T_a < T_{-b}) = \frac{b}{a+b}; \qquad \mathbb{P}(T_{-b} < T_a) = \frac{a}{a+b}. \qquad (4.5.18)$$

Furthermore,

$$\mathbb{E}(T_a \wedge T_{-b}) = ab. \qquad (4.5.19)$$

The larger b is, the more likely the Brownian motion is to exit through a from the interval $]-b,a[$.

Proof Since $\mathbb{P}(T < +\infty) = 1$, we know that

$$\mathbb{P}(T_a < T_{-b}) + \mathbb{P}(T_{-b} < T_a) = 1.$$

Moreover, the stopped martingale $M_t = B_{T\wedge t}$ is bounded by $\sup(a,b)$, and we can apply (4.5.10) to time T. This implies that

$$0 = \mathbb{E}(M_0) = \mathbb{E}(M_T) = a\,\mathbb{P}(T = T_a) - b\,\mathbb{P}(T = T_{-b}).$$

But $\{T = T_a\} = \{T_a < T_{-b}\}$ and $\{T = T_{-b}\} = \{T_{-b} < T_a\}$ so that finally, we can compute the probabilities of exiting through one or the other bound of the interval $]-b,a[$ and obtain (4.5.18). To show (4.5.19), we will use a similar argument using the martingale $M_t = B_t^2 - t$. Indeed, $\mathbb{E}(M_{T\wedge t}) = 0$ for all $t \geq 0$. Moreover we have

$$\mathbb{E}(M_{T\wedge t}) = a^2\,\mathbb{P}(B_T = a, T \leq t) + b^2\,\mathbb{P}(B_T = -b, T \leq t) + \mathbb{E}(B_t^2 \mathbb{1}_{T>t}) - \mathbb{E}(T \wedge t).$$

By dominated convergence and monotone convergence as t tends to infinity, we deduce that

$$\mathbb{E}(M_T) = 0 = a^2\,\mathbb{P}(B_T = a) + b^2\,\mathbb{P}(B_T = -b) - \mathbb{E}(T),$$

from which the result follows. □

D – Applications to barrier hitting times

As in the case of random walks, we will be interested in absorbing or reflecting barriers.

Absorbing barrier at 0

Imagine a plankton particle originating from $x > 0$ (the position 0 is the sea floor) and diffusing vertically according to a Brownian motion. We assume that the surface is at a height a from the bottom, and initially we will assume that $a = +\infty$, that is, the depth is sufficiently large compared to the size of the cell for its movement to be considered possible over all $\mathbb{R}_+$. The sea floor is covered with predators, and the particle is absorbed as soon as it touches the floor (as in the discrete case studied previously). Before being absorbed, the particle's position at time t is a Brownian motion starting from x, defined by $X_t = x + B_t$. Let us find the average capture time of a particle originating from a depth $x > 0$. This time is given by $m(x) = \mathbb{E}(T_0 | X_0 = x)$, where $T_0 = \inf\{t : X_t = 0\}$.

By adapting the proof of (4.5.15), we immediately obtain that $m(x) = +\infty$.

Now suppose that the sea surface is at height $a > 0$ and that the particle is also absorbed at the surface, if it reaches it (oil slick, pollution, whale). This cell will then move according to a Brownian motion originating from $x \in [0, a]$, absorbed at 0 and a. Thus, in this case,

$$m(x) = \mathbb{E}(T_0 \wedge T_a | X_0 = x) = \mathbb{E}_x(T_0 \wedge T_a).$$

We can then apply the previous reasoning by replacing b with $-x$ and a with $a-x$. We find that

$$\mathbb{P}_x(T_a < T_0) = \frac{x}{a};$$
$$\mathbb{P}_x(T_0 < T_a) = 1 - \frac{x}{a},$$

and then that $m(x) = x(a-x)$.

Reflecting barrier at a

Now suppose that the sea surface is at height a and that the plankton cell naturally reflects at the surface. It will then move according to a Brownian motion originating from x, absorbed at 0 and reflected at a. If the Brownian motion reaches a before 0, it will behave after reflection as a new Brownian motion originating from a and independent of the previous one. For this second Brownian motion, the probability of reaching 0 starting from a will then be the same as the probability of reaching $2a$ starting from a for the first one (strong Markov property). We will adapt the computations of the previous case by replacing a with $2a$. Thus, in this case,

$$m(x) = \mathbb{E}(T_0 \wedge T_{2a} | X_0 = x) = x(2a - x),$$

where $0 < x < a$.

We have therefore shown the following results.

Proposition 4.5.25

(i) *The expected exit time from* $]0, a[$ *by a Brownian motion originating from* x *is given by*
$$\mathbb{E}_x(T_0 \wedge T_a) = x(a-x).$$

(ii) *The expected first hitting time of* 0 *by a Brownian motion originating from* x *and reflected at* a *is given by*
$$\mathbb{E}_x(T_0 \wedge T_{2a}) = x(2a-x).$$

4.6 Stochastic Integrals and Stochastic Differential Equations

We saw in Remark 4.1.2 2) that certain normalizations of random walks can lead to processes of the type $t \longrightarrow b\,t + \sigma B_t$, the sum of a deterministic function and a random process. If $b \neq 0$, the first term describes the central tendency and the second term is a random perturbation. This process can model the dynamics of a particle whose speed is b on average, with random fluctuations of variance $\sigma^2 t$. In this section, we will generalize this type of process.

Suppose, for example, that the speed of movement of a plankton particle depends on the heterogeneity of the water column in which it evolves. For example, the water temperature or pollution can influence the speed (the plankton can move faster in warm waters and slow down if the temperature drops). This heterogeneity will be reflected in the fact that the coefficients b and σ will depend on the spatial position where the plankton is located. We will develop this idea.

Usually, when random fluctuations are nonexistent or extremely small, the description of the movement of a particle is described by the solution of an ordinary differential equation, of the type

$$\mathrm{d}x_t = b(x_t)\,\mathrm{d}t,$$

where the function $b(x)$ models the spatial dependence of the speed. Suppose that the movement is disturbed by a random component. As we have seen above, we can model the randomness of the fluctuations of a particle during an infinitesimal time $\mathrm{d}t$ by a random variable of centered normal law of variance $\sigma^2\,\mathrm{d}t$, which we can think of as "$\sigma\,\mathrm{d}B_t$" (since B_t has variance t). If we further assume that the standard deviation of fluctuations depends on the position of the particle at time t, the random movement of the particle will be a random function $t \longrightarrow X_t$, a solution of an equation that could have the form

$$\mathrm{d}X_t = b(X_t)\,\mathrm{d}t + \sigma(X_t)\,\mathrm{d}B_t. \tag{4.6.1}$$

An equation of this type is called a *stochastic differential equation*. One might think that this is a generalization of an ordinary differential equation but in fact, it cannot have the same meaning, since Brownian trajectories are nowhere differentiable (cf. Proposition 4.3.6).

The only meaning given to this equation will be an integral one, defined as follows: if we assume that the initial condition is a random variable X_0, a solution of (4.6.1) is by definition a solution of the integral equation

$$X_t = X_0 + \int_0^t b(X_s)\,\mathrm{d}s + \int_0^t \sigma(X_s)\,\mathrm{d}B_s.$$

A major problem is therefore to give meaning to the integral $\displaystyle\int_0^t \sigma(X_s)\,\mathrm{d}B_s$, which will be an integral of a different nature than a usual integral. This integral, called a stochastic integral, was constructed by Itô in 1944.

4.6.1 Stochastic Integrals

We wish to give meaning to the integral $\int_0^t H_s \, \mathrm{d}B_s$ for B a Brownian motion and $H = (H_t)_{t \geq 0}$ a process almost surely with continuous and bounded trajectories (these assumptions could be somewhat relaxed). Such an integral will be called a *stochastic integral*.

As usual, the natural approach is to obtain $\int_0^t H_s \, \mathrm{d}B_s$ as the limit of discrete sums

$$I(H,n)_t = \sum_{i=1}^{[nt]} H_{(i-1)/n}(B_{i/n} - B_{(i-1)/n}). \tag{4.6.2}$$

Heuristically, the variable $B_{i/n} - B_{(i-1)/n}$ is of order $1/\sqrt{n}$, as it is centered and of variance $1/n$, and therefore $H_{(i-1)/n}(B_{i/n} - B_{(i-1)/n})$ is also of order $1/\sqrt{n}$. The variable $I(H,n)_t$ is therefore of order $\sqrt{n}$. This heuristic reasoning indicates that the sequence of random variables $(I(H,n)_t)_n$ will generally not converge, at least almost surely. (This is consistent with the fact that the trajectories of B are of infinite variation.)

However, a sort of "miracle" occurs, if we assume further that the process H is adapted to the filtration $(\mathcal{F}_t)$ of the Brownian motion when we compute the $\mathbb{L}^2$-norm of $I(H,n)_t$.

As H is adapted to the filtration $(\mathcal{F}_t)$, $H_{(i-1)/n}$ is $\mathcal{F}_{(i-1)/n}$-measurable and the random variable

$$Y(n,i) = H_{(i-1)/n}(B_{i/n} - B_{(i-1)/n})$$

satisfies

$$\mathbb{E}(Y(n,i)|\mathcal{F}_{(i-1)/n}) = H_{(i-1)/n}\mathbb{E}(B_{i/n} - B_{(i-1)/n}|\mathcal{F}_{(i-1)/n}) = 0,$$

since $B_{i/n} - B_{(i-1)/n}$ is independent of $\mathcal{F}_{(i-1)/n}$ and centered. Assume now that the process H under consideration is bounded by a constant C. By a similar argument, the conditional second-order moment of $Y(n,i)$ satisfies

$$\mathbb{E}(Y(n,i)^2|\mathcal{F}_{(i-1)/n}) = H^2_{(i-1)/n}\frac{1}{n} \leq \frac{C^2}{n},$$

since $\mathbb{E}((B_{i/n} - B_{(i-1)/n})^2) = 1/n$. Furthermore, for $i < j$,

$$\begin{aligned}
&\mathbb{E}(Y(n,i)Y(n,j)) \\
&\quad = \mathbb{E}\Big(\mathbb{E}(H_{(i-1)/n}(B_{i/n} - B_{(i-1)/n})H_{(j-1)/n}(B_{j/n} - B_{(j-1)/n})\,|\,\mathcal{F}_{\frac{j-1}{n}})\Big) \\
&\quad = \mathbb{E}\Big(H_{(i-1)/n}(B_{i/n} - B_{(i-1)/n})H_{(j-1)/n})\mathbb{E}(B_{j/n} - B_{(j-1)/n})\Big) = 0,
\end{aligned} \tag{4.6.3}$$

by the independence of the increments of the Brownian motion. Thus in this case, the cross terms when computing the $\mathbb{L}^2$-norm of $I(H,n)_t$, defined in (4.6.2), cancel out. This kind of reasoning has already appeared in the proof of Proposition 4.3.5, when $H = 1$.

We deduce that $I(H,n)_t = \sum_{i=1}^{[nt]} Y(n,i)$ is a centered and square integrable random variable with variance

$$\sum_{i=1}^{[nt]} \mathbb{E}(Y(n,i)^2) \leq C^2 t.$$

We can show much more, namely the following theorem.

Theorem 4.6.1 *Let H be a bounded, continuous process adapted to the filtration of a Brownian motion B. Then, the sequence*

$$I(H,n)_t = \sum_{i=1}^{[nt]} H_{(i-1)/n}(B_{i/n} - B_{(i-1)/n})$$

converges in $\mathbb{L}^2$*, as n tends to infinity, towards a limit, denoted by*

$$\int_0^t H_s \, \mathrm{d}B_s.$$

Definition 4.6.2 The square-integrable random variable $\int_0^t H_s \, \mathrm{d}B_s$ is called the *stochastic integral of H with respect to B over the interval* $[0,t]$.

Remark 4.6.3 1) It can be shown that for any increasing partition $(t_n)_n$ of $[0,t]$ with step tending to 0, an approximation similar to (4.6.2) obtained by replacing $\frac{i}{n}$ by t_n also converges in $\mathbb{L}^2$ towards the same limit. This is where Theorem 4.6.1 takes its full strength.

2) The convergence of the Riemann sums stated in Theorem 4.6.1 is related to the adaptation of the process (H_t). If $H_{(i-1)/n}(B_{i/n} - B_{(i-1)/n})$ is replaced by $H_{t(n,i)}(B_{i/n} - B_{(i-1)/n})$, with $(i-1)/n < t(n,i) \leq i/n$, as it is possible to do for the approximations by Riemann sums for usual integrals, the associated sequence $I(H,n)_t$ does not necessarily converge, and if it does converge, then the limit can be different from $\int_0^t H_s \, \mathrm{d}B_s$.

3) Let us emphasize that the stochastic integral is not a usual integral, taken for each value of ω, but is defined as an $\mathbb{L}^2$ limit and therefore only makes sense almost surely. As we have seen previously, the object $\mathrm{d}B_t$ is just a notation and is in no way a measure defined ω by ω.

Proof We show that, for a fixed t, the sequence $(I(H,n)_t)_n$ is a Cauchy sequence in $\mathbb{L}^2$. Let us first show that for a fixed $m \in \mathbb{N}^*$, the sequence $\mathbb{E}\big((I(H,n)_t - I(H,mn)_t)^2\big)$ tends to 0 as n tends to infinity. Each interval $(\frac{i-1}{n}, \frac{i}{n}]$ of amplitude $\frac{1}{n}$ is composed of m subintervals $(\frac{i-1}{n}, \frac{i-1}{n} + \frac{1}{mn}], \ldots, (\frac{i-1}{n} + \frac{m-1}{mn}, \frac{i}{n}]$. We then have

$$\begin{aligned}
I(H,n)_t &- I(H,mn)_t \\
&= \sum_{i=1}^{[nt]} H_{\frac{i-1}{n}} (B_{i/n} - B_{(i-1)/n}) - \sum_{i=1}^{[nt]} \sum_{j=1}^{m} H_{\frac{i-1}{n}+\frac{j-1}{mn}} (B_{\frac{i-1}{n}+\frac{j}{mn}} - B_{\frac{i-1}{n}+\frac{j-1}{mn}}) \\
&= \sum_{i=1}^{[nt]} \sum_{j=1}^{m} (B_{\frac{i-1}{n}+\frac{j}{mn}} - B_{\frac{i-1}{n}+\frac{j-1}{mn}}) (H_{\frac{i-1}{n}} - H_{\frac{i-1}{n}+\frac{j-1}{mn}}).
\end{aligned}$$

By the same reasoning as previously in (4.6.3) (by conditioning), it is easy to show that

$$\begin{aligned}
\mathbb{E}\big((I(H,n)_t - I(H,mn)_t)^2\big) &= \sum_{i=1}^{[nt]} \sum_{j=1}^{m} \mathbb{E}\Big((B_{\frac{i-1}{n}+\frac{j}{mn}} - B_{\frac{i-1}{n}+\frac{j-1}{mn}})^2 (H_{\frac{i-1}{n}} - H_{\frac{i-1}{n}+\frac{j-1}{mn}})^2\Big) \\
&= \frac{1}{mn} \sum_{i=1}^{[nt]} \sum_{j=1}^{m} \mathbb{E}\Big((H_{\frac{i-1}{n}} - H_{\frac{i-1}{n}+\frac{j-1}{mn}})^2\Big).
\end{aligned}$$

From the uniform continuity of H on $[0,t]$ and thanks to the dominated convergence theorem, we deduce that this quantity tends to 0 as n tends to infinity. By a similar argument for $\mathbb{E}((I(H,m)_t - I(H,mn)_t)^2)$ as n tends to infinity, we can finally deduce that the sequence $(I(H,n)_t)_n$ is a Cauchy sequence in $\mathbb{L}^2$. This implies the convergence of the sequence $(I(H,n)_t)_n$ in $\mathbb{L}^2$. □

Properties of the stochastic integral

The essential properties of the stochastic integral are deduced from the properties of $I(H,n)_t$ by passing to the limit.

Theorem 4.6.4 *Let H and K be bounded, continuous processes adapted to the filtration of a Brownian motion B.*

1. *For all real numbers α and β, $\int_0^t (\alpha H_s + \beta K_s)\,\mathrm{d}B_s = \alpha \int_0^t H_s\,\mathrm{d}B_s + \beta \int_0^t K_s\,\mathrm{d}B_s$.*
2. *The process M defined by $M_t = \int_0^t H_s\,\mathrm{d}B_s$ is a continuous square-integrable martingale, starting at 0, and therefore of zero expectation.*
3. *The process $(M_t^2 - \int_0^t H_s\,\mathrm{d}s)_t$ is a continuous martingale and the following fundamental isometry is satisfied:*

$$\mathbb{E}(M_t^2) = \mathbb{E}\Big(\Big(\int_0^t H_s\,\mathrm{d}B_s\Big)^2\Big) = \mathbb{E}\Big(\int_0^t H_s^2\,\mathrm{d}s\Big). \tag{4.6.4}$$

If in addition $N_t = \int_0^t K_s\,\mathrm{d}B_s$, the process MN defined by

$$M_t N_t \;-\; \int_0^t H_s K_s\,\mathrm{d}s$$

is a continuous martingale starting at 0. *It follows that*

$$\mathbb{E}(M_t N_t) = \mathbb{E}\Big(\int_0^t H_s K_s \,\mathrm{d}s\Big). \tag{4.6.5}$$

The isometry formula (4.6.4) is fundamental since it reduces the computation of the second moment of a stochastic integral to a classical integral computation.

Proof 1. This is obvious by passing to the limit.

2. By construction, M is adapted and square integrable. If we take the partition $t_i = s + \frac{(t-s)i}{n}$ of $[s,t]$, we also have $\int_s^t H_u \,\mathrm{d}B_u = \lim_n \sum_i H_{t_{i-1}}(B_{t_i} - B_{t_{i-1}})$, where the limit is taken in $\mathbb{L}^2$. Note that this convergence implies the convergence of moments of order 1 and order 2, which will be useful below. Moreover,

$$\begin{aligned}\mathbb{E}\Big(\int_s^t H_u \,\mathrm{d}B_u|\mathcal{F}_s\Big) &= \lim_n \sum_i \mathbb{E}\left(H_{t_{i-1}}(B_{t_i} - B_{t_{i-1}})|\mathcal{F}_s\right)\\ &= \lim_n \sum_i \mathbb{E}\left(H_{t_{i-1}}\mathbb{E}(B_{t_i} - B_{t_{i-1}}|\mathcal{F}_{t_{i-1}})|\mathcal{F}_s\right) = 0.\end{aligned}$$

We deduce from this that M is a martingale. Note that $I(H,n)_t$ can still be written as

$$I(H,n)_t = \sum_{i=1}^{[nt]} H_{(i-1)/n}(B_{t\wedge i/n} - B_{t\wedge(i-1)/n})$$

and is a continuous-time martingale. We can apply Doob's inequality (4.5.5) to the martingale $M_t - I(H,n)_t$. We then have for $\varepsilon > 0$

$$\mathbb{P}(\sup_{t\in[0,T]} |M_t - I(H,n)_t| > \varepsilon) \le \frac{4}{\varepsilon^2}\mathbb{E}(|M_T - I(H,n)_T|^2),$$

which tends to 0 as n tends to infinity. Thus, the sequence $I(H,n)$ converges in probability to M in the set of continuous functions on $[0,T]$ equipped with the uniform norm. We can then extract a subsequence such that $I(H,n_k)$ converges almost surely to M, for the topology of uniform convergence. We deduce in particular that M is almost surely continuous on $[0,T]$.

3. Let us show that $M_t^2 - \int_0^t H_s \,\mathrm{d}s$ is a martingale. As M is a martingale, we have for $s < t$,

$$\mathbb{E}(M_t^2 - M_s^2|\mathcal{F}_s) = \mathbb{E}((M_t - M_s)^2|\mathcal{F}_s)$$

and it is enough to check that

$$\mathbb{E}((M_t - M_s)^2|\mathcal{F}_s) = \mathbb{E}\Big(\Big(\int_s^t H_u \,\mathrm{d}B_u\Big)^2|\mathcal{F}_s\Big) = \mathbb{E}\Big(\int_s^t H_u^2 \,\mathrm{d}u|\mathcal{F}_s\Big).$$

By arguments similar to those invoked for (4.6.3), we have

$$\begin{aligned}
&\mathbb{E}\Big(\Big(\int_s^t H_u \,\mathrm{d}B_u\Big)^2 |\mathcal{F}_s\Big)\\
&= \mathbb{E}\Big(\lim_n \Big(\sum_i H_{t_{i-1}}(B_{t_i}-B_{t_{i-1}})\Big)^2 |\mathcal{F}_s\Big) = \lim_n \mathbb{E}\Big(\Big(\sum_i H_{t_{i-1}}(B_{t_i}-B_{t_{i-1}})\Big)^2 |\mathcal{F}_s\Big)\\
&= \lim_n \Big[\sum_i \mathbb{E}\Big(\mathbb{E}(H^2_{t_{i-1}}(B_{t_i}-B_{t_{i-1}})^2 |\mathcal{F}_{t_{i-1}})|\mathcal{F}_s\Big)\\
&\qquad +2\sum_{i<j}\mathbb{E}\Big(\mathbb{E}(H_{t_{i-1}}H_{t_{j-1}}(B_{t_i}-B_{t_{i-1}})(B_{t_j}-B_{t_{j-1}})\,|\mathcal{F}_{t_{j-1}})|\mathcal{F}_s\Big)\Big]\\
&= \lim_n \sum_i \mathbb{E}\Big(H^2_{t_{i-1}}(t_i-t_{i-1})\,|\mathcal{F}_s\Big) = \mathbb{E}\Big(\int_s^t H_u^2\, du|\mathcal{F}_s\Big).
\end{aligned}$$

The limits are justified by the convergence of partial sums in $\mathbb{L}^2$ and the last limit is a convergence of Riemann sums.

The last assertion and Formula (4.6.5) can be easily deduced by studying $(M+N)^2$ and $(M-N)^2$. □

Remark 4.6.5 The stochastic integral can be generalized over $[0,T]$ to any integrand $(H_t, t\in[0,T])$ adapted and continuous such that $\mathbb{E}\Big(\int_0^T H_s^2\,\mathrm{d}s\Big) < +\infty$. Moreover, the isometry (4.6.4) is preserved.

The proof of this generalization is more delicate and will not be developed here.

4.6.2 Stochastic Differential Equations (SDEs)

We are now able to define what a solution of a stochastic differential equation is.

Definition 4.6.6 On a probability space $(\Omega,\mathcal{F},\mathbb{P})$, consider a Brownian motion $(B_t, t\geq 0)$ and a real random variable X_0 independent of B. Let $(\mathcal{F}_t)_{t\geq 0}$ be the filtration generated by the Brownian motion B and by X_0. Also consider measurable functions b and σ, defined on $\mathbb{R}$. A *solution of the stochastic differential equation (SDE)*

$$\mathrm{d}X_t = b(X_t)\,\mathrm{d}t + \sigma(X_t)\,\mathrm{d}B_t;\ \ X_0 \tag{4.6.6}$$

is any process $(X_t, t\geq 0)$ continuous, adapted to the filtration $(\mathcal{F}_t)_{t\geq 0}$ and satisfying for all $t\geq 0$,

$$X_t = X_0 + \int_0^t b(X_s)\,\mathrm{d}s + \int_0^t \sigma(X_s)\,\mathrm{d}B_s. \tag{4.6.7}$$

Note that the adaptation of X is necessary for the stochastic integral in (4.6.7) to make sense.

We will look for sufficient conditions for the existence and uniqueness of a solution. In the purely deterministic case of an ordinary differential equation, of the form

$$\mathrm{d}X_t = b(X_t)\,\mathrm{d}t, \qquad X_0 = x_0, \tag{4.6.8}$$

where the function b and the initial condition x_0 are given, a classical result (the Cauchy–Lipschitz Theorem) states that (4.6.8) admits one and only one solution as soon as b is Lipschitz continuous. For the stochastic differential equation (4.6.6), we can prove a similar result in the set of adapted processes, almost surely with continuous trajectories and square integrable uniformly on $[0,T]$:

$$\mathbb{L}_T^2 = \Big\{X \text{ adapted continuous process with real values}, \mathbb{E}\Big(\sup_{t\le T}|X_t|^2\Big) < \infty\Big\}.$$

Theorem 4.6.7 *Suppose that the coefficients b and σ are Lipschitz continuous and that the initial condition X_0 belongs to $\mathbb{L}^2$. Then for all $T > 0$, there exists a unique (up to indistinguishability) solution $(X_t, t \in [0,T])$ in $\mathbb{L}_T^2$ to the stochastic differential equation* (4.6.7)

$$X_t = X_0 + \int_0^t b(X_s)\,\mathrm{d}s + \int_0^t \sigma(X_s)\,\mathrm{d}B_s.$$

Definition 4.6.8 The process X is often called a *diffusion process*. The functions $\sigma(x)$ and $b(x)$ are called, respectively, the *diffusion coefficient* and *drift coefficient* of X.

Proof The argument of the proof essentially consists of a fixed point theorem in $\mathbb{L}_T^2$. Let us denote by K_b and K_σ the Lipschitz constants of b and σ. Note that in this case, both functions b and σ are sublinear, in the sense that there exists a constant $\tilde{K}_b > 0$ such that $|b(x)|^2 \le \tilde{K}_b(1+|x|^2)$ and the same goes for σ.

(i) Uniqueness. Suppose there exist two solutions X and Y of (4.6.7) belonging to $\mathbb{L}_T^2$. Note that in this case, $\mathbb{E}(\sup_{t\le T}|b(X_t)|^2) < +\infty$, and the same will be true when replacing X with Y or b with σ. We then have

$$\begin{aligned}
\mathbb{E}(|X_t - Y_t|^2) &\le 2\mathbb{E}\Big(\Big(\int_0^t (b(X_s) - b(Y_s))\,\mathrm{d}s\Big)^2 + \Big(\int_0^t (\sigma(X_s) - \sigma(Y_s))\,\mathrm{d}B_s\Big)^2\Big)\\
&\le 2\Big(T\int_0^t \mathbb{E}(|b(X_s) - b(Y_s)|^2)\,\mathrm{d}s + \int_0^t \mathbb{E}(|\sigma(X_s) - \sigma(Y_s)|^2)\,\mathrm{d}s\Big)\\
&\le 2(TK_b^2 + K_\sigma^2)\int_0^t \mathbb{E}(|X_s - Y_s|^2)\,\mathrm{d}s.
\end{aligned}$$

We used the Cauchy–Schwarz inequality for the first term of the inequality and the isometry (4.6.4) for the second term, then the Lipschitz continuity of σ and b.

The function $s \mapsto \mathbb{E}(|X_s - Y_s|^2)$ is measurable and bounded by hypothesis because $X - Y$ belongs to $\mathbb{L}_T^2$. We can then apply Gronwall's Lemma to this function (see Lemma 4.6.10 below) and thus deduce that for all t, $X_t = Y_t$ almost surely. Thus for all rational t there exists a negligible set N_t such that if $\omega \notin N_t$, $X_t(\omega) = Y_t(\omega)$. But then, $N = \cup_{t \in \mathbb{Q} \cap [0,T]} N_t$ is a negligible set as a countable union of negligible sets. Moreover, for all $\omega \notin N$, for all $t \in \mathbb{Q} \cap [0,T]$, $X_t(\omega) = Y_t(\omega)$. The continuity of X and Y allows us to conclude that the trajectories of X and Y coincide almost surely and that these two processes are therefore indistinguishable.

Note that another proof of this result uses Hölder's inequality for the drift term and Doob's inequality (4.5.5), applied to the martingale $\int_0^t \sigma(X_s)\,\mathrm{d}B_s$ with $p = 2$. Indeed, for all $u \leq T$,

$$\begin{aligned}
&\mathbb{E}\Big(\sup_{t\leq u}|X_t - Y_t|^2\Big)\\
&\quad\leq 2\mathbb{E}\Big(\sup_{t\leq u}\Big(\int_0^t (b(X_s)-b(Y_s))\,\mathrm{d}s\Big)^2 + \sup_{t\leq u}\Big(\int_0^t (\sigma(X_s)-\sigma(Y_s))\,\mathrm{d}B_s\Big)^2\Big)\\
&\quad\leq 2\Big(T\int_0^u \mathbb{E}(|b(X_s)-b(Y_s)|^2)\,\mathrm{d}s + 4\int_0^u \mathbb{E}(|\sigma(X_s)-\sigma(Y_s)|^2)\,\mathrm{d}s\Big)\\
&\quad\leq 2\Big(T\int_0^u \mathbb{E}(\sup_{v\leq s}|b(X_v)-b(Y_v)|^2)\,\mathrm{d}s + 4\int_0^u \mathbb{E}(\sup_{v\leq s}|\sigma(X_v)-\sigma(Y_v)|^2)\,\mathrm{d}s\Big)\\
&\quad\leq 2(TK_b^2 + 4K_\sigma^2)\int_0^u \mathbb{E}(\sup_{v\leq s}|X_v - Y_v|^2)\,\mathrm{d}s,
\end{aligned}$$

which also gives the result since in this case we immediately obtain from Gronwall's Lemma that the processes X and Y are indistinguishable on $[0,T]$.

(ii) Existence. The idea is to iteratively define the sequence of processes $(X^n)_n$ on $[0,T]$ by $X_t^0 = X_0$ and for all n, by

$$X_t^n = X_0 + \int_0^t b(X_s^{n-1})\,\mathrm{d}s + \int_0^t \sigma(X_s^{n-1})\,\mathrm{d}B_s.$$

Let us first show that the processes X^n belong to $\mathbb{L}_T^2$, by induction on n. Let us fix n and assume that $X^{n-1} \in \mathbb{L}_T^2$. Introducing for $k \in \mathbb{N}^*$ the stopping times

$$T_k^{(n-1)} = \inf\{t > 0, |X_t^{n-1}| \geq k\}; \quad \inf\emptyset = +\infty,$$

we have already seen that the sequence $(T_k^{(n-1)})_k$ converges almost surely to infinity.

For a fixed k, the random variables $X^{n-1}_{t\wedge T_k^{n-1}}$ are bounded, just as $b(X^{n-1}_{t\wedge T_k^{n-1}})$ and $\sigma(X^{n-1}_{t\wedge T_k^{n-1}})$, since b and σ are Lipschitz continuous. In particular,

$$\int_0^{t\wedge T_k^{n-1}} \sigma(X_s^{n-1})\,\mathrm{d}B_s$$

is a martingale.

By applying Hölder's inequality to the drift term and Doob's inequality (4.5.5) to the stochastic integral with $p = 2$, we have for all $u \leq T$,

$$\begin{aligned}
&\mathbb{E}\Big(\sup_{t\le u}|X^n_{t\wedge T^{n-1}_k}|^2\Big)\\
&\quad\le 4\mathbb{E}\Big(X_0^2+\sup_{t\le u}\Big(\int_0^{t\wedge T^{n-1}_k} b(X^{n-1}_s)\,\mathrm{d}s\Big)^2+\sup_{t\le u}\Big(\int_0^{t\wedge T^{n-1}_k}\sigma(X^{n-1}_s)\,\mathrm{d}B_s\Big)^2\Big)\\
&\quad\le 4\Big(\mathbb{E}(X_0^2)+T\int_0^u \mathbb{E}(\sup_{v\le s}|b(X^{n-1}_{v\wedge T^{n-1}_k})|^2)\,\mathrm{d}s+4\int_0^u \mathbb{E}(\sup_{v\le s}|\sigma(X^{n-1}_{v\wedge T^{n-1}_k})|^2)\,\mathrm{d}s\Big)\\
&\quad\le \widetilde{C}(b,\sigma,T)\Big(\mathbb{E}(X_0^2)+\int_0^u \mathbb{E}(\sup_{v\le s}|X^{n-1}_{v\wedge T^{n-1}_k}|^2)\,\mathrm{d}s\Big),
\end{aligned}$$

where $\widetilde{C}(b,\sigma,T)$ is a constant independent of k. By making k tend to infinity and since $\mathbb{E}(\sup_{v\le T}|X^{n-1}_v|^2)<+\infty$, we deduce from this that $\mathbb{E}\big(\sup_{t\le T}(X^n_t)^2\big)<+\infty$.

By a similar computation, we can prove that there exists a constant $C>0$ such that for all n,

$$\mathbb{E}\Big(\sup_{s\le t}|X^n_s-X^{n-1}_s|^2\Big)\le C\int_0^t E\big(\sup_{u\le s}|X^{n-1}_u-X^{n-2}_u|^2\big)\,\mathrm{d}s.$$

This inequality can be iterated, which gives

$$\mathbb{E}\Big(\sup_{t\le T}|X^n_t-X^{n-1}_t|^2\Big)\le C^*\frac{(CT)^{n-1}}{(n-1)!},$$

where $C^*=\mathbb{E}(\sup_{t\le T}|X^1_t-X_0|^2)<+\infty$. But then

$$\mathbb{P}\Big(\sup_{t\le T}|X^n_t-X^{n-1}_t|>\frac{1}{2^n}\Big)\le 4C^*\frac{(4CT)^{n-1}}{(n-1)!},$$

which is the general term of a convergent series. We deduce from the Borel–Cantelli Lemma that the series of general term $\sup_{t\le T}|X^n_t-X^{n-1}_t|$ converges almost surely, and thus, the sequence $(X^n)_n$ converges almost surely and uniformly on any bounded compact set towards a continuous process X. It is easy to prove that this process is a solution of the stochastic differential equation (4.6.6) and that it belongs to $\mathbb{L}^2_T$ for all $T>0$. □

Remark 4.6.9 1) The fact that we obtain a solution in the space $\mathbb{L}^2_T$ is not surprising: it is indeed in this framework that we have developed the stochastic calculus.

2) We have defined the solution of the SDE (4.6.6) on a fixed probability space, given a Brownian motion B and an initial condition X_0 on this space. Another (weaker) approach consists in defining a solution given a probability space, a Brownian motion, an initial condition with the same law as X_0 and a process X satisfying (4.6.7). This leads to another notion of uniqueness, called weak uniqueness, in the case where the laws of two solutions in the previous sense (called weak solutions) coincide. This notion is weaker than trajectory uniqueness. (See, for example, Revuz–Yor [71, Theorem 1.7].)

The next lemma is commonly called Gronwall's Lemma.

Lemma 4.6.10 *Let $T > 0$ and f be a positive and bounded Borel function on $[0,T]$. Suppose further that there exist two positive constants α and β such that for all $t \in [0,T]$,*

$$f(t) \le \alpha + \beta \int_0^t f(s)\,\mathrm{d}s. \tag{4.6.9}$$

Then for all $t \in [0,T]$,

$$f(t) \le \alpha\, \mathrm{e}^{\beta t}.$$

In particular, if $\alpha = 0$, the function f is identically zero.

Proof Iterating (4.6.9), we obtain

$$\begin{aligned} f(t) \le \alpha + \alpha(\beta t) + \alpha\frac{(\beta t)^2}{2} + \cdots + \alpha\frac{(\beta t)^n}{n!} \\ + \beta^{n+1}\int_0^t \mathrm{d}s_1 \int_0^{s_1} \mathrm{d}s_2 \cdots \int_0^{s_n} f(s_{n+1})\,\mathrm{d}s_{n+1}. \end{aligned}$$

If f is bounded by A, the last term is bounded by $A\frac{(\beta t)^{n+1}}{(n+1)!}$ and tends to 0 as n tends to infinity. Thus the left-hand term is bounded by α times the expansion of the exponential function, which concludes the lemma. □

There is an existence and pathwise uniqueness result for solutions of the SDE (4.6.7) which does not require the Lipschitz continuity of the diffusion coefficient σ. It is only true in this one-dimensional framework.

Theorem 4.6.11 (Yamada–Watanabe Theorem) *Let $X_0 \in \mathbb{L}^2$. We assume that the functions b and σ are measurable and satisfy*

$$|b(x) - b(y)| + |\sigma(x) - \sigma(y)|^2 \le C|x - y|. \tag{4.6.10}$$

In this case, there exists on any time interval $[0,T]$ a pathwise unique (up to indistinguishability) solution to the SDE (4.6.7) starting from X_0.

Proof We will only provide a sketch of the proof here. For the complete proof, we refer the reader to [48, Section 5.5] or to [42, chap. IV, Theorems 2.2 and 3.3].

Theorem 3.2 of [42] proves the pathwise uniqueness of a solution to this equation. Theorem 2.2 of [42] shows that as soon as the coefficients b and σ are continuous and bounded on $\mathbb{R}$, there exists a solution (X,B) to (4.6.7), obtained as the limit (in law) of discretized (in time) processes. Let us apply this result to our case. Let us introduce for all $n \in \mathbb{N}^*$ the stopping time $T_n = \inf\{t \ge 0, X_t = n\}$. The restrictions of the coefficients b and σ to the interval $[0,n]$ being bounded, it is easy to prove (by stochastic calculus arguments that we have already used) that the solution $X_{\cdot \wedge T_n}$ is well defined for all n and moreover,

$$\mathbb{E}\big(\sup_{s \le t}(X_{s\wedge T_n})^2\big) \le \widetilde{C}\left(1 + \int_0^t \mathbb{E}\big(\sup_{u \le s}(X_{u\wedge T_n})^2\big)\,\mathrm{d}s\right),$$

where $\widetilde{C}$ is a positive constant independent of n. Since the process $X_{\cdot\wedge T_n}$ is bounded by n, Gronwall's Lemma implies that $\sup_n \mathbb{E}\big(\sup_{t\leq T\wedge T_n}(X_t)^2\big) < +\infty$. By the arguments used in the proof of Theorem 4.6.7, we deduce that the sequence $(T_n)_n$ converges a.s. to infinity, implying that a solution is well defined on any time interval $[0,T]$ and that it also satisfies $\mathbb{E}\big(\sup_{t\leq T}(X_t)^2\big) < +\infty$. □

The assumption (4.6.10) allows σ to be only Hölder continuous with coefficient $1/2$. In the following, we will use this result with $\sigma(x) = \sqrt{x}$ in the case of Feller diffusion (4.7.1) or $\sigma(x) = \sqrt{x(1-x)}$ in the case of Wright–Fisher diffusion (4.7.4). These processes are classical in population dynamics and in population genetics and their justification from discrete models will be developed in Chapters 5 and 6. We will provide some properties of the Feller equation and the Wright–Fisher equation at the end of this chapter.

4.6.3 Stochastic Differential Systems

We have defined a stochastic differential equation in dimension 1, but this notion can be generalized to any dimension d, as soon as a vector $b(\cdot) = (b_i(\cdot))_{1\leq i\leq d}$, composed of d measurable functions, a matrix $\sigma(\cdot) = ((\sigma_{i,j}(\cdot)))_{1\leq i\leq d, 1\leq j\leq m}$ composed of dm measurable functions and $B = (B^j)_{1\leq j\leq m}$, an m-dimensional Brownian motion, are given. In this case, we can define the solution of a system of stochastic differential equations as follows.

Definition 4.6.12 Given an m-dimensional Brownian motion B and a random vector $X_0 \in \mathbb{R}^d$ independent of B, a *solution of the stochastic differential system*

$$\mathrm{d}X_t = \sigma(X_t)\cdot \mathrm{d}B_t + b(X_t)\,\mathrm{d}t$$

is a continuous adapted process $X = (X^i)_{1\leq i\leq d}$ with values in $\mathbb{R}^d$ such that

$$X_t = X_0 + \int_0^t \sigma(X_s)\cdot \mathrm{d}B_s + \int_0^t b(X_s)\,\mathrm{d}s,$$

in the sense that for all $1 \leq i \leq d$,

$$X_t^i = X_0^i + \sum_{j=1}^m \int_0^t \sigma_{ij}(X_s)\,\mathrm{d}B_s^j + \int_0^t b_i(X_s)\,\mathrm{d}s. \tag{4.6.11}$$

By a proof similar to that of Theorem 4.6.7, replacing the absolute value with the Euclidean norm in $\mathbb{R}^d$, it is easy to show that for Lipschitz continuous coefficients b_i and $\sigma_{i,j}$ (for all i and j) and a square integrable random vector X_0, there exists a pathwise unique solution to (4.6.11) that is square integrable uniformly on $[0,T]$.

4.6.4 The Markov Property for a Solution to an SDE

In the following, we will assume the existence and pathwise uniqueness of a solution to the stochastic differential equation (4.6.6) over all $\mathbb{R}_+$, in the sense that the solution exists and is unique on any closed bounded time interval of $\mathbb{R}_+$.

Theorem 4.6.13 *The solution X to the stochastic differential equation* (4.6.6) *is a homogeneous Markov process. It is even a strong Markov process.*

Proof Let us fix $T > 0$. We then consider the following stochastic differential equation,

$$\mathrm{d}X'_t = b(X'_t)\,\mathrm{d}t + \sigma(X'_t)\,\mathrm{d}B^T_t, \qquad X'_0 = X_T, \tag{4.6.12}$$

where $B^T_t = B_{T+t} - B_T$. The process B^T is still a Brownian motion, independent of the past before T, and therefore independent of X_T. The coefficients in the equation (4.6.12) being Lipschitz continuous, the solution X' exists and is unique.

We now define a new process X'' by $X''_t = X_t$ if $t < T$ and $X''_t = X'_{t-T}$ if $t \geq T$. Clearly, in view of (4.6.12), X'' satisfies the equation (4.6.7). Thus, by uniqueness, we obtain $X'' = X$, and $X_{T+t} = X'_t$ if $t \geq 0$.

It follows that the random variable X_{T+t} depends only on X_T and B^T, and the conditional law of X_{T+t} given $X_T = x$ is equal to the law of X_t if $X_0 = x$. We then deduce that X is a homogeneous Markov process.

By repeating the previous argument for a finite stopping time T instead of a fixed time, we also prove the strong Markov property for X. □

In the following, we will denote by $(P_t)_{t\geq 0}$ the transition semigroup of the solution X of the stochastic differential equation (4.6.6), defined for any bounded continuous function f by

$$P_t f(x) = \mathbb{E}(f(X_t)|X_0 = x). \tag{4.6.13}$$

By the Markov property, it is possible to show the semigroup property, as in the Brownian case:

$$P_{t+s} f = P_t(P_s f) = P_s(P_t f). \tag{4.6.14}$$

4.6.5 Itô's Formula

When $t \mapsto x(t)$ is a real function, continuous and of finite variation, the integration by parts formula implies that for any continuously differentiable function f, we have

$$f(x(t)) = f(x(0)) + \int_0^t f'(x(s))\,\mathrm{d}x(s). \tag{4.6.15}$$

This becomes false when the function x is replaced by a Brownian motion B. Indeed, if (4.6.15) were true, by taking $f(x) = x^2$, we would get $B_t^2 = 2\int_0^t B_s\,\mathrm{d}B_s$

and, since the process defined by the stochastic integral is still a martingale, we could deduce that B_t^2 is a martingale. However, we have seen in Proposition 4.5.3 that the process $B_t^2 - t$ is also a martingale. By difference, the "process" t should be a martingale, which is obviously false. Thus (4.6.15) is false for Brownian motion. The proper formula needs an additional term and more regularity on f.

Theorem 4.6.14 (Itô's Formula) *Consider X as the solution of the stochastic differential equation* (4.6.7). *Let f be a function of class C^2, bounded with bounded derivatives, then we have*

$$f(X_t) = f(X_0) + \int_0^t f'(X_s)b(X_s)\,\mathrm{d}s + \int_0^t f'(X_s)\sigma(X_s)\,\mathrm{d}B_s + \frac{1}{2}\int_0^t f''(X_s)\sigma^2(X_s)\,\mathrm{d}s, \qquad (4.6.16)$$

where f' is the first and f'' the second derivative of f.

Proof We shall only give an idea of the proof when $X = B$. We can already see that for $t_i = \frac{it}{n}$,

$$\begin{aligned} B_t^2 &= \lim_n \sum_{i=1}^n (B_{t_i}^2 - B_{t_{i-1}}^2) \\ &= 2\lim_n \sum_{i=1}^n B_{t_{i-1}}(B_{t_i} - B_{t_{i-1}}) + \lim_n \sum_{i=1}^n (B_{t_i} - B_{t_{i-1}})^2. \end{aligned}$$

By virtue of Theorem 4.6.1, we know that the first term on the right-hand side converges to $2\int_0^t B_s\,\mathrm{d}B_s$ and by Proposition 4.3.5, the second term converges to t. Finally, this gives

$$B_t^2 = 2\int_0^t B_s\,\mathrm{d}B_s + t,$$

and we thus find that $B_t^2 - t$ is a martingale.

Let us now consider a function $f \in C_b^2$, in the sense that f, f', f'' are continuous and bounded. A first-order expansion gives

$$f(B_t) = f(B_0) + \sum_{i=1}^n f'(B_{\theta_i})(B_{t_i} - B_{t_{i-1}}),$$

with $\theta_i \in]t_{i-1}, t_i[$. As $f'(B_{\theta_i})$ is not $\mathcal{F}_{t_{i-1}}$-measurable, we do not know how the right-hand side term converges when n tends to infinity. We are therefore led to consider a second-order Taylor expansion of $f(B_t)$. We write

$$f(B_t) = f(B_0) + \sum_{i=1}^n f'(B_{t_{i-1}})(B_{t_i} - B_{t_{i-1}}) + \frac{1}{2}\sum_{i=1}^n f''(B_{\theta_i})(B_{t_i} - B_{t_{i-1}})^2.$$

In this case, the first sum on the right-hand side converges in $\mathbb{L}^2$ to $\int_0^t f'(B_s)\,\mathrm{d}B_s$. Let us show that the second term converges to $\frac{1}{2}\int_0^t f''(B_s)\,\mathrm{d}s$. First, let us note that by convergence of Riemann sums and the dominated convergence theorem,

$$\lim_n \mathbb{E}\Big(\Big(\sum_{i=1}^n f''(B_{t_{i-1}})(t_i - t_{i-1}) - \int_0^t f''(B_s)\,\mathrm{d}s\Big)^2\Big) = 0.$$

Furthermore, we have

$$\begin{aligned}&\mathbb{E}\Big(\Big(\sum_{i=1}^n f''(B_{\theta_i})(B_{t_i} - B_{t_{i-1}})^2 - \sum_{i=1}^n f''(B_{t_{i-1}})(B_{t_i} - B_{t_{i-1}})^2\Big)^2\Big)\\ &\leq \mathbb{E}\Big(\sup_i |f''(B_{\theta_i}) - f''(B_{t_{i-1}})|^2\Big(\sum_{i=1}^n (B_{t_i} - B_{t_{i-1}})^2\Big)^2\Big).\end{aligned}$$

Since $|\theta_i - t_{i-1}| \leq \frac{t}{n}$, and $t \to B_t$ is almost surely uniformly continuous on $[0,T]$, $(\sup_i |f''(B_{\theta_i}) - f''(B_{t_{i-1}})|^2)$ tends almost surely to 0 as n tends to infinity. As f'' is continuous and bounded,

$$\sup_i |f''(B_{\theta_i}) - f''(B_{t_{i-1}})|^2\Big(\sum_{i=1}^n (B_{t_i} - B_{t_{i-1}})^2\Big)^2 \leq 2\|f''\|_\infty^2\Big(\sum_{i=1}^n (B_{t_i} - B_{t_{i-1}})^2\Big)^2,$$

and $\mathbb{E}\left(\left(\sum_{i=1}^n (B_{t_i} - B_{t_{i-1}})^2\right)^2\right)$ is bounded by $C^{te}\,t^2$ (as seen in Proposition 4.3.5). We deduce from the dominated convergence theorem that

$$\sum_{i=1}^n f''(B_{\theta_i})(B_{t_i} - B_{t_{i-1}})^2 - \sum_{i=1}^n f''(B_{t_{i-1}})(B_{t_i} - B_{t_{i-1}})^2$$

converges to 0 in $\mathbb{L}^2$.

Moreover, by now familiar reasoning (introducing the appropriate conditioning, cf. Proposition 4.3.5), we have

$$\begin{aligned}\mathbb{E}\Big(\Big(\sum_{i=1}^n f''(B_{t_{i-1}})(B_{t_i} - B_{t_{i-1}})^2 &- \sum_{i=1}^n f''(B_{t_{i-1}})(t_i - t_{i-1})\Big)^2\Big)\\ &\leq \|f''\|_\infty^2 \sum_{i=1}^n \mathbb{E}\Big(\Big((B_{t_i} - B_{t_{i-1}})^2 - (t_i - t_{i-1})\Big)^2\Big)\\ &\leq 2\,\|f''\|_\infty^2 \sum_{i=1}^n (t_i - t_{i-1})^2\\ &\leq \frac{2t^2}{n}\,\|f''\|_\infty^2,\end{aligned}$$

which tends to 0 as n tends to infinity. By combining all these remarks, we finally obtain the Itô formula for $f(B_t)$. We could easily generalize this argument to any solution X of an SDE. □

Remark 4.6.15 1) If f is of class C^2, the formula is still true as long as each term of the Itô formula makes sense. Recall that the stochastic term will be well defined as soon as $\mathbb{E}\left(\int_0^t (f'(B_s))^2\,\mathrm{d}s\right) < \infty$.

2) As X is the solution of the SDE (4.6.7), it is the sum of a martingale and a finite variation term. In this case, the process $(X_t, t \geq 0)$ is called a semi-martingale. If f is of class C^2, the process $f(X)$ keeps this semi-martingale structure.

Example 4.6.16 The Itô formula applied to $f(x) = x^2$ and to the solution X of the SDE (4.6.7) gives

$$X_t^2 = X_0^2 + 2\int_0^t X_s\,\sigma(X_s)\,\mathrm{d}B_s + \int_0^t (2X_s\,b(X_s) + \sigma^2(X_s))\,\mathrm{d}s.$$

There exists a multi-dimensional version of the Itô formula, which can be proved by analogous arguments.

Theorem 4.6.17 *Consider a solution X of* (4.6.11)*: for all $1 \leq i \leq d$,*

$$X_t^i = X_0^i + \sum_{j=1}^m \int_0^t \sigma_{i,j}(X_s)\,\mathrm{d}B_s^j + \int_0^t b_i(X_s)\,\mathrm{d}s.$$

Then for any function ϕ of class C_b^2 from $\mathbb{R}^d$ to $\mathbb{R}$, we have

$$\phi(X_t) = \phi(X_0) + \int_0^t \sum_{i=1}^d \partial_i\phi(X_s) b_i(X_s)\,\mathrm{d}s + \sum_{j=1}^m \int_0^t \sum_{i=1}^d \partial_i\phi(X_s)\sigma_{i,j}(X_s)\,\mathrm{d}B_s^j$$
$$+ \frac{1}{2}\int_0^t \sum_{i,k=1}^d (\sigma\sigma^t)_{i,k}(X_s)\partial_{ik}^2\phi(X_s)\,\mathrm{d}s.$$

Corollary 4.6.18 *If X is a solution of the one-dimensional SDE* (4.6.7) *and if $\psi(t,x)$ is a function defined on $\mathbb{R}_+ \times \mathbb{R}$ of class C_b^1 in t and C_b^2 in x, then*

$$\psi(t, X_t) = \psi(0, X_0) + \int_0^t \partial_x\psi(s, X_s)\sigma(X_s)\,\mathrm{d}B_s$$
$$+ \int_0^t \left(\partial_t\psi(s, X_s) + \partial_x\psi(s, X_s) b(X_s) + \frac{1}{2}\partial_x^2\psi(s, X_s)\sigma^2(X_s)\right)\mathrm{d}s.$$

Itô's Formula is extremely useful. We will see many applications in the rest of the book. In the following subsection we will reveal a link between the solution of a stochastic differential equation and the solution of a partial differential equation, which will be useful for making explicit computations related to these diffusion processes.

4.6.6 Generators – A Link With Partial Differential Equations

Let $(X_t, t \le T)$ be the solution of the stochastic differential equation (4.6.7). Suppose that the functions σ and b are Lipschitz continuous and bounded functions. For any bounded function f of class C^2 with bounded derivatives, the process $\int_0^t f'(X_s)\sigma(X_s)\,\mathrm{d}B_s$ is a martingale starting at zero, since the integrand is adapted, continuous and bounded. Taking the expectation of each term in (4.6.16), we obtain

$$\mathbb{E}(f(X_t)) = \mathbb{E}(f(X_0)) + \int_0^t \mathbb{E}\Big(f'(X_s)b(X_s) + \frac{1}{2}\sigma^2(X_s)f''(X_s)\Big)\,\mathrm{d}s.$$

Therefore, the family of operators (P_t) defined in (4.6.13) satisfies for any x and t,

$$P_t f(x) = f(x) + \int_0^t P_s\Big(f'b + \frac{1}{2}\sigma^2 f''\Big)(x)\,\mathrm{d}s.$$

We deduce from the dominated convergence theorem the differentiability at 0 of the function $t \mapsto P_t f(x)$ for each f in C_b^2 and $x \in \mathbb{R}$. More generally, we can define the infinitesimal generator A of $(X_t, t \le T)$ as follows:

Definition 4.6.19 The *infinitesimal generator* of the Markov process $(X_t, t \le T)$ is defined on its domain by

$$Af(x) = \lim_{t \to 0, t > 0} \frac{P_t f(x) - f(x)}{t},$$

under simple bounded convergence. In particular, the functions $f \in C_b^2$ belong to the domain of A, and

$$Af(x) = b(x)f'(x) + \frac{1}{2}\sigma^2(x)f''(x). \tag{4.6.17}$$

Thus, for all $x \in \mathbb{R}$, we can write

$$P_t f(x) = f(x) + \int_0^t P_s(Af)(x)\,\mathrm{d}s,$$

which leads to the Kolmogorov forward equation:

$$\frac{\partial}{\partial t} P_t f(x) = P_t(Af)(x).$$

Furthermore, if the law of X_t starting from x admits a density $p_t(x, y)$, we then have $P_t f(x) = \int_{\mathbb{R}} f(y)\,p_t(x, y)\,\mathrm{d}y$ and for a C^∞-function f with compact support,

$$
\begin{aligned}
&\int_{\mathbb{R}} f(y)p_t(x,y)\,\mathrm{d}y \\
&\quad = f(x) + \int_0^t \int_{\mathbb{R}} \left(f'(y)b(y) + \frac{1}{2}\sigma^2(y)f''(y) \right) p_s(x,y)\,\mathrm{d}y\mathrm{d}s, \\
&\quad = f(x) + \int_{\mathbb{R}} f(y) \int_0^t \left(-\frac{\partial}{\partial y}(b(y)p_s(x,y)) + \frac{1}{2}\frac{\partial^2}{\partial y^2}(\sigma^2(y)p_s(x,y)) \right) \mathrm{d}s\mathrm{d}y,
\end{aligned}
$$

using integration by parts and Fubini's Theorem. We deduce that for all x, $(p_t(x,\cdot))$ is a (weak) solution of the partial differential equation

$$
\frac{\partial}{\partial t}(p_t(x,y)) = -\frac{\partial}{\partial y}(b(y)p_t(x,y)) + \frac{1}{2}\frac{\partial^2}{\partial y^2}(\sigma^2(y)p_t(x,y)). \tag{4.6.18}
$$

This equation is called the *Fokker–Planck equation*.

If the coefficients σ and b are smooth and the density $p_t(x,y)$ is a smooth function of (t,y), this density is a strong solution of the equation (4.6.18).

Differentiating the second equality of (4.6.14) with respect to s, at the point $s=0$, we deduce a second equation, called the *Kolmogorov backward equation*:

$$
\frac{\partial}{\partial t}P_t f(x) = A(P_t f)(x). \tag{4.6.19}
$$

In the case where the law of X_t starting from x admits a density $p_t(x,y)$ twice differentiable in x, we deduce that

$$
\begin{aligned}
P_t f(x) &= f(x) + \int_0^t A(P_s f)(x)\,\mathrm{d}s \\
&= f(x) \\
&+ \int_0^t \left\{ b(x)\,\frac{\partial}{\partial x}\left(\int p_s(x,y)f(y)\,\mathrm{d}y \right) + \frac{1}{2}\sigma^2(x)\,\frac{\partial^2}{\partial x^2}\left(\int p_s(x,y)f(y)\,\mathrm{d}y \right) \right\} \mathrm{d}s.
\end{aligned}
$$

Finally, we obtain for all y,

$$
\frac{\partial}{\partial t}(p_t(x,y)) = b(x)\,\frac{\partial}{\partial x}p_t(x,y) + \frac{1}{2}\sigma^2(x)\,\frac{\partial^2}{\partial x^2}p_t(x,y).
$$

We have thus exhibited solutions of partial differential equations from probabilistic objects. These equations model macroscopically (on average) a phenomenon whose random fluctuations are described by the underlying diffusion process.

Under certain assumptions, it is possible to prove that the law of X_t given that $X_0 = x$ admits a density. For example, we have the following

Theorem 4.6.20 *If the function b is bounded Lipschitz continuous and if σ is of class C_b^1 with $\sigma(x) \geq \sigma^* > 0$ for all x, then for all $t > 0$, there exists a function $(x,y) \mapsto p_t(x,y)$ on $\mathbb{R}\times\mathbb{R}$, such that $P_t f(x) = \int p_t(x,y)f(y)\,\mathrm{d}y$ (including the case when the initial condition is deterministic).*

Proof This result shows the regularization effect due to Brownian motion. Let us give a brief idea of the proof. Thanks to the properties of σ, the function $f(x) = \int_0^x \frac{1}{\sigma(u)}\,\mathrm{d}u$ is bijective and of class C^2. We can apply the Itô formula,

$$f(X_t) = f(X_0) + B_t + \int_0^t \left(\frac{b(X_s)}{\sigma(X_s)} - \sigma'(X_s)\right) \mathrm{d}s.$$

The process $(f(X_t))_t$ is therefore the sum of a Brownian motion and a drift term with bounded coefficient. We then use a very well-known theorem called Girsanov's Theorem (which is beyond the scope of this course, see for example [48, Chap. 3]), which shows in particular that the law of the process $f(X)$ is equivalent to that of a Brownian motion. We deduce that for all $t > 0$, the random variable $f(X_t)$ admits a density with respect to Lebesgue measure and there exists a measurable positive function h_t with integral 1 such that for any bounded continuous function g,

$$\mathbb{E}(g \circ f(X_t)) = \int_{\mathbb{R}} g(z) h_t(z)\,\mathrm{d}z.$$

The bijectivity of f completes the proof. □

Proving in a more general framework the existence of density for the marginals of the solution of a stochastic differential equation and showing its regularity is the subject of Malliavin calculus. Many results exist, particularly in the multi-dimensional case, depending on the properties of the matrix σ (see for example [66]).

4.6.7 Applications to Hitting Times of Barriers

From the point of view of applications it is interesting to know how a diffusion process behaves when it evolves in a bounded interval and in particular its behavior at the boundary of this interval. We will therefore revisit in this context the problems of hitting absorbing and reflecting barriers that we have studied for Brownian motion.

In this subsection, we consider a solution X (a diffusion process) of the stochastic differential equation (4.6.6) with Lipschitz continuous coefficients b and σ and a diffusion coefficient σ bounded below by a strictly positive number: for all x, $\sigma(x) \geq \sigma^* > 0$. We are given $a < c$, two real numbers.

Absorbing barriers

Suppose that the diffusion process X starts from $x_0 \in]a,c[$. Since the functions σ and b are Lipschitz continuous, they are bounded on $[a,c]$. As before, we denote by T_a and T_c the respective hitting times of a and c for the diffusion process X and

$$T_{a,c} = T_a \wedge T_c.$$

Note that $T_{a,c}$ is a stopping time. We introduce for $x \in [a,c]$ the functions

$$
\begin{aligned}
Q(x) &= -\int_a^x \frac{2b(z)}{\sigma^2(z)}\,\mathrm{d}z,\\
f(x) &= \int_a^x \exp\left(-\int_a^y \frac{2b(z)}{\sigma^2(z)}\,\mathrm{d}z\right)\mathrm{d}y = \int_a^x \mathrm{e}^{Q(y)}\,\mathrm{d}y
\end{aligned}
\tag{4.6.20}
$$

and

$$
\begin{aligned}
g(x) =& -\int_a^x \mathrm{e}^{Q(y)}\left(\int_a^y \frac{2\mathrm{e}^{-Q(z)}}{\sigma^2(z)}\,\mathrm{d}z\right)\mathrm{d}y\\
&+\frac{\int_a^x \mathrm{e}^{Q(y)}\,\mathrm{d}y}{\int_a^c \mathrm{e}^{Q(y)}\,\mathrm{d}y}\int_a^c \mathrm{e}^{Q(y)}\left(\int_a^y \frac{2\mathrm{e}^{-Q(z)}}{\sigma^2(z)}\,\mathrm{d}z\right)\mathrm{d}y.
\end{aligned}
$$

One can easily prove that the functions f and g are C_b^2 functions, respectively solutions of $f(a)=0$, and for all x,

$$
Af(x) = \frac{1}{2}\sigma^2(x)f''(x) + b(x)f'(x) = 0,
$$

and $g(a)=g(c)=0$ and for all x,

$$
Ag(x) = \frac{1}{2}\sigma^2(x)g''(x) + b(x)g'(x) = -1.
$$

Proposition 4.6.21 *For all* $x_0 \in]a,c[$,

$$
\mathbb{E}_{x_0}(T_{a,c}) = g(x_0) < +\infty, \qquad \mathbb{P}_{x_0}(T_a < T_c) = \frac{f(c)-f(x_0)}{f(c)}.
$$

Proof We apply Itô's formula to the function g, which gives

$$
\begin{aligned}
g(X_{t\wedge T_{a,c}}) &= g(X_0) + \int_0^{t\wedge T_{a,c}} g'(X_s)\sigma(X_s)\,\mathrm{d}B_s\\
&\quad + \int_0^{t\wedge T_{a,c}} \left(\frac{1}{2}\sigma^2(X_s)g''(X_s) + b(X_s)g'(X_s)\right)\mathrm{d}s\\
&= g(x_0) + \int_0^{t\wedge T_{a,c}} g'(X_s)\sigma(X_s)\,\mathrm{d}B_s - t\wedge T_{a,c}.
\end{aligned}
$$

The process $\int_0^{t\wedge T_{a,c}} g'(X_s)\sigma(X_s)\,\mathrm{d}B_s$ is a stopped martingale. Thus, for all $t>0$,

$$
\mathbb{E}_{x_0}(g(X_{t\wedge T_{a,c}})) = g(x_0) - \mathbb{E}_{x_0}(t\wedge T_{a,c}).
$$

We deduce that for all $t>0$,

$$
\mathbb{E}_{x_0}(t\wedge T_{a,c}) \le 2\|g\|_\infty,
$$

where $\|g\|_\infty$ is the supremum of g on $[a,c]$, then by letting t tend to infinity, that

$$\mathbb{E}_{x_0}(T_{a,c}) < +\infty.$$

In particular, $T_{a,c}$ is finite almost surely and $\mathbb{E}_{x_0}(T_{a,c}) = g(x_0)$ because $g(a) = g(c) = 0$.

We now apply Itô's formula to the function f,

$$\begin{aligned} f(X_t) &= f(X_0) + \int_0^t f'(X_s)\sigma(X_s)\,\mathrm{d}B_s + \int_0^t \left(\frac{1}{2}\sigma^2(X_s)f''(X_s) + b(X_s)f'(X_s)\right)\mathrm{d}s \\ &= f(x_0) + \int_0^t f'(X_s)\sigma(X_s)\,\mathrm{d}B_s. \end{aligned}$$

Thus $(f(X_{t\wedge T_{a,c}}), t \geq 0)$ is a bounded martingale (because $X_t \in [a,c]$) and $\mathbb{E}(f(X_t)) = \mathbb{E}(f(X_0))$. We can therefore apply the stopping theorem to the stopping time $T_a \wedge T_c$. Thus

$$\mathbb{E}_{x_0}(f(X_{T_a\wedge T_c})) = f(a)\mathbb{P}_{x_0}(T_a < T_c) + f(c)\mathbb{P}_{x_0}(T_c < T_a) = f(x_0),$$

and as we have seen above,

$$\mathbb{P}_{x_0}(T_a < T_c) + \mathbb{P}_{x_0}(T_c < T_a) = 1.$$

We thus deduce that for all $x_0 \in (a,c)$,

$$\mathbb{P}_{x_0}(T_a < T_c) = \frac{f(c) - f(x_0)}{f(c)}; \quad \mathbb{P}_{x_0}(T_c < T_a) = \frac{f(x_0) - f(a)}{f(c)}. \qquad \square$$

Note that we considered f satisfying $f(a) = 0$, but for the previous formula, the function f can be chosen up to an additional constant. For example, it may be more appropriate to choose a function f vanishing at x_0.

Reflecting barriers

We now assume that the process starts from $x \in [a,c]$, evolves like a diffusion inside the interval and is instantly reflected when it reaches the edge. As in the case of the random walk, we have the intuition that the process will evolve indefinitely in the interval, stabilizing as time tends to infinity. We want to look for the invariant law. This will be the limit of the law of X_t as t tends towards infinity. By adapting the results from Subsection 4.6.6, we can show that the law of the process reflected at a and at c is a solution to the Fokker–Planck equation (4.6.18)

$$\frac{\partial}{\partial t}(p_t(x,y)) = -\frac{\partial}{\partial y}(b(y)p_t(x,y)) + \frac{1}{2}\frac{\partial^2}{\partial y^2}(\sigma^2(y)p_t(x,y))$$

with the Neumann boundary conditions $\frac{\partial}{\partial y}p_t(x,a) = \frac{\partial}{\partial y}p_t(x,c) = 0$.

If we let t tend towards infinity, we find that the limiting density q is a solution to the ordinary differential equation

$$\frac{\partial}{\partial y}(b(y)q(y)) = \frac{1}{2}\frac{\partial^2}{\partial y^2}(\sigma^2(y)q(y)),$$

with the boundary conditions $q'(a) = q'(c) = 0$.

We can then solve this equation and obtain the invariant measure of the process.

4.7 Stochastic Differential Equations for Population Studies

4.7.1 Feller's Diffusion Equation

Feller's diffusion equation, named after the mathematician W. Feller who introduced it (see [37]), models the size of a population under the assumption that genealogies of distinct individuals are independent. This equation will be studied in more detail in Chapters 5 and 7, where we will justify it as an approximation of a discrete process that describes the size of a population.

This equation is given by

$$dZ_t = rZ_t\,dt + \sqrt{\gamma Z_t}\,dB_t; \qquad Z_0 = z_0 > 0, \tag{4.7.1}$$

where $(B_t, t \geq 0)$ is a standard Brownian motion, $r \in \mathbb{R}$ is often called the growth rate and $\gamma > 0$.

The infinitesimal terms of drift and variance are proportional to Z. By Theorem 4.6.11, we have existence and pathwise uniqueness of a solution that belongs to $\mathbb{L}^2_T$ for all $T > 0$. The pathwise uniqueness implies that the solution stays nonnegative. Indeed if it attains 0, then by uniqueness it stays equal to 0 (since the constant process equal to 0 is the solution of the equation with initial condition 0).

By introducing, as in the proof of Theorem 4.6.11, the increasing sequence of hitting times T_n, we can prove that the sequence $(T_n)_n$ converges a.s. to infinity, and that for all $t > 0$, $\mathbb{E}(Z_t) = z_0 + r\int_0^t \mathbb{E}(Z_s)\,ds$, leading to

$$\mathbb{E}(Z_t) = z_0 e^{rt}.$$

Thus, as t tends to infinity, $\mathbb{E}(Z_t)$ converges to $+\infty$ when $r > 0$ and to 0 when $r < 0$, while it remains equal to z_0 when $r = 0$.

Let us study more finely the behavior of the process Z at the boundary of the interval $(0, +\infty)$. As $z_0 > 0$, $z_0 \in (\frac{1}{n}, n)$, for n a non-zero well chosen integer. The function Q defined by (4.6.20) is here $Q(x) = -\dfrac{2rx}{\gamma}$ if $r \neq 0$ and $Q(x) = 0$ if $r = 0$ (up to a multiplicative constant). The computations developed in Subsection 4.6.7 then show that if $r \neq 0$,

$$\mathbb{P}_{z_0}(T_{\frac{1}{n}} < T_n) = \frac{e^{-\frac{2rn}{\gamma}} - e^{-\frac{2rz_0}{\gamma}}}{e^{-\frac{2rn}{\gamma}} - e^{-\frac{2r\frac{1}{n}}{\gamma}}}.$$

As n tends to infinity, we see that if $r < 0$, $\mathbb{P}_{z_0}(T_0 < \infty) = 1$. We can show the same result for the critical case $r = 0$ since if r is close to 0, an equivalent of $\mathbb{P}_{z_0}(T_{\frac{1}{n}} < T_n)$ is $\frac{n-z_0}{n-1/n}$. Thus for $r \leq 0$, the process tends to 0 almost surely. Note that when $r = 0$, the process Z tends almost surely to 0 as t tends to infinity while the expectation of Z_t is constant. (The hypotheses of the dominated convergence theorem are not satisfied.)

In the case where $r > 0$, we obtain that

$$\mathbb{P}_{z_0}(T_0 < \infty) = e^{-\frac{2rz_0}{\gamma}} = \left(e^{-\frac{2r}{\gamma}}\right)^{z_0}.$$

The last equality shows that the probability of reaching 0 starting from the initial condition z_0 is equal to the probability of reaching 0 starting from the initial condition 1, to the power z_0. This peculiarity is related to the branching property, which we will prove below. Recall that this property was introduced in the context of the study of the Galton–Watson process, Proposition 3.2.1. The branching property means that the law of the sum of two independent solutions of (4.7.1) arising from two distinct initial conditions (modeling two disjoint populations with the same dynamics) is that of a solution of (4.7.1) arising from the sum of the initial conditions. Thus the effect of the initial conditions is additive.

To include the initial condition, we will denote the process Z with initial condition z_0 by $Z^{(z_0)}$.

Proposition 4.7.1 *The process Z satisfies the branching property, i.e.*

$$Z^{(z_0+z_0')} \stackrel{d}{=} Z^{(z_0)} + Z'^{(z_0')} \qquad (z_0, z_0' \in \mathbb{R}_+),$$

where $Z^{(z_0)}$ and $Z'^{(z_0')}$ are two solutions of the Feller equation (4.7.1) *with the two initial conditions z_0 and z_0' and driven by two independent Brownian motions B and B'.*

Proof To simplify the notation, let us write $X_t = Z_t + Z_t'$, where $Z_0 = z_0$, $Z_0' = z_0'$ and Z and Z' are two independent solutions of (4.7.1). The process X satisfies

$$X_t = z_0 + z_0' + \int_0^t rX_s\,ds + \int_0^t \sqrt{\gamma Z_s}\,dB_s + \sqrt{\gamma Z_s'}\,dB_s', \qquad (4.7.2)$$

where B and B' are two independent Brownian motions.

Let us introduce the process defined by

$$\int_0^t \frac{1}{\sqrt{\gamma X_s}}(\sqrt{\gamma Z_s}\,dB_s + \sqrt{\gamma Z_s'}\,dB_s').$$

As long as X does not cancel out (the nullity of X implies the nullity of Z and Z'), the integrands are bounded and this process is a continuous martingale. To overcome the fact that X can cancel out, we introduce a Brownian motion $\widetilde{B}$ independent of B and B', constructed on an auxiliary space, and define the process

$$B''_t = \int_0^t \frac{1}{\sqrt{\gamma X_s}} \mathbb{1}_{\{X_s \neq 0\}} (\sqrt{\gamma Z_s}\, \mathrm{d}B_s + \sqrt{\gamma Z'_s}\, \mathrm{d}B'_s) + \int_0^t \mathbb{1}_{\{X_s = 0\}}\, \mathrm{d}\widetilde{B}_s,$$

which is a continuous martingale. We can easily show that

$$\mathbb{E}((B''_t)^2) = \mathbb{E}\left(\int_0^t \frac{\gamma(Z_s + Z'_s)}{\gamma X_s} \mathbb{1}_{\{X_s \neq 0\}}\, \mathrm{d}s + \int_0^t \mathbb{1}_{\{X_s = 0\}}\, \mathrm{d}s \right) = t.$$

This computation can be extended to any bounded stopping time. This implies by Theorem 4.5.15 that $(B''_t)^2 - t$ defines a martingale and Proposition 4.5.4 implies that B'' is a Brownian motion. Equation (4.7.2) can then be written as

$$X_t = z_0 + z'_0 + \int_0^t rX_s\, \mathrm{d}s + \int_0^t \sqrt{\gamma X_s} \mathrm{d}B''_s.$$

The process X is therefore a solution to (4.7.1) with the initial condition $z_0 + z'_0$. □

The branching property implies in particular that for all $\lambda > 0$,

$$\mathbb{E}_{z_0 + z'_0}\Big(\exp(-\lambda Z_t) \Big) = \mathbb{E}_{z_0}\Big(\exp(-\lambda Z_t) \Big) \mathbb{E}_{z'_0}\Big(\exp(-\lambda Z_t) \Big).$$

The function $g_t(z) = \mathbb{E}_z\Big(\exp(-\lambda Z_t) \Big)$ therefore satisfies $g_t(z + z') = g_t(z) g_t(z')$. Moreover, the function g_t is decreasing in z. Indeed, if two solutions of (4.7.1) are constructed on the same probability space with $z \le z'$, then almost surely for all t, $Z_t^{(z)} \le Z_t^{(z')}$. To prove this, suppose the contrary. This implies that if T is the first time when the two trajectories meet, necessarily $\mathbb{P}(T < +\infty) > 0$. But then, the strong Markov property and the strong uniqueness of solutions imply that the two solutions coincide after the stopping time T. We therefore still have the inequality.

We finally deduce that there exists a positive real number $u_t(\lambda)$, called the *Laplace exponent* of Z, such that

$$g_t(z) = \mathrm{e}^{-u_t(\lambda) z}.$$

Let us apply (4.6.19) to $z \to f_\lambda(z) = \exp(-\lambda z)$ and for $z = 1$: $\frac{\partial}{\partial t} P_t f_\lambda(1) = A P_t f_\lambda(1)$, where A is the infinitesimal generator of Z, defined by $Af(x) = \frac{\gamma}{2} x f''(x) + r x f'(x)$ (see (4.6.17)).

Since $P_t f_\lambda(z) = g_t(z) = \exp(-z u_t(\lambda))$, this gives

$$-\frac{\partial u_t(\lambda)}{\partial t} \exp(-u_t(\lambda)) = (r u_t(\lambda) - \frac{\gamma}{2} (u_t(\lambda))^2) \exp(-u_t(\lambda)).$$

We deduce that the Laplace exponent $u_t(\lambda)$ is a solution of

$$\frac{\partial}{\partial t}u_t(\lambda) = ru_t(\lambda) - \frac{\gamma}{2}(u_t(\lambda))^2, \qquad u_0(\lambda) = \lambda.$$

The (elementary) solution of the equation gives that for all t,

$$u_t(\lambda) = \frac{2r\lambda e^{rt}}{2r - \gamma\lambda(1 - e^{rt})}.$$

This implies in particular that

$$\mathbb{P}_1(T_0 < t) = \lim_{\lambda\to\infty} \mathbb{E}_1\Big(\exp(-\lambda Z_t)\Big) = \lim_{\lambda\to\infty} \exp(-u_t(\lambda)) = \exp\left(-\frac{2re^{rt}}{\gamma(e^{rt} - 1)}\right),$$

which characterizes the law of T_0. By letting t tend to infinity, we can find the extinction probability: if $r \le 0$, $\mathbb{P}_1(T_0 < \infty) = 1$ and if $r > 0$, $\mathbb{P}_1(T_0 < \infty) = e^{-\frac{2r}{\gamma}}$.

4.7.2 The Logistic Feller Equation

If we wish to take into account population regulation effects due to ecological interactions, the Feller equation is not suitable. For that purpose, we will consider the following stochastic differential equation, called the *logistic Feller equation*,

$$dY_t = (rY_t - cY_t^2)\,dt + \sqrt{\gamma Y_t}\,dB_t, \tag{4.7.3}$$

with $r \in \mathbb{R}$, $c > 0$, $\gamma > 0$ and $Y_0 > 0$ a.s. A process solution to Equation (4.7.3) can be obtained by constructing the limit of normalized birth and death processes in the case where competition between individuals is modeled through the quadratic term and the competition parameter c (see Chapter 5, Section 5.6.2).

We can prove that a solution to this equation does not explode by comparing it to a solution of the Feller equation (4.7.1). This is based on the following comparison lemma.

Lemma 4.7.2 *Let b_1 and b_2 be two bounded Borel functions such that $b_1 \ge b_2$ and b_1 is Lipschitz continuous. Consider solutions X^1 and X^2 of the SDE* (4.6.6) *with σ and b_1 (resp. b_2) defined on the same probability space with the same Brownian motion.*

Then, if we assume that $X_0^1 \ge X_0^2$, we will have

$$\mathbb{P}(X_t^1 \ge X_t^2, \text{ for all } t) = 1.$$

The proof of this lemma uses stochastic calculus tools that we have not covered in this course. We refer, for example, to Theorem 3.7 in [42].

We can deduce that a solution to (4.7.3) will almost surely be smaller than the solution to the Feller equation ($c = 0$) starting from the same initial condition and driven by the same Brownian motion. This ensures that the solution doesn't explode and is well defined on $\mathbb{R}_+$.

To study it further, we introduce the change of variable $x = 2\sqrt{\frac{y}{\gamma}}$ and apply Itô's formula. We find that the process $X = 2\sqrt{\frac{Y}{\gamma}}$ is a solution to the SDE

$$dX_t = dB_t - q(X_t)\,dt,$$

where

$$q(x) = \frac{1}{2x} - \frac{rx}{2} + \frac{c\gamma x^3}{8}.$$

The function q is defined and of class C^1 on $(0,+\infty)$ and the process X is uniquely defined until the explosion time $T_0 \wedge T_\infty$.

To determine the behavior of the process X, and therefore of Y, as t tends to infinity, we introduce as in Section 4.6.6 the functions Q and f defined by

$$Q(x) = 2\int_1^x q(y)\,dy; \qquad f(x) = \int_1^x e^{Q(y)}\,dy.$$

A simple computation gives that for $x > 0$,

$$Q(x) = \log x - \frac{r}{2}(x^2 - 1) + \frac{c\gamma}{16}(x^4 - 1).$$

Considering $[a,b] \subset (0,+\infty)$ and $x \in [a,b]$, we have seen that $\mathbb{E}_x(T_a \wedge T_b) < +\infty$ and that $\mathbb{P}_x(T_a < T_b) = \frac{f(b) - f(x)}{f(b) - f(a)}$. Note that $f(b)$ tends to infinity as b tends to $+\infty$ and that $f(0^+) < \infty$. Moreover, we have seen that $T_\infty = +\infty$ a.s. We deduce that $\mathbb{P}_x(T_0 < +\infty) = 1$. The process X reaches 0 almost surely. The same is true for Y, regardless of the growth rate r. Then, the solution of the logistic Feller equation goes to 0 as t tends to infinity, whatever the sign of r is. The interesting case is when $r > 0$, for which, without competition ($c = 0$), the population can explode with positive probability (see the previous subsection). Competition between individuals regulates population growth and leads it almost surely to extinction, regardless of the initial condition. We can thus measure, on this very simple model, the impact that the environment can have on the survival of the population.

4.7.3 Ornstein–Uhlenbeck Processes

We wish to model the movement of an animal influenced by the presence of a home (its den, its food source, etc.). Suppose the animal moves along a road, a river, a blade of grass, or any one-dimensional space: the state space of positions is $\mathbb{R}$. We also assume that its home is located at 0. We denote by X_t the position of the animal at time t. This position is defined as the solution of the equation

$$\mathrm{d}X_t = -X_t\,\mathrm{d}t + \sigma\,\mathrm{d}B_t;\; X_0 = x \in \mathbb{R},$$

where B is a Brownian motion and σ a real number. Such a process is called an Ornstein–Uhlenbeck process. The existence and uniqueness of the trajectory are immediate since the drift coefficient is linear and the diffusion coefficient is constant. The coefficient σ quantifies the random fluctuations of the animal, whose central tendency is to return to its home. We are interested in the distance the animal can travel during an excursion away from home. Let $x > 0$. Here again we will use Section 4.6.6 to compute $\mathbb{P}_x(T_0 < T_L)$, for all $L > 0$ (we have seen that these hitting times are finite almost surely). We obtain, by setting $f(x) = \int_1^x \mathrm{e}^{\frac{y^2}{\sigma^2}}\,\mathrm{d}y$, that

$$\mathbb{P}_x(T_L < T_0) = \frac{f(x) - f(0)}{f(L) - f(0)}.$$

Using integration by parts, we can write

$$f(x) = \int_1^x \mathrm{e}^{\frac{y^2}{\sigma^2}}\,\mathrm{d}y = \frac{\sigma^2}{2}\frac{1}{x}\mathrm{e}^{\frac{x^2}{\sigma^2}} + \frac{\sigma^2}{2}\int_1^x \frac{1}{y^2}\mathrm{e}^{\frac{y^2}{\sigma^2}}\,\mathrm{d}y + C^{te},$$

and deduce that

$$\lim_{x\to\infty} x\mathrm{e}^{-\frac{x^2}{\sigma^2}} f(x) = \frac{\sigma^2}{2}.$$

Indeed, let $\varepsilon > 0$. Write

$$\begin{aligned} x\mathrm{e}^{-\frac{x^2}{\sigma^2}}\int_1^x \frac{1}{y^2}\mathrm{e}^{\frac{y^2}{\sigma^2}}\,\mathrm{d}y &= x\mathrm{e}^{-\frac{x^2}{\sigma^2}}\int_1^{x-\varepsilon} \frac{1}{y^2}\mathrm{e}^{\frac{y^2}{\sigma^2}}\,\mathrm{d}y + x\mathrm{e}^{-\frac{x^2}{\sigma^2}}\int_{x-\varepsilon}^x \frac{1}{y^2}\mathrm{e}^{\frac{y^2}{\sigma^2}}\,\mathrm{d}y \\ &\le x\mathrm{e}^{-\frac{x^2}{\sigma^2}}\int_1^{x-\varepsilon} \mathrm{e}^{\frac{y^2}{\sigma^2}}\,\mathrm{d}y + x\int_{x-\varepsilon}^x \frac{1}{y^2}\,\mathrm{d}y \\ &\le x(x-\varepsilon-1)\mathrm{e}^{-(x^2-(x-\varepsilon)^2)} + \frac{\varepsilon x}{x(x-\varepsilon)}. \end{aligned}$$

The two terms on the right-hand side tend to 0 as x tends to infinity.

We deduce that

$$\mathbb{P}_x(T_L < T_0) \sim_{L\to\infty} \frac{2f(x)}{\sigma^2} L\mathrm{e}^{-\frac{L^2}{\sigma^2}}.$$

This implies that

$$\sum_{n=1}^{\infty} \mathbb{P}_x(T_n < T_0) < +\infty.$$

Applying the Borel–Cantelli Lemma, we deduce that, almost surely, there are only a finite number of realized sets $\{T_n < T_0\}$. Thus, almost surely, there exists a distance D (the maximum of these integers n) that the animal will not exceed during a given excursion away from its home.

4.7.4 Other Examples of Spatial Movements

Here are some examples of specific spatial movements modeled by stochastic differential equations. These examples are by no means exhaustive but illustrate some biological contexts. The reader can further study them in the corresponding publications.

1. *Evolution of a virus in a cell* (Lagache, Holcman [55]). This model is used to understand the early stages of viral infection where the virus enters the cell and infiltrates the nucleus. The movement of the virus involves a succession of ballistic phases along microtubules (the cell's highways) and diffusion phases to move from one microtubule to another. The virus is absorbed as soon as it arrives in the nucleus. The dynamics of the virus can be summarized by the bi-dimensional diffusion equation

$$\mathrm{d}X_t = b(X_t)\,\mathrm{d}t + \sqrt{2D}\,\mathrm{d}B_t,$$

to which is added a term for the degradation of the virus over time.

2. *Movement of a fish* (Cattiaux, Chafaï, Motsch [16]). In this model, the sea is identified as a surface. The movement of the fish is described using three variables: the position $X \in \mathbb{R}^2$, the angular velocity $\theta \in \mathbb{R}$ and the curvature $K \in \mathbb{R}$. The dynamics are random and governed by the following system of differential equations

$$\begin{aligned}\mathrm{d}X_t &= \tau(\theta_t, K_t)\,\mathrm{d}t,\\ \mathrm{d}\theta_t &= K_t\,\mathrm{d}t,\\ \mathrm{d}K_t &= -K_t\,\mathrm{d}t + \sqrt{2\alpha}\,\mathrm{d}B_t.\end{aligned}$$

Simulations of this model yield Figure 4.5, which represents the trajectory $t \to (X_t^1, X_t^2)$ with $\alpha = 1$ and $\tau(\theta, K) = \frac{1}{1+2|K|}(\cos\theta, \sin\theta)$.

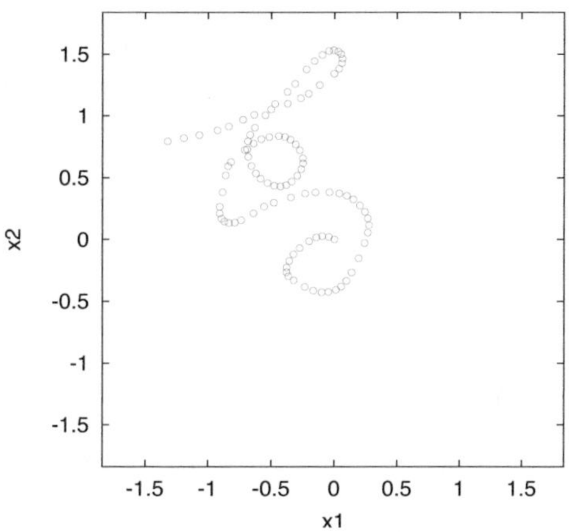

Fig. 4.5 Simulated trajectory of a fish.

3. *Ant aggregation* (Boi, Capasso, Morale [12], Capasso, Oelschläger [62]). Aggregation plays a key role in explaining population dynamics. It concerns populations of all types: ant columns, fish schools, cancer cell clusters, social interactions. The authors describe the (random) position of each ant and model the entire movement of the ants by a system of stochastic differential equations. The N ants are at respective positions $X_t^1, \ldots, X_t^N$, at time t. The position of the k-th ant satisfies the SDE

$$dX_t^k = H(X_t^1, \ldots, X_t^N)\, dt + \sigma_N(X_t^1, \ldots, X_t^N)\, dB_t^k.$$

The Brownian motions B^k are independent. The function H models the strong chemical attraction that pushes the ants to form columns.

4.7.5 The Wright–Fisher Diffusion Process

In Chapter 6, we will introduce and study the classic models of population genetics. These models describe the evolution of the proportion of individuals carrying an allele A in a population of constant size, where individuals carry one of the two alleles A or a. The most famous model in continuous time is the Wright–Fisher diffusion process, justified in Theorem 6.3.1. This diffusion process is defined as the solution to the stochastic differential equation

$$Z_t = z + \int_0^t \sqrt{Z_s(1 - Z_s)}\, dB_s, \tag{4.7.4}$$

where $(B_t, t \geq 0)$ is a standard Brownian motion and $z \in]0,1[$. Theorem 4.6.11 has established the existence and uniqueness of a solution. As in Section 4.7.1, pathwise uniqueness ensures that the solution stays between 0 and 1. Moreover, the points 0 and 1 are absorbing. These points correspond to states where the population is composed of individuals all of the same type: either of allele A ($Z = 1$), or of allele a ($Z = 0$). In this case, we say that allele A (resp. allele a) has become fixed in the population. Let us study the fixation time of one of the alleles, namely

$$T^* = \inf\{t > 0, Z_t = 0 \text{ or } 1\},$$

with the convention $\inf \emptyset = +\infty$.

The infinitesimal generator of the Wright–Fisher process is given for $h \in C_b^2$ by

$$Lh(z) = \frac{1}{2} z(1 - z)\, h''(z).$$

Let us apply the computations from Section 4.6.7 (with the same notation) with $\sigma(z) = z(1 - z)$ and $b(z) = 0$. We get $Q(z) = 0$, $f(z) = z$. We know that $\mathbb{E}_z(T^*) = g(z)$, where g is the solution of $\frac{1}{2} z(1 - z)\, g''(z) = -1$, with $g(0) = g(1) = 0$. It is easy to see that $g(z) = -2(z \log z + (1 - z) \log(1 - z))$. Therefore we obtain

$$\mathbb{E}_z(T^*) = -2(z\log z + (1-z)\log(1-z)); \ \mathbb{P}_z(T_0 < T_1) = 1-z. \tag{4.7.5}$$

4.8 Exercises

Exercise 4.8.1
Let $(B_t, t \geq 0)$ be a Brownian motion.

1. Show that the process B', defined by $B'_t = -B_t$ for all $t > 0$, is a Brownian motion.
2. Show that the process B^c, defined by $B_t^{(c)} = \frac{1}{c}B_{c^2t}$ for all $c > 0$, is a Brownian motion.

Exercise 4.8.2 (Dispersion of plants or animals (cf. Renshaw [70]))
We study the dispersion over time of a population, starting from the origin $(0,0)$ in a 2-dimensional space. For each individual, it is modeled by $(X_t, Y_t) = D(B_t^1, B_t^2)$, where (B_t^1, B_t^2) is a bi-dimensional Brownian motion. For each t, the dispersion can also be measured by the polar coordinates (R_t, θ_t).

1. Give the law of (R_t, θ_t) for $t > 0$. What can be deduced about the random variables R_t and θ_t? What is the law of R_t?
   What is the law of the area S_t swept by R_t? Prove that on average, these areas form expanding concentric circles over time.
2. Assume that the population is constant and of size N. The wave front at time t is the number $F(t) \in \mathbb{R}_+$ such that, almost surely, one can find one and only one individual at a distance $r > F(t)$. Give the value of $F(t)$ as a function of N and t.
3. Now suppose that N has a deterministic exponential growth as a function of t, with rate $\lambda > 0$. Prove that in this case, the front $F(t)$ advances with constant speed.
4. We apply this model to the study of the dispersion of certain oaks that colonized England after the last glaciation. These oak trees begin to produce acorns at around 60 years old, and continue to do so for several hundred years. These acorns will disperse (particularly thanks to birds), allowing for the spatial propagation of the species. We will therefore assume that the unit of time is one generation (60 years). We want to measure this dispersion under the assumption that it follows the model above. We assume that the first oak trees appeared at the end of the Ice Age 18,000 years ago and we estimate that a mature oak tree produces 9×10^6 acorns in one generation.
   Knowing that the wavefront has increased by 600 miles, find the dispersion coefficient D per generation for this species.

Exercise 4.8.3
The process $(X_t)_{t\geq 0}$ is called a *submartingale* if for each $t \geq 0$, X_t is integrable and if for $0 \leq s \leq t$, $\mathbb{E}(X_t|\mathcal{F}_s) \geq X_s$, where $\mathcal{F}_s = \sigma(X_u, u \leq s)$.

Let $(X_t)_{t\geq 0}$ be a positive submartingale. Show that for all $c > 0$, we have

$$\mathbb{P}(\sup_{s\leq t} X_s \geq c) \leq \frac{1}{c}\mathbb{E}(X_t).$$

From this, deduce that for a Brownian motion $(B_t)_{t\geq 0}$ and for all $c>0$,

$$\mathbb{P}(\sup_{s\leq t} B_s \geq c) \leq \mathrm{e}^{-\frac{c^2}{2t}}.$$

Hint: you may want to study $\mathrm{e}^{\theta B_t}$, *for* $\theta > 0$.

Exercise 4.8.4 (Law of the iterated logarithm)
In this exercise, we will prove Proposition 4.3.2. For large t, let us introduce the function $h(t) = \sqrt{2t \log\log t}$.

1. Let $K > 1$ and $\delta > 0$. For all $n \in \mathbb{N}$, define $C_n = \sqrt{2(1+\delta)K^n \log\log(K^n)}$.
   (a) Use Exercise 4.8.3 and the Borel–Cantelli Lemma to show that
   $$\mathbb{P}(\limsup_{n\to\infty}\{\sup_{s\leq K^n} B_s \geq C_n\}) = 0.$$
   (b) Deduce that, with probability 1, $\limsup_{t\to\infty} \frac{B_t}{h(t)} \leq 1$.
2. Let $N \in \mathbb{N}$ and $\varepsilon \in]0,1[$. For all $n \in \mathbb{N}$, we introduce the set
   $$F_n = \{B_{N^{n+1}} - B_{N^n} > (1-\varepsilon)\, h(N^{n+1} - N^n)\}.$$
   (a) Let φ be the density of the standard normal distribution. Show that
   $$\frac{\varphi(x)}{x} = \int_x^\infty \left(1+\frac{1}{y^2}\right)\varphi(y)\,\mathrm{d}y \leq \left(1+\frac{1}{x^2}\right)\int_x^\infty \varphi(y)\,\mathrm{d}y.$$
   (b) Deduce that
   $$\mathbb{P}(F_n) \geq \frac{1}{\sqrt{2\pi}}\exp\left(\frac{-y^2}{2}\right)\left(\frac{y}{1+y^2}\right),$$
   where $y = (1-\varepsilon)\sqrt{2\log\log(N^{n+1} - N^n)}$.
   (c) Using the Borel–Cantelli Lemma again, show that
   $$\mathbb{P}(\limsup_{n\to\infty} F_n) = 1.$$
   (d) Conclude.

Exercise 4.8.5 (Reproduce before being devoured)
Consider a water spider moving along a river as a one-dimensional Brownian motion $(B_t : t \geq 0)$ and expected at a point $a > 0$ by a band of frogs that will immediately snap it up. Suppose that the spider reproduces after an exponential time of parameter $\mu > 0$, independent of its movement. We seek to know if and when the spider will die and if it will have time to reproduce beforehand.

1. Biologically interpret the hypothesis of independence between the spider's movement and reproduction.
2. Let $\lambda > 0$. Show that $\exp(\lambda B_t - \lambda^2 t/2)$ is a martingale.

3. Denote by T_a the time the spider reaches a. Show that

$$\exp(\lambda B_{t\wedge T_a} - \lambda^2(t\wedge T_a)/2)$$

is a bounded martingale.
4. Show that the probability p that the spider dies without offspring is given by

$$p = \exp\big(-a\sqrt{2\mu}\big).$$

5. There is now another colony of frogs in position $-a$.
Denote by $T_a^\star := T_a \wedge T_{-a}$ the time the spider reaches one of the frog zones. Find a martingale that allows you to compute the new probability p of death without reproduction and prove that

$$p = \frac{1}{\cosh(a\sqrt{2\mu})}.$$

6. Using the previous result, prove that

$$\mathbb{E}\big(T_a^\star\, \mathrm{e}^{-\mu T_a^\star}\big) = \frac{a}{\sqrt{2\mu}}\,\frac{\sinh(a\sqrt{2\mu})}{\cosh^2(a\sqrt{2\mu})}.$$

7. Deduce rigorously that $\mathbb{E}(T_a^\star) = a^2$ and recover this result using a very simple martingale.

Exercise 4.8.6 (Feller's diffusion process (cf. Feller [37]))
Let $T > 0$. Let $(Z_t, t \geq 0)$ be the diffusion process defined as the unique solution (in the set of processes $\{(Y_t, t \geq 0);\ E(\sup_{t\leq T}|Y_t|^2) < +\infty\}$) of the following stochastic differential equation

$$\mathrm{d}Z_t = rZ_t\,\mathrm{d}t + \sqrt{\sigma Z_t}\,\mathrm{d}B_t, \qquad Z_0 = z_0 > 0,$$

for $r \in \mathbb{R}$ and $\sigma > 0$. It is called *Feller's diffusion process*.

1. Compute the mean and variance of this process.
2. Recall Itô's formula for this process.
3. Find a function f of class C^2, bounded and with bounded first and second derivatives, defined from $\mathbb{R}_+$ to $\mathbb{R}_+$ such that $f(Z_t)$ is a martingale for the filtration generated by the Brownian motion.
4. Deduce the extinction probability of this diffusion as a function of the initial condition z_0.

Exercise 4.8.7 (The logistic Feller diffusion process)
We are interested in the Markovian process Z that is the solution of the stochastic differential equation

$$\mathrm{d}Z_t = (rZ_t - cZ_t^2)\,\mathrm{d}t + \sqrt{Z_t}\,\mathrm{d}B_t, \qquad Z_0 = z_0 > 0,$$

where $r, c > 0$ and $(B_t)_{t\geq 0}$ is a standard Brownian motion. We call this solution the *logistic Feller diffusion process.*

1. Prove that when $c = 0$, the solution Z_t of this stochastic differential equation exists for all $t \geq 0$ and is unique.
2. Provide an interpretation of the term $\sqrt{Z_t}\,\mathrm{d}B_t$ and the term $(rZ_t - cZ_t^2)\,\mathrm{d}t$ in terms of the demographic parameters of the model.
3. In order to study the long-term behavior of Z, we introduce the change of variable $X_t = 2\sqrt{Z_t}$.

   (a) What is peculiar about the point 0?

   (b) Show that as long as Z has not reached 0, X is the solution of the stochastic differential equation

   $$\mathrm{d}X_t = q(X_t)\,\mathrm{d}t + \mathrm{d}B_t, \qquad \text{where} \quad q(x) = -\frac{1}{2x} + \frac{rx}{2} - \frac{cx^3}{8},$$

   with initial value $x_0 = 2\sqrt{z_0}$.

4. (a) Define

   $$Q(x) := \int_1^x 2q(y)\,\mathrm{d}y,$$

   and $\Lambda(x) := \int_1^x \mathrm{e}^{-Q(y)}\,\mathrm{d}y$. Verify that $\Lambda'(x)q(x) + \frac{1}{2}\Lambda''(x) = 0$.

   (b) Deduce that for all $0 < a < x_0 < b$, the process $(\Lambda(X_{t\wedge T_a\wedge T_b}))_{t\geq 0}$ is a martingale, and that

   $$\mathbb{P}_{x_0}[T_a < T_b] = \frac{\Lambda(b) - \Lambda(x_0)}{\Lambda(b) - \Lambda(a)}.$$

Chapter 5
Continuous Time Population Processes

He now needed a wide variety of models, perhaps transformable into each other, according to a combinatorial procedure, to find the one that suited a reality which, in turn, was always made up of several different realities, in time as well as in space.

Italo Calvino, *Palomar*

In Chapter 3, we considered branching processes indexed by a discrete time, which models, for example, the number of generations or the number of time periods in a cyclical reproduction process (seasonal, annual reproduction) or a discretized physical time. In this chapter, we model population dynamics whose individuals reproduce and die at random times, continuously in time, allowing overlapping generations characteristic of certain species. The dynamics of these populations will depend on the physical time $t \in \mathbb{R}_+$.

We define the random process $(X_t, t \geq 0)$ as modeling the population size dynamics. At each birth time, the process increases by 1, or more if there is multiple reproduction, and it decreases by 1 at each death of an individual. This process is therefore piecewise constant and jumps at random birth and death times. We assume that the random process $(X_t, t \geq 0)$ is a Markov process, that is, its future random behaviour depends on the past only through the information given by its current state (we will see a precise definition below). This assumption actually implies a non-aging property, which is a strong biological assumption. (Problems such as the accumulation of defects or senescence, for example, cannot be studied in this framework.) However, the Markov assumption provides a mathematical framework that allows for rigorous results and quantitative answers to a number of questions. We will systematically study such Markov processes, the prototype of which is the Poisson process. This process, like Brownian motion, is a process with independent and stationary increments. We will show that the durations between two jumps of this process are independent random variables with an exponential distribution and that

S. Méléard, *Random Models in Biology, Ecology and Evolution*, Texts in Applied Mathematics 87, https://doi.org/10.1007/978-3-662-73483-4_5

the law of the process at a fixed instant is a Poisson law. We will show that a similar structure exists in a more general framework where the law of the jump times and the reproduction law can depend on the instantaneous state of the population, thus describing situations such as density-dependence. We will show that the dynamics of the law of the process at each instant follows a matrix differential equation (called Kolmogorov's equation) driven by a specific matrix defined as the infinitesimal generator of the process. Thanks to the Markov property, the generator and the law of the initial condition are sufficient to characterize the law of the process. The specific cases of continuous time branching processes and birth and death processes will be studied, with a particular emphasis on the study of the extinction probability and the law of the extinction time.

5.1 Markov Pure Jump Processes

Definition 5.1.1 A *pure jump process* $(X_t, t \geq 0)$ is a process with values in a finite or countable space E whose trajectories are almost surely right-continuous and left-limited and constant between isolated jump instants. The knowledge of the process $(X_t, t \geq 0)$ is equivalent to the knowledge of the sequence of random variables (T_n, Z_n), $n \geq 0$, where $T_n \in \mathbb{R} \cup \{+\infty\}$ is the n-th jump time and $Z_n \in E$ is the position of the process just before the jump T_{n+1}. Moreover, $T_0 = 0$ and the jump times $T_n \in \mathbb{R} \cup \{+\infty\}$ form an increasing sequence such that

$$T_n(\omega) < T_{n+1}(\omega) \text{ if } T_n(\omega) < +\infty.$$

We also assume that the jump times do not accumulate in a finite time:

$$\lim_{n \to \infty} T_n(\omega) = +\infty. \tag{5.1.1}$$

In this case, we have

$$X_t(\omega) = \sum_{n \geq 0, T_n(\omega) < +\infty} Z_n(\omega) \mathbf{1}_{[T_n(\omega), T_{n+1}(\omega)[}(t).$$

A typical trajectory is represented in Figure 5.1.

Remark 5.1.2 Our motivations mainly concern population sizes and we will limit our study to the case of processes with integer values, i.e. $E = \mathbb{N}$

The pure jump processes we are interested in are Markovian, in the following sense.

Definition 5.1.3 An $\mathbb{N}$-valued pure jump process $(X_t, t \geq 0)$ is a *Markov pure jump process* (also called a *continuous time Markov chain*) if, for all $0 < s < t$, the conditional law of the random variable X_t given $(X_u, u \leq s)$ only depends on the law of X_s: for all $n \in \mathbb{N}$, for all $0 \leq t_0 < t_1 < \cdots < t_n < s$ and the states $i_0, i_1, \ldots, i_n, i, j \in \mathbb{N}$,

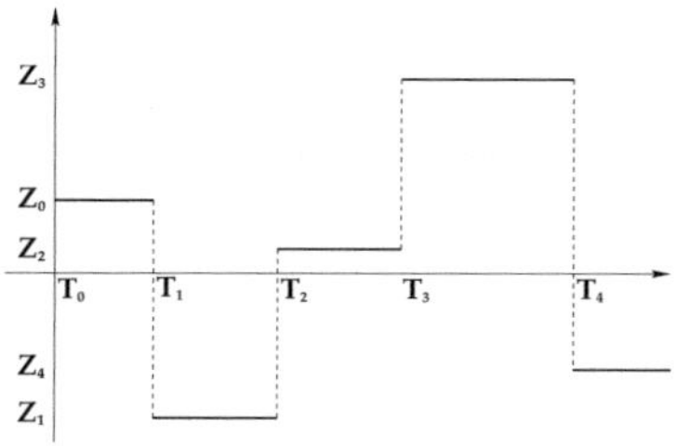

Fig. 5.1 Trajectory of a pure jump process.

we have

$$\mathbb{P}\left(X_t = j \mid X_{t_0} = i_0, X_{t_1} = i_1, \ldots, X_{t_n} = i_n, X_s = i\right) = \mathbb{P}\left(X_t = j \mid X_s = i\right).$$

We say that this Markov process is *homogeneous* if the probability $\mathbb{P}(X_t = j \mid X_s = i)$ depends on s and t only through the difference $t-s$.

In the following, we shall study these homogeneous Markov processes. We use the notation

$$\mathbb{P}(X_t = j \mid X_s = i) = P_{i,j}(t-s).$$

For all $t > 0$, the infinite size matrix $P(t) = (P_{i,j}(t))_{i,j\in\mathbb{N}}$ is a Markov matrix on $\mathbb{N}\times\mathbb{N}$: $\forall i, j \in \mathbb{N}$, $\forall t > 0$,

$$P_{i,j}(t) \geq 0, \quad \sum_j P_{i,j}(t) = 1.$$

The matrix $P(t)$ is called the *transition matrix*. It can also be considered as an operator on sequences with the immediate property that for a sequence $U = (u_n)_n$, the sequence $P(t)U$ is the vector obtained by the formulas of the matrix product. We note that

$$P(0) = \mathrm{Id}.$$

We denote by $\mu(t)$ the law of X_t and by $\mu(0) = \mu$ the initial law of the process.

We identify a function $g : \mathbb{N} \to \mathbb{R}$ with the vector $(g_j)_{j\in\mathbb{N}}$, where $g_j = g(j)$, and a measure μ on $\mathbb{N}$ with the vector (μ_i), where $\mu_i = \mu(\{i\})$.

Proposition 5.1.4 *Let $(X_t)_t$ be a homogeneous Markov pure jump process with initial law μ and matrix $(P(t), t > 0)$. For all $n \in \mathbb{N}$, for all $0 < t_1 < \cdots < t_n$, the law of the random vector $(X_0, X_{t_1}, \ldots, X_{t_n})$ is given by: for all $i_0, i_1, \ldots, i_n \in \mathbb{N}$,*

$$\mathbb{P}\left(X_0 = i_0, X_{t_1} = i_1, \ldots, X_{t_n} = i_n\right) = \mu_{i_0} P_{i_0,i_1}(t_1) P_{i_1,i_2}(t_2 - t_1) \times \cdots \times P_{i_{n-1},i_n}(t_n - t_{n-1}).$$

Proof It immediately follows from conditional probabilities and the Markov property that

$$\begin{aligned}
&\mathbb{P}\left(X_0 = i_0, X_{t_1} = i_1, \cdots, X_{t_n} = i_n\right)\\
&= \mathbb{P}(X_0 = i_0)\,\mathbb{P}(X_{t_1} = i_1 | X_0 = i_0)\,\mathbb{P}\left(X_{t_2} = i_2 | X_0 = i_0, X_{t_1} = i_1\right)\\
&\qquad\qquad \times \cdots \times \mathbb{P}\left(X_{t_n} = i_n | X_0 = i_0, \ldots, X_{t_{n-1}} = i_{n-1}\right)\\
&= \mu_{i_0}\, P_{i_0,i_1}(t_1)\, P_{i_1,i_2}(t_2 - t_1) \times \cdots \times P_{i_{n-1},i_n}(t_n - t_{n-1}).
\end{aligned}$$

□

This proposition is fundamental because it shows that the law μ and the transition matrices $(P(t), t > 0)$ are sufficient to characterize the law of the process. Indeed, we know from Theorem 4.2.2 that the law of a process with values in the set of functions from $\mathbb{R}_+$ to $\mathbb{R}$, equipped with the product σ-field, is characterized by the laws of the finite-dimensional marginals $(X_{t_1}, \ldots, X_{t_n})$, for all $0 \le t_1 < \cdots < t_n$ and for all $n \in \mathbb{N}^*$.

In particular, for all $t > 0$, for any positive or bounded function $g : \mathbb{N} \to \mathbb{R}$, we have

$$\mathbb{E}(g(X_t) |\, X_0 = i) = (P(t)g)_i = \sum_{j \in \mathbb{N}} P_{i,j}(t) g_j.$$

The law $\mu(t)$ of X_t will be given by

$$\mu(t) = \mu P(t) \ :\ \forall j \in \mathbb{N}\,,\ \mu_j(t) = \sum_{i \in \mathbb{N}} \mu_i P_{i,j}(t).$$

(Multiplication on the left by a measure is dual to multiplication on the right by a sequence.)

Moreover,

Proposition 5.1.5 *The matrices* $(P(t), t > 0)$ *satisfy* $P(0) = \mathrm{Id}$ *and the semi-group relation, called the* Chapman–Kolmogorov equation*:*

$$P(t+s) = P(t)P(s) = P(s)P(t). \tag{5.1.2}$$

The product of two transition matrices commutes.

Remark 5.1.6 This property justifies calling the family of operators $(P(t), t \ge 0)$ the *transition semi-group* of the process.

Proof Let $i, k \in \mathbb{N}$ and $s, t > 0$. We have

$$\begin{aligned}
P_{i,k}(t+s) &= \mathbb{P}(X_{t+s} = k | X_0 = i) = \sum_j P(X_t = j, X_{t+s} = k | X_0 = i)\\
&= \sum_j \mathbb{P}(X_{t+s} = k | X_t = j)\mathbb{P}(X_t = j | X_0 = i) = \sum_j P_{i,j}(t) P_{j,k}(s)
\end{aligned}$$

by the Markov property, hence $P(t+s) = P(t)P(s)$. □

5.2 A Prototype: The Poisson Process

5.2.1 Definition of a Poisson Process

We introduce the prototype of pure jump Markov processes, which we assume to have independent and stationary increments. This process models the times of successive occurrences of random events.

Definition 5.2.1 A *point process* on $\mathbb{R}_+$ is a pure jump process with jumps of amplitude 1. It is characterised by an (almost surely) increasing sequence of random times

$$0 < T_1 < T_2 < \cdots < T_n < \cdots,$$

defined on a probability space $(\Omega, \mathcal{F}, \mathbb{P})$, with values in $\mathbb{R}_+$ and satisfying

$$T_n \to +\infty \text{ almost surely, as } n \text{ tends to infinity.}$$

The random variables T_n model the moments when the events occur.

The random variables

$$S_1 = T_1,\ S_2 = T_2 - T_1,\ \ldots,\ S_n = T_n - T_{n-1},\ \ldots$$

model the lengths of the intervals or waiting times between two successive events.

Definition 5.2.2 The *random counting function* $(N_t)_{t\geq 0}$ associated with the point process $\{T_n, n \in \mathbb{N}\}$ is defined by

$$N_t = \sup\{n,\ T_n \leq t\} = \sum_{j\geq 1} \mathbf{1}_{\{T_j \leq t\}}.$$

N_t is therefore the number of events that have occurred before time t and $N_{T_n} = n$ for all n.

Let us note that $N_0 = 0$ since $T_1 > 0$ and that for all t, $N_t < +\infty$, since the sequence $(T_n)_n$ tends to infinity. For $0 \leq s < t$, $N_t - N_s$ is the number of events that have occurred during the time interval $]s,t]$.

The trajectories $t \to N_t(\omega)$ of such a process are a.s. right-continuous and left-limited, by definition.

Remark 5.2.3 The law of the point process and the law of the random counting function can be deduced from each other. Indeed, we have

$$\begin{aligned}
\{N_t \geq n\} &= \{T_n \leq t\} \\
\{N_t = n\} &= \{T_n \leq t < T_{n+1}\} \\
\{N_t \geq n > N_s\} &= \{s < T_n \leq t\}.
\end{aligned}$$

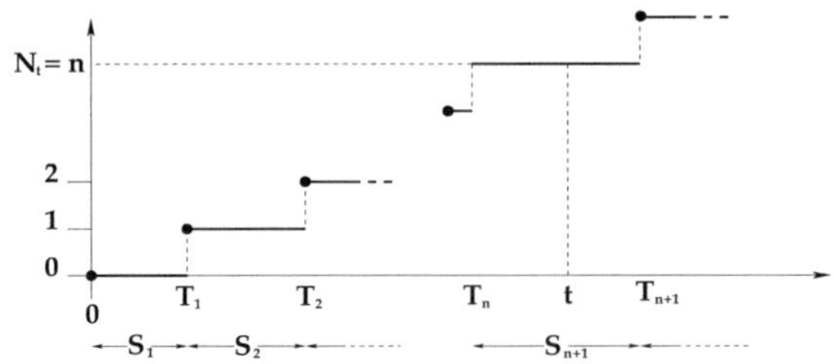

Fig. 5.2 Trajectory of a point process.

Definition 5.2.4 The point process (T_n) or equivalently the random process $(N_t, t \geq 0)$ is called a *Poisson process* when $(N_t, t \geq 0)$ has independent and stationary increments, that is to say when

1) For all $t_0 < t_1 < \cdots < t_n$ in $\mathbb{R}_+$, the increments $\left(N_{t_j} - N_{t_{j-1}},\ 1 \leq j \leq n\right)$ are independent random variables.
2) For $0 \leq s < t$, the law of $N_t - N_s$ depends on s and t only by the difference $t - s$. It is therefore equal to the law of N_{t-s}.

Property (2) is called the *stationarity of increments*.

The name Poisson process is justified by the following property:

Proposition 5.2.5 *Let $(N_t, t \geq 0)$ be a Poisson process. Then there exists $\lambda > 0$ such that for all $t > 0$, N_t is a Poisson random variable with parameter λt:*

$$\mathbb{P}(N_t = k) = \mathrm{e}^{-\lambda t} \frac{(\lambda t)^k}{k!}, \ \forall k \in \mathbb{N}.$$

Definition 5.2.6 The parameter λ is called the *intensity* of the Poisson process. It is equal to the average number of events that occur during a unit of time, since

$$\mathbb{E}(N_{t+1} - N_t) = \mathbb{E}(N_1) = \lambda.$$

We also say that the events *occur at rate λ*.

Proof Let g_t be the generating function of N_t. We have, for $u \in [0, 1]$,

$$g_t(u) = \mathbb{E}(u^{N_t}) = \sum_{k \geq 0} \mathbb{P}(N_t = k)\, u^k.$$

We want to prove that g_t is the generating function of a Poisson random variable with parameter λt, that is to say,

$$g_t(u) = \exp(-(\lambda t(1-u)).$$

By independence of increments and for $s < t$, we have that $g_t(u) = g_{t-s}(u)g_s(u)$, and more generally we can prove that $g_t(u) = (g_1(u))^t$. (We prove it for integers, then for rationals, and we conclude by using that $t \to g_t(u)$ is decreasing.) Furthermore, $g_t(u) \geq \mathbb{P}(N_t = 0) = \mathbb{P}(T_1 > t)$, which tends to 1 as $t \to 0$, implying that $g_1(u)$ is not null. As, moreover, $g_1(u) \leq 1$, there exists $\lambda(u) > 0$ such that

$$g_t(u) = \mathrm{e}^{-\lambda(u)t}.$$

Let us show that $\lambda(u)$ has the form $\lambda \times (1-u)$, with λ constant.
We have

$$\lambda(u) = \lim_{t\downarrow 0} \frac{1}{t}(1 - g_t(u)) = \lim_{t\downarrow 0} \frac{1}{t} \sum_{k\geq 1} \mathbb{P}(N_t = k)(1-u^k).$$

Since $u \leq 1$, we deduce that for all t,

$$0 \leq \frac{1}{t}(1-g_t(u)) - \frac{1}{t}\mathbb{P}(N_t = 1)(1-u) = \frac{1}{t}\sum_{k\geq 2}\mathbb{P}(N_t = k)(1-u^k) \leq \frac{1}{t}\mathbb{P}(N_t \geq 2).$$

Assume first that $\frac{1}{t}\mathbb{P}(N_t \geq 2)$ tends to 0 as $t \to 0$. Then

$$\lambda(u) = \lim_{t\downarrow 0} \frac{1}{t}\mathbb{P}(N_t = 1)(1-u)$$

exists and $\lambda(u) = (1-u)\lambda(0)$, with $\lambda(0) = \lim_{t\downarrow 0} \frac{1}{t}\mathbb{P}(N_t = 1)$. We will therefore have proved Proposition 5.2.5 with $\lambda = \lambda(0)$.

To study the asymptotic behavior of $\mathbb{P}(N_2 \geq t)$, note that

$$\cup_n \{N_{nt} = 0, N_{(n+1)t} \geq 2\} \subset \{T_2 < T_1 + t\},$$

and that by the independence and stationarity of increments,

$$\mathbb{P}(\cup_n \{N_{nt} = 0, N_{(n+1)t} \geq 2\}) = \sum_n \mathbb{P}(N_{nt} = 0)\mathbb{P}(N_t \geq 2) \leq \mathbb{P}(T_2 < T_1 + t).$$

We deduce that

$$\sum_n \mathrm{e}^{-\lambda(0)nt}\,\mathbb{P}(N_t \geq 2) = (1-\mathrm{e}^{-\lambda(0)t})^{-1}\mathbb{P}(N_t \geq 2) \leq \mathbb{P}(T_2 < T_1 + t).$$

As t tends to 0, this last quantity equals $\mathbb{P}(T_2 \leq T_1) = 0$. As for t sufficiently small, we also have that $(\lambda(0)t)^{-1} \leq (1-\mathrm{e}^{-\lambda(0)t})^{-1}$, we finally deduce that $\dfrac{1}{t}\mathbb{P}(N_t \geq 2)$ tends to 0 as $t \to 0$. □

Remark 5.2.7 We can give an intuitive interpretation of this result. It results from the proof above that

$$\begin{aligned}\mathbb{P}(N_{t+h}-N_t=0) &= 1-\lambda h+o(h)\\ \mathbb{P}(N_{t+h}-N_t=1) &= \lambda h+o(h)\\ \mathbb{P}(N_{t+h}-N_t\geq 2) &= o(h).\end{aligned}$$

So up to $o(h)$, $N_{t+h}-N_t$ is a Bernoulli random variable taking the value 0 with probability $1-\lambda h$ and the value 1 with probability λh. This property combined with the independence of increments and the formula

$$N_{t+s}-N_t=\sum_{j=1}^{n}(N_{t+jh}-N_{t+(j-1)h})\ ,\quad \text{with } h=\frac{s}{n},$$

implies that $N_{t+s}-N_t$ follows approximately a binomial law with parameter $(n,\lambda s/n)$. It can be easily shown that as n tends to infinity, this sequence of distributions weakly converges to a Poisson distribution with parameter λs.

Remark 5.2.8 Observe that Proposition 5.2.5 and the property of independent increments yield the law of any vector $(N_{t_1},\ldots,N_{t_d})$, for $t_1<\cdots<t_d$.

We can also deduce from Proposition 5.2.5 the law of the first jump time of the process (N_t).

Corollary 5.2.9 *The law of the first jump time T_1 is an exponential law with parameter λ. Similarly, for any $s>0$, the law of the first event after s, i.e., $T_{N_s+1}-s$, is an exponential law with parameter λ.*

Proof For $t>0$, we have $\mathbb{P}(T_1>t)=\mathbb{P}(N_t=0)=\mathrm{e}^{-\lambda t}$. Similarly,

$$\mathbb{P}(T_{N_s+1}-s>t)=\mathbb{P}(N_{s+t}-N_s=0)=\mathbb{P}(N_t=0). \qquad \square$$

Let us recall that an exponential random variable is memoryless, in the sense that for all $t,s>0$,

$$\mathbb{P}(T_1>t+s\,|\,T_1>t)=\mathbb{P}(T_1>s).$$

We will also find exponential variables in the general structure of a Markov jump process. This is strongly related to the Markov property.

We finish this section by exhibiting a martingale related to the Poisson process. Observe that the distribution of $(N_s,s\leq t)$ is equivalent to that of $(N_t,T_1,T_2,\ldots,T_{N_t})$. The σ-field $\mathcal{F}_t^N=\sigma(N_s,s\leq t)$ is the σ-field generated by these variables and therefore describes the information given by the process $(N_t)_t$ up to time t.

Proposition 5.2.10 *Let* $(N_t, t \geq 0)$ *be a Poisson process with parameter* λ*. Then the process* $(M_t, t \geq 0)$ *defined by* $M_t = N_t - \lambda t$ *is an* $\mathcal{F}_t^N$*-martingale.*

Proof For each $t > 0$, N_t is a random variable with finite expectation and therefore the same is true for M_t. Moreover, for $s < t$,

$$\mathbb{E}(N_t|\mathcal{F}_s^N) = \mathbb{E}(N_t - N_s + N_s|\mathcal{F}_s^N) = \mathbb{E}(N_t - N_s) + N_s = \lambda(t-s) + N_s.$$

We used the independence of the random variables $N_t - N_s$ and N_s, and Proposition 5.2.5. □

5.2.2 The Strong Markov Property

Let $(N_t, t \geq 0)$ be a Poisson process with intensity λ. For any $s > 0$, we introduce the process $(N_t^s, t \geq 0)$ defined by

$$N_t^s = N_{t+s} - N_s.$$

This process counts the number of events in the time interval $]s,t]$. It is easy to see that it is also a Poisson process with intensity λ, independent of $(N_u, u \leq s)$. In particular, this implies that $(N_t, t \geq 0)$ satisfies the Markov property and from Proposition 5.2.5, we deduce the following

Proposition 5.2.11 *The Poisson process with intensity* λ *is a Markov process. Its transition matrix is given by*

$$P_{i,j}(t) = \mathrm{e}^{-\lambda t}\,\frac{(\lambda t)^{j-i}}{(j-i)!}, \quad \textit{if } j \geq i,$$

and $P_{i,j}(t) = 0$ *otherwise.*

We will generalize this property to $\mathcal{F}_t^N$-stopping times. Let us note that for all n, the jump time T_n is an $\mathcal{F}_t^N$-stopping time, since $\{T_n \leq t\} = \{N_t \geq n\}$. However, if $t < s$, T_{N_s} is not because

$$\{T_{N_s} \leq t\} = \{N_s - N_t = 0\} \notin \mathcal{F}_t^N.$$

Proposition 5.2.12 *Let* $(N_t, t \geq 0)$ *be a Poisson process with intensity* λ *and* S *an* $\mathcal{F}_t^N$*-stopping time. On the event* $\{S < +\infty\}$*, we set for* $t \geq 0$

$$N_t^S = N_{S+t} - N_S.$$

Conditionally on $\{S < +\infty\}$*, the process* $(N_t^S, t \geq 0)$ *is a Poisson process with intensity* λ*, independent of the* σ*-field generated by the trajectory of* N *up to* S*.*

Proof We already know that the result is true if S is constant. Suppose that S takes its values in an increasing sequence of positive real numbers $(s_j)_j$. As S is a stopping time,

$$\{S = s_j\} = \{S \le s_j\} \backslash \{S \le s_{j-1}\} \in \mathcal{F}^N_{s_j}.$$

Let $0 = t_0 < t_1 < t_2 < \cdots < t_k$, $n_1, n_2, \ldots, n_k \in \mathbb{N}$ and A be an event from the σ-field generated by the trajectory of N up to S. We have $A \cap \{S \le t\} \in \mathcal{F}^N_t$. Then

$$\begin{aligned}
&\mathbb{P}\left(A \cap_{i=1}^k \{N^S_{t_i} - N^S_{t_{i-1}} = n_i\}\right) \\
&= \sum_j \mathbb{P}\left(\{S = s_j\} \cap A \cap_{i=1}^k \{N_{s_j+t_i} - N_{s_j} - N_{s_j+t_{i-1}} + N_{s_j} = n_i\}\right) \\
&= \sum_j \mathbb{P}\left(\{S = s_j\} \cap A\right) \prod_{i=1}^k \mathbb{P}\left(N_{s_j+t_i} - N_{s_j+t_{i-1}} = n_i\right) \\
&= \mathbb{P}(A) \prod_{i=1}^k \mathbb{P}\left(N_{t_i - t_{i-1}} = n_i\right). \qquad (5.2.1)
\end{aligned}$$

Here we have used the fact that the process $(N_t)_t$ has independent and stationary increments.

The result is therefore established if the sequence of values of S is discrete. Suppose now that S is any stopping time. We introduce the sequence $(R_n)_n$ defined by

$$R_n = \sum_{k \in \mathbb{N}^*} k 2^{-n} \mathbb{1}_{\{(k-1)2^{-n} < S \le k 2^{-n}\}}.$$

It is easy to see that the R_n are stopping times that converge in decreasing order to S. The equality (5.2.1) is true for each R_n and we can easily justify a passage to the limit, due to the right-continuity of the trajectories of $(N_t)_t$. This concludes the proof. □

We deduce the following fundamental result.

Proposition 5.2.13 *A Poisson process $(N_t, t \ge 0)$ with intensity λ is a strong Markov process. Let us denote as before by $(S_n)_n$ the sequence of waiting times between jumps. Then the variables S_i are independent and follow an exponential law with parameter λ.*

Proof The strong Markov property is an immediate corollary of Proposition 5.2.12. We already know, by Corollary 5.2.9, that $S_1 = T_1$ follows an exponential law with parameter λ. Let us apply Proposition 5.2.12 with $S = T_n$. Thus $S_{n+1} = T_{n+1} - T_n$ is the first jump time of the Poisson process $(N_{T_n+t} - N_{T_n}, t \ge 0)$ with intensity λ and independent of $T_1, \ldots, T_n$, therefore also of $S_1, \ldots, S_n$. The result follows. □

The following proposition provides a constructive proof of the existence of a Poisson process and a simulation algorithm.

Proposition 5.2.14 *Let $(S_n)_n$ be a sequence of independent random variables following an exponential law with parameter λ. For all $n \geq 1$ and $t > 0$, let us define $T_n = S_1 + \cdots + S_n$ and $N_t = \sup\{n, T_n \leq t\} = \mathrm{Card}\{n, T_n \leq t\}$. Then $(N_t, t \geq 0)$ is a Poisson process of intensity λ.*

5.2.3 Asymptotic Behavior of a Poisson Process

Let $(N_t, t \geq 0)$ be a Poisson process with intensity λ. We then have

$$\mathbb{E}(N_t) = \lambda t, \quad \mathrm{Var}(N_t) = \lambda t.$$

Thus,

$$\mathbb{E}(t^{-1} N_t) = \lambda, \quad \mathrm{Var}(t^{-1} N_t) = \frac{\lambda}{t},$$

so $\dfrac{N_t}{t}$ converges in mean square to λ, as t tends to infinity.

In fact, we also have a strong version of this law of large numbers.

Proposition 5.2.15 *Let $(N_t, t \geq 0)$ be a Poisson process of intensity λ. Then $\frac{N_t}{t}$ almost surely converges to λ, as t tends to infinity.*

Proof Let us first note that $N_n = \sum_{i=1}^{n}(N_i - N_{i-1})$ is the sum of independent random variables with the same Poisson distribution of parameter λ. It then follows from the strong law of large numbers that $\dfrac{N_n}{n}$ almost surely converges to λ, as n tends to infinity. We can then write

$$\frac{N_t}{t} = \frac{N_{[t]}}{[t]}\frac{[t]}{t} + \frac{N_t - N_{[t]}}{t}.$$

It is then sufficient to show that $\displaystyle\sup_{n \leq t < n+1} \frac{N_t - N_n}{n}$ tends almost surely to 0 as n tends to infinity. Let us set

$$\xi_n = \sup_{n \leq t < n+1} (N_t - N_n) = N_{n+1} - N_n.$$

The random variables ξ_n are independent and have the same Poisson distribution with parameter λ. The law of large numbers then implies that $\frac{\xi_1 + \cdots + \xi_n}{n}$ converges to λ almost surely and therefore that $\frac{\xi_n}{n}$ converges almost surely to 0. □

We also have a central limit theorem.

Theorem 5.2.16 *Let $(N_t, t \geq 0)$ be a Poisson process with intensity λ. Then the process $\left(\frac{N_t - \lambda t}{\sqrt{\lambda t}}, t \geq 0\right)$ converges in distribution, as t tends to infinity, to a standard normal random variable.*

Proof We will reason as in the previous proof. By the central limit theorem, we know that the sequence $\left(\frac{N_n-\lambda n}{\sqrt{\lambda n}}\right)_n$ converges in distribution to Z as n tends to infinity, where Z is a standard normal random variable. Moreover, $\frac{N_t-N_{[t]}}{\sqrt{\lambda t}} \le \frac{\xi_{[t]}}{\sqrt{\lambda [t]}}$. However, $\mathbb{P}(\frac{\xi_n}{\sqrt{n}} > \varepsilon) = \mathbb{P}(\xi_n > \varepsilon\sqrt{n}) = \mathbb{P}(\xi_1 > \varepsilon\sqrt{n})$, which tends to 0 as n tends to infinity. Therefore $\frac{N_t-N_{[t]}}{\sqrt{\lambda t}}$ tends in probability to 0. Finally, we have

$$\frac{N_t-\lambda t}{\sqrt{\lambda t}} = \frac{N_{[t]}-\lambda[t]}{\sqrt{\lambda[t]}} \times \sqrt{\frac{[t]}{t}} + \frac{N_t-N_{[t]}}{\sqrt{\lambda t}} + \sqrt{\lambda}\frac{[t]-t}{\sqrt{t}}.$$

Knowing that $N_t - N_{[t]}$ and $N_{[t]}$ are independent and using characteristic functions, as in Section 4.2, we conclude the proof. □

We can in fact also establish a functional central limit theorem, which gives the convergence of the random function

$$t \longrightarrow Y_t^u = \frac{N_{ut}-\lambda tu}{\sqrt{\lambda u}}$$

to a Brownian motion, as u tends to infinity. The proof is inspired by the one we have already seen in Section 4.2. Indeed, we have shown that for fixed t, Y_t^u converges in distribution towards a centred random variable B_t, with variance t.

Similarly, we can study the distribution behavior of $(Y_{t_1}^u, Y_{t_2}^u - Y_{t_1}^u, \ldots, Y_{t_k}^u - Y_{t_{k-1}}^u)$, for $t_1 < \cdots < t_k$, and show that this vector converges to $(Z_{t_1}, Z_{t_2-t_1}, \ldots, Z_{t_k-t_{k-1}})$, where the random variables $Z_{t_i-t_{i-1}}$ are independent and follow Gaussian distributions with respective variances $t_i - t_{i-1}$. Note also that the amplitude of the jumps of Y_t^u is $\frac{1}{\sqrt{\lambda u}}$, which tends to 0 as u tends to infinity. We have therefore proved that the process $(Y_t^u, t \ge 0)$ converges, in the sense of finite-dimensional marginals, towards a process $(B_t, t \ge 0)$ which has independent and stationary increments, with continuous trajectories and such that for each $t > 0$ the random variable B_t is a centred normal variable with variance t. This is a Brownian motion (see Definition 4.2.5).

Remark 5.2.17 We have seen in Proposition 5.2.10 that the process $N_t - \lambda t$ is a martingale. But (4.5.6) is not satisfied. Indeed, by taking $t = n/\lambda$, we can show that

$$\mathbb{E}(|N_t - \lambda t|) = 2\left(n - n\mathrm{e}^{-n}\sum_{k=n+1}^{\infty}\frac{n^k}{k!} - \mathrm{e}^{-n}\sum_{k=0}^{n} k\frac{n^k}{k!}\right),$$

which tends to $+\infty$ as n tends to infinity. Thus, we cannot use Theorem 4.5.6 to conclude the convergence of N_t/t.

5.2.4 Compound Poisson Processes

Given a Poisson process, we can construct more complex pure jump Markov processes by assuming the jump amplitudes are random and independent of the Poisson process.

Example 5.2.18 Let $(N_t, t \geq 0)$ be a Poisson process with intensity λ and jump times $(T_n)_n$ and let $(Z_n)_n$ be a Markov chain with values in $\mathbb{Z}$, independent of $(N_t)_t$ and with transition matrix $M_{i,j}$. Then

$$X_t = \sum_{n=0}^{\infty} Z_n \mathbb{1}_{[T_n, T_{n+1}[}(t)$$

is a Markov pure jump process, called a *compound Poisson process*, with transition matrix

$$P_{i,j}(t) = \sum_n \mathbb{P}(T_n \leq t < T_{n+1}, Z_n = j | Z_0 = i) = \sum_n \mathbb{P}(N_t = n)\mathbb{P}(Z_n = j | Z_0 = i),$$

by independence. Hence

$$P_{i,j}(t) = \mathrm{e}^{-\lambda t} \sum_n \frac{(\lambda t)^n}{n!} M_{i,j}^{(n)}.$$

We will see in the next chapter that the distribution of a pure jump Markov process can be described by generalizing this decomposition: a separation between the distribution of the time between jumps and distribution of the amplitude of jumps.

5.3 The Generator of a Pure Jump Markov Process

5.3.1 The Infinitesimal Generator

We return to the general framework of pure jump Markov processes. We have seen that the law of such a process is characterized by its semi-group as soon as we know its initial condition (Proposition 5.1.4). The semi-group is given by the set of matrices $P(t), t > 0$. We will see in this subsection that it is enough to know a single matrix that describes the infinitesimal behavior of $P(t)$ in the neighborhood of 0. This is due to the Chapman–Kolmogorov property (Proposition 5.1.5). This matrix, called the *infinitesimal generator* of the process, allows us to describe the structure of the process.

Let us first note that a pure jump Markov process satisfies the strong Markov property.

Theorem 5.3.1 *Let $(X_t, t \geq 0)$ be a Markov jump process and S an $\mathcal{F}_t^X$-stopping time. Conditionally on $\{S < +\infty\}$ and on $\{X_S = i\}$, the process $(X_{S+t}, t \geq 0)$ is independent of the σ-field generated by X up to time S and its law is that of $(X_t, t \geq 0)$ starting from $X_0 = i$.*

The proof is an adaptation of that given for the Poisson process (Proposition 5.2.12) and is left to the reader.

The semi-group property (5.1.2) implies that the knowledge of $P(t)$ for all t can be deduced from the knowledge of $P(t)$ for small t. In fact, we will prove that it is sufficient to know its right derivative at 0.

Theorem 5.3.2 *Let $P(t)$, $t > 0$, be the semi-group of transition matrices of a pure jump Markov process $(X_t)_t$. There exists a matrix $(Q_{i,j}, i, j \in \mathbb{N})$, called the* infinitesimal generator *of the semi-group $(P(t))_t$ or of the Markov process $(X_t)_t$, which satisfies*

$$Q_{i,j} \geq 0 \text{ for } i \neq j \ ; \ Q_{i,i} = -\sum_{j \in \mathbb{N} \setminus \{i\}} Q_{i,j} \leq 0,$$

this last inequality being strict except if the state i is absorbing. Let $q_i = -Q_{i,i}$.

When $h \downarrow 0$,

$$\begin{aligned} P_{i,j}(h) &= hQ_{i,j} + o(h) \text{ for } i \neq j, \\ P_{i,i}(h) &= 1 - hq_i + o(h). \end{aligned} \tag{5.3.1}$$

Moreover, conditionally on $X_0 = i$ and if $q_i \neq 0$, the first jump T_1 and the position $Z_1 = X_{T_1}$ after this jump are independent, with T_1 of exponential law with parameter q_i and Z_1 of law given by $\left(\frac{Q_{i,j}}{q_i}, j \neq i\right)$. If $q_i = 0$, the process remains absorbed at i.

Remark 5.3.3 As in the case of a diffusion process (cf. Definition 4.6.19), there is a functional definition of the infinitesimal generator. For any bounded function f on $\mathbb{N}$, we set

$$Qf(i) = \lim_{t \to 0, t > 0} \frac{P_t f(i) - f(i)}{t}.$$

By (5.3.1), we have for t near 0,

$$\begin{aligned} P_t f(i) &= \sum_j P_{i,j}(t) f(j) = P_{i,i}(t) f(i) + \sum_{j \neq i} P_{i,j}(t) f(j) \\ &= \Big(1 + tQ_{i,i} + o(t)\Big) f(i) + \sum_{j \neq i} \Big(t Q_{i,j} + o(t)\Big) f(j) \\ &= f(i) + t \left(\sum_{j \neq i} Q_{i,j}(f(j) - f(i)) \right) + o(t), \end{aligned}$$

and therefore

$$Qf(i) = \sum_{j \neq i} Q_{i,j}\Big(f(j) - f(i)\Big). \tag{5.3.2}$$

Definition 5.3.4 We call q_i the *jump rate* of the process starting from i. We thus have

$$\mathbb{P}(T_1 > t | X_0 = i) = \mathrm{e}^{-q_i t}.$$

The number $Q_{i,j}$ will be called the *transition rate* from i to j.

Thanks to the strong Markov property, we deduce the following result.

Corollary 5.3.5 *For all $n \in \mathbb{N}^*$, conditionally on the σ-field generated by the process up to time T_{n-1}, the random variable $T_n - T_{n-1}$ is independent of $Z_n = X_{T_n}$. Moreover, the conditional law of $T_n - T_{n-1}$ is an exponential law with parameter $q_{Z_{n-1}}$, and the conditional law of Z_n is given by $(\frac{Q_{Z_{n-1},j}}{q_{Z_{n-1}}}, j \in \mathbb{N})$.*

Proof of Theorem 5.3.2 We want to compute $\mathbb{P}(T_1 > t | X_0 = i)$. Consider $n \in \mathbb{N}$ and $h > 0$ and suppose that $h \to 0$ and $n \to \infty$ in such a way that $nh \uparrow t$. We use the semi-group property. For this, first observe that

$$\{T_1 > nh\} \subset \{X_0 = X_h = \cdots = X_{nh}\} \subset \{T_1 > nh\} \cup \{T_2 - T_1 \leq h\}.$$

As $\mathbb{P}(T_2 - T_1 \leq h) \to 0$ when $h \to 0$, we have

$$\begin{aligned}
\mathbb{P}(T_1 > t | X_0 = i) &= \lim_{h \to 0, nh \to t} \mathbb{P}(X_0 = X_1 = \cdots = X_{nh} \,|\, X_0 = i) \\
&= \lim_{h \to 0, nh \to t} \big(P_{i,i}(h)\big)^n \\
&= \lim_{h \to 0, nh \to t} \mathrm{e}^{n \ln P_{i,i}(h)}, \text{ by the Markov property.}
\end{aligned}$$

As $P_{i,i}(h)$ tends to 1 when $h \to 0$, we know that $\log P_{i,i}(h) \sim P_{i,i}(h) - 1$ in a neighborhood of 0. As, moreover, n is of order $\frac{t}{h}$, we deduce from the existence of the previous limit that there exists $q_i \in [0, +\infty]$ such that $\lim_{h \to 0} \frac{1}{h}(1 - P_{i,i}(h)) = q_i$. Thus,

$$\mathbb{P}(T_1 > t | X_0 = i) = \mathrm{e}^{-q_i t}.$$

This last property implies that $q_i < +\infty$ and that $q_i = 0$ if and only if i is an absorbing state. We then set $Q_{i,i} = -q_i$.

The proof of the existence of a limit to $\dfrac{P_{i,j}(h)}{h}$, for $i \neq j$, is done in a similar way. We have

$$\{T_1 \leq t, Z_0 = i, Z_1 = j\} = \lim_{h \to 0, nh \to t} \cup_{0 \leq m \leq n} \{X_0 = X_h = \cdots = X_{(m-1)h} = i, X_{mh} = j\},$$

from which

$$\begin{aligned}
\mathbb{P}(T_1 \le t, Z_1 = j | X_0 = i) &= \lim_{h\to 0, nh\to t} \sum_{m=0}^{n-1} \left(P_{i,i}(h)\right)^m P_{i,j}(h) \\
&= \lim_{h\to 0, nh\to t} \frac{1-\left(P_{i,i}(h)\right)^n}{1-P_{i,i}(h)} P_{i,j}(h) \\
&= \lim_{h\to 0, nh\to t} \left(1-\left(P_{i,i}(h)\right)^n\right) \frac{h}{1-P_{i,i}(h)} \frac{1}{h} P_{i,j}(h) \\
&= \frac{1-\mathrm{e}^{-q_i t}}{q_i} \lim_{h\to 0} \frac{1}{h} P_{i,j}(h).
\end{aligned}$$

Thus, $Q_{i,j} = \lim_{h\to 0} \frac{1}{h} P_{i,j}(h)$ exists for $i \neq j$ and

$$\mathbb{P}(T_1 \le t, Z_1 = j \,|\, X_0 = i) = (1-\mathrm{e}^{-q_i t}) \frac{Q_{i,j}}{q_i},$$

from which $\mathbb{P}(T_1 \le t, Z_1 = j \,|\, X_0 = i) = \mathbb{P}(T_1 \le t \,|\, X_0 = i)\mathbb{P}(Z_1 = j \,|\, X_0 = i)$, which implies the independence of T_1 and Z_1 conditionally on $X_0 = i$. Moreover,

$$\mathbb{P}(Z_1 = j \,|\, X_0 = i) = \frac{Q_{i,j}}{q_i}. \tag{5.3.3}$$

We deduce from this that

$$\sum_{j\neq i} Q_{i,j} = q_i = -Q_{i,i}. \qquad \square$$

This theorem allows us to obtain the Kolmogorov equations, which are fundamental in practice. These equations describe the time dynamics of the marginal distributions from the rate matrix $(Q_{i,j}))_{i,j}$. Let I be the identity matrix on $\mathbb{N}$.

Theorem 5.3.6 *Under Assumption* (5.1.1),

1) $(P(t), t \ge 0)$ *is the unique solution of the backward Kolmogorov equation*

$$\frac{\mathrm{d}P}{\mathrm{d}t}(t) = QP(t), \ \textit{for} \ t > 0; \ P(0) = I,$$

that is, for all $i, j \in \mathbb{N}$,

$$\frac{\mathrm{d}P_{i,j}}{\mathrm{d}t}(t) = \sum_{k\in\mathbb{N}} Q_{i,k} P_{k,j}(t).$$

For all i *in* $\mathbb{N}$ *and any function* g, $u(t,i) = \mathbb{E}(g(X_t)|X_0 = i)$ *is a solution of*

$$\frac{\partial u}{\partial t}(t,i) = \sum_{k\in\mathbb{N}} Q_{i,k}\, u(t,k),\ t > 0, \ ; \ u(0,i) = g(i),\ i \in \mathbb{N}.$$

2) $(P(t), t \geq 0)$ *is the unique solution of the forward Kolmogorov equation*

$$\frac{\mathrm{d}P}{\mathrm{d}t}(t) = P(t)Q, \text{ for } t > 0; \; P(0) = I,$$

that is, for all $i, j \in \mathbb{N}$,

$$\frac{\mathrm{d}P_{i,j}}{\mathrm{d}t}(t) = \sum_{k \in \mathbb{N}} P_{i,k}(t) Q_{k,j}.$$

Furthermore, the family of marginal distributions $\mu(t)$ *of* $(X_t, t \geq 0)$ *satisfies the Fokker–Planck equation*

$$\frac{\partial \mu_j(t)}{\partial t} = \sum_{k \in \mathbb{N}} \mu_k(t) Q_{k,j}, \text{ for } t > 0, j \in \mathbb{N}.$$

Proof The idea of the proof is as follows. To establish the backward Kolmogorov equation, it is sufficient to differentiate $P(t+h)$ at $h=0$ using the semi-group property $P(t+h) = P(h)P(t)$. The equation for u is deduced by multiplying on the right by the column vector (g_j). The forward equation is obtained in the same way, by writing $P(t+h) = P(t)P(h)$. The Fokker–Planck equation is then immediately deduced by multiplying on the left by the row vector $(\mu_i(0))$. □

5.3.2 Embedded Markov Chains

Let $(X_t)_t$ be a pure jump Markov process associated with (T_n, Z_n), $n \geq 0$. The sequence $Z_n = X_{T_n}$ is a discrete-time Markov chain. This is a consequence of the strong Markov property of $(X_t)_t$. It is called the *embedded Markov chain* and satisfies $Z_{n+1} \neq Z_n$, almost surely, for all n. Its transition matrix is easily computed in terms of the generator Q of X_t, thanks to (5.3.3):

$$\tilde{P}_{i,j} = \begin{cases} \frac{Q_{ij}}{q_i} & \text{if } j \neq i \\ 0 & \text{if } j = i. \end{cases}$$

If we set

$$S_n = q_{Z_{n-1}}(T_n - T_{n-1}),$$

where $q_i = \sum_{i \neq j} Q_{i,j}$, and for all $t \geq 0$, $N_t = \sup\{n, \sum_{k=1}^{n} S_k \leq t\}$, then the process $(N_t)_t$ is a Poisson process with intensity 1. Indeed, it is sufficient to apply Corollary 5.3.5: the conditional law of $T_n - T_{n-1}$ given Z_{n-1} is an exponential law with parameter $q_{Z_{n-1}}$, and therefore the law of S_n is an exponential law with parameter 1. This is based on the fact that if U is an exponential random variable with parameter λ, then λU is an exponential random variable with parameter 1.

Conversely, it is possible to define a continuous-time jump Markov process with values in $\mathbb{N}$ from its infinitesimal generator. Let Q be a rate matrix, that is, a matrix indexed by $\mathbb{N}$ such that for all $i, j \in \mathbb{N}$,

$$Q_{i,j} \geq 0 \text{ if } j \neq i \, ; \, Q_{i,i} = -\sum_{j \neq i} Q_{i,j} \leq 0. \tag{5.3.4}$$

Let us set $q_i = -Q_{i,i}$ and define the transition matrix $\tilde{P}$ by $\tilde{P}_{i,j} = \frac{Q_{i,j}}{q_i}$ if $i \neq j$ and $q_i \neq 0$, and $\tilde{P}_{i,j} = 0$ if $j = i$, with the convention $\tilde{P}_{i,j} = 0 \, \forall j$ if $q_i = 0$. To every $i \in \mathbb{N}$, we associate the chain $(Z_n)_n$ with transition matrix $\tilde{P}$ and we consider a Poisson process (N_t) of intensity 1 independent of $(Z_n)_n$. The sequence of jump times of (N_t) is denoted $(T'_n)_n$ and for $n \geq 1$, we define

$$U_n = \frac{T'_n - T'_{n-1}}{q_{Z_{n-1}}}; \qquad T_n = U_1 + \cdots + U_n.$$

If the non-explosion condition (5.1.1) is satisfied by $(T_n)_n$, then

$$X_t = \sum_{n \geq 0} Z_n \mathbf{1}_{[T_n, T_{n+1}[}(t), \; t \geq 0 \tag{5.3.5}$$

is a pure jump Markov process with infinitesimal generator Q.

Algorithmic construction of $(X_t)_t$

It is easy to construct the process $(X_t, t \geq 0)$ starting from state i by iterating the following procedure.

- Start from $X_0 = i$ and wait an exponential time S_1 with parameter q_i. The process remains constant equal to i until time S_1.
- At time S_1, the process jumps from state i to state j with probability $\frac{Q_{i,j}}{q_i}$.
- Repeat the procedure (wait an exponential time S_2 with parameter q_j and independent of $S_1, \ldots$).

The exponential lifetimes $U_1, U_2, \ldots$ are called *sojourn times* in the respective states $Z_1, Z_2, \ldots$. The process has a finite lifetime if

$$T_\infty = \sum_{k \geq 1} U_k = \lim_n T_n < +\infty.$$

Note that $T_\infty < +\infty$ obviously implies that $\lim_{t \uparrow T_\infty} X_t = +\infty$; in this case, we say that the process *explodes* in finite time.

Let us now study the non-explosion condition (5.1.1) for $(T_n)_n$, which ensures that the process is defined by (5.3.5) for all $t \in \mathbb{R}_+$, and thus proves the existence of a pure jump Markov process with infinitesimal generator Q.

Proposition 5.3.7 *The non-explosion condition* $\lim_n T_n = +\infty$*, almost surely, is satisfied if and only if*

$$\sum_{n\geq 0} \frac{1}{q_{Z_n}} = +\infty \quad \textit{almost surely.}$$

Two simple sufficient conditions are given in the next corollary.

Corollary 5.3.8 *For an infinitesimal generator* Q *to be the infinitesimal generator of a non-exploding pure jump Markov process satisfying* (5.3.4), *it is sufficient to check one of the following two conditions:*

(i) $\sup_{i\in E} q_i < +\infty$.
(ii) *The Markov chain* $(Z_n)_n$ *with transition matrix* $\tilde{P}$ *is recurrent.*

Proof of Proposition 5.3.7 We have seen that $T_n - T_{n-1} = \frac{S_n}{q_{Z_{n-1}}}$ and the random times S_n are independent. Moreover, S_n follows an exponential distribution with parameter 1, so conditionally on Z_{n-1}, the random variable $\frac{S_n}{q_{Z_{n-1}}}$ follows an exponential distribution with parameter $q_{Z_{n-1}}$.

We prove that for a sequence (e_i) of independent exponential random variables with parameters q_i, we have

$$\mathcal{E} = \sum_i e_i = +\infty \Longleftrightarrow \sum_i \frac{1}{q_i} = +\infty.$$

Indeed, as $\mathbb{E}(e_i) = \frac{1}{q_i}$, it is clear that if $\sum_i \frac{1}{q_i} < +\infty$, then $\mathcal{E}$ is finite almost surely.

Let's now study the case where $\sum_i \frac{1}{q_i} = +\infty$. The Laplace transform of $\mathcal{E}$ is given by

$$\mathbb{E}(\mathrm{e}^{-\lambda\mathcal{E}}) = \prod_i \mathbb{E}(\mathrm{e}^{-\lambda e_i}) = \prod_i \frac{q_i}{\lambda + q_i}.$$

It is null if and only if $\sum_i e_i = +\infty$ with probability 1.

The product above is null if and only if the sum $\sum_i \log\left(1 - \frac{\lambda}{\lambda+q_i}\right)$ equals $-\infty$. However, for sufficiently small λ, this series behaves as the series $-\sum_i \frac{1}{q_i}$, which completes the proof. □

5.4 Continuous Time Branching Processes

5.4.1 Definition and Branching Property

Consider a process $(X_t)_{t\geq 0}$ describing the following population dynamics:

- At time $t = 0$, there is a random number X_0 of individuals.
- Each individual has a random lifetime that follows an exponential distribution with parameter $a > 0$.

- At the end of this time, the individual reproduces according to the reproduction law $(p_i)_{i\in\mathbb{N}}$. The probability that its lineage stops is therefore p_0. Since we are interested in the population size dynamics we will avoid the trivial cases and assume that
$$0 < p_0 < 1; \;\; p_1 = 0.$$
- The lifetimes and the numbers of children of each individual are independent of each other.

Definition 5.4.1 The process $(X_t)_{t\geq 0}$ defined above is called a *continuous time branching process*. For any t, X_t represents the number of individuals at time t.

When $p_0 + p_2 = 1$, the process is called a *binary branching process* or *linear birth and death process*. Such a process models, for example, the mechanism of cell division or death.

If, in addition, $p_0 = 0$ (a lineage never disappears), the process is called a *binary fission process* or *Yule process*.

We introduce the expectation of the reproduction law $m = \sum_{k\geq 0} k\, p_k$ and for $s \in [0,1]$, its generating function $g(s) = \sum_{k\geq 0} p_k\, s^k$. Recall that if $m < +\infty$, the function g is differentiable at 1 and $g'(1) = m$. Let s_0 be the smallest root of the equation $g(s) = s$.

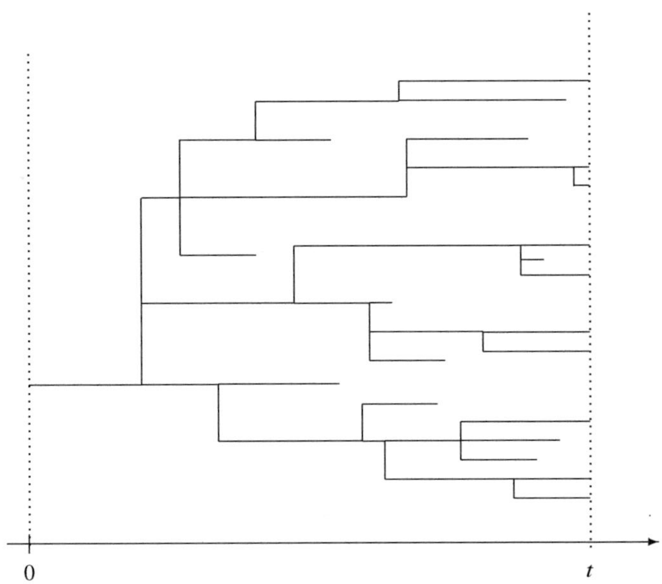

Fig. 5.3 A trajectory of a continuous time branching process.

In Figure 5.3, the branches represent the lineages of individuals. The figure models the evolution over time of these lineages.

Remark 5.4.2 The modeling of the lifetime of individuals by an exponential law needs to be discussed. Indeed, an exponential law has the property of non-aging, which does not necessarily represent reality. However, some species do not age. The hydra, a small freshwater polyp of a few millimeters, is an example, as well as some turtles, some mollusks and corals, and some fish like the sturgeon (see Tully [77] on this subject). This hypothesis of an exponential law is linked to the Markov property for the jump process $(X_t)_{t\geq 0}$, as we saw in the previous section. This makes computations easy. There are works that deviate from this hypothesis and markovianity, to take into account other population dynamics, such as age-dependent branching processes (cf. Kaj–Sagitov [47]) or branching trees ("splitting trees") (cf. Lambert [56]), for which the lifetimes of individuals follow any laws.

Theorem 5.4.3 *The branching process $(X_t)_{t\geq 0}$ is a pure jump Markov process. The process $(Z_n)_n = (X_{T_n})_n$ is a Markov chain with transition matrix P such that $P_{i,j} = p_{j-i+1}$ if $j \geq i-1$ and $P_{i,j} = 0$ otherwise. The random times $T_{n+1} - T_n$ are independent conditionally on $Z_n = i$ and follow an exponential law with parameter ia.*

Fundamental property: From the definition of the process, we can see that if the initial population is composed of i individuals, the sub-processes resulting from these i ancestors are independent and have the same law. The process $(X_t)_t$ can therefore be written as the sum

$$X_t = Z_t^1 + \cdots + Z_t^i,$$

where the processes $\big((Z_t^k, t \geq 0), k = 1, \dots, i\big)$ are independent and identically distributed with the same law as X but starting from a single individual. One says that the process X satisfies the *branching property*. In particular, for any time $t > 0$, the generating function of X_t is equal to the product of the generating functions of the Z_t^k, which is written: for all $i \geq 0$ and $s \in [0, 1]$,

$$\sum_{j=0}^{\infty} P_{i,j}(t) s^j = \left(\sum_{j=0}^{\infty} P_{1,j}(t) s^j \right)^i, \tag{5.4.1}$$

with $(P(t), t \geq 0)$ the transition semi-group of the process.

The proof of Theorem 5.4.3 follows from Corollary 5.3.5 and the following proposition, which gives the infinitesimal generator of the process.

Proposition 5.4.4 *The generator of the continuous-time branching process is given by*

1) *$q_i = ia$. In particular, $q_0 = 0$ and the point 0 is absorbing.*
2)

$$\forall\, i \neq j\,,\; Q_{i,j} = \begin{cases} ia p_{j-i+1}, & \textit{if } j \geq i-1, \\ 0 & \textit{otherwise.} \end{cases}$$

Proof 1) We know that $q_i = \lim_{t\to 0} \dfrac{1-P_{ii}(t)}{t}$ and that $1-P_{i,i}(t) = \mathbb{P}_i(X(t)\neq i)$. Let us prove that this probability is close to $\mathbb{P}_i(T_1 < t) = 1-\mathrm{e}^{-iat}$ for small t, where T_1 is the first jump time. Indeed,

$$\begin{aligned}&\mathbb{P}_i(T_1 < t) - \mathbb{P}_i(X(t)\neq i)\\&\quad = \mathbb{P}_i(\text{there are at least 2 jumps before } t, \text{ one to leave } i \text{ and one to return})\\&\quad \le \mathbb{P}_i(T_2 < t).\end{aligned}$$

However, $\mathbb{P}_i(T_2 < t) = \mathbb{P}_i(T_1+T_2-T_1 < t) \le \mathbb{P}_i(T_1 < t)\,\mathbb{P}_i(T_2-T_1 < t)$ since by construction, T_1 and T_2-T_1 are independent. We deduce that $\mathbb{P}_i(T_2 < t)$ is of order t^2, negligible compared to $\mathbb{P}_i(T_1 < t)$ for small t.

2) By a similar argument, we also have that for $i\neq j$,

$$P_{i,j}(t) = \mathbb{P}(X_t = j\,|\,X_0 = i) = \mathbb{P}(T_1 < t, X_{T_1} = j\,|\,X_0 = i) + o(t).$$

Thus, $Q_{i,j} = \lim_{t\to 0}\frac{P_{i,j}(t)}{t} = \lim_{t\to 0}\frac{1}{t}\mathbb{P}_i(T_1 < t)\,\mathbb{P}_i(Z_1 = j) = aip_{j-i+1}$. □

We deduce from these computations that

$$\begin{aligned}&P_{i,j}(h) = iap_{j-i+1}h + o(h), \qquad \text{for } j \ge i-1,\\&P_{i,i}(h) = 1 - iah + o(h).\end{aligned}$$

Using the previous section, we can easily write the Kolmogorov equations for the process Z

$$\frac{\mathrm{d}}{\mathrm{d}t}P_{i,j}(t) = (PQ)_{ij}(t) = -j\,a\,P_{i,j}(t) + a \sum_{1\le k\le j+1, k\neq i} k\,p_{j-k+1}\,P_{i,k}(t), \quad \text{(forward)}$$

$$\frac{\mathrm{d}}{\mathrm{d}t}P_{i,j}(t) = (QP)_{ij}(t) = -i\,a\,P_{i,j}(t) + i\,a \sum_{k\ge i-1, k\neq i} p_{k-i+1}\,P_{k,j}(t) \quad \text{(backward)},$$

with the initial conditions

$$P_{i,j}(0+) = \begin{cases} 1 & \text{for } i = j,\\ 0 & \text{for } i \neq j.\end{cases}$$

For example, in the case of critical binary reproduction where $p_0 = p_2 = \frac{1}{2}$, the backward Kolmogorov equation becomes for all i, j

$$\frac{\mathrm{d}}{\mathrm{d}t}P_{i,j}(t) = \frac{ia}{2}\Big(P_{i+1,j}(t) + P_{i-1,j}(t) - 2P_{i,j}(t)\Big),$$

which gives a recurrence equation that can be solved. However, these Kolmogorov equations are not the most suitable tool for studying the law of a continuous-time branching process and it is more judicious to study the dynamic equation satisfied by the generating functions, as soon as the process is defined at all times.

5.4.2 Equation for the Generating Function

We have seen that, thanks to the branching property, the generating functions satisfy the property (5.4.1) for all time t, allowing us to reduce the initial condition to $X_0 = 1$, as we will now assume.

For $s \in [0, 1]$, we set

$$F(s,t) = \mathbb{E}(s^{X_t} \,|\, X_0 = 1) = \sum_{j\geq 0} \mathbb{P}(X_t = j \,|\, X_0 = 1)\, s^j = \sum_{j\geq 0} P_{1,j}(t)\, s^j .$$

We then have

$$\begin{aligned}
\frac{\partial}{\partial t} F(s,t) &= \sum_{j\geq 0} \frac{\partial}{\partial t} P_{1,j}(t)\, s^j \\
&= \sum_{j\geq 0} s^j \sum_{k\geq 0} Q_{1,k} P_{k,j}(t) \qquad \text{(backward Kolmogorov Equation)} \\
&= \sum_{k\geq 0} Q_{1,k} \sum_{j\geq 0} s^j\, P_{k,j}(t) \\
&= \sum_{k\geq 0} Q_{1,k}\, (F(s,t))^k \qquad \text{(branching property (5.4.1))}.
\end{aligned}$$

But $Q_{1,1} = -a$ and $Q_{1,j} = a\, p_j$ if $j \neq 1$. Hence

$$\frac{\partial}{\partial t} F(s,t) = a \left(\sum_{j\geq 0} p_j\, (F(s,t))^j - F(s,t) \right) = a\, (g(F(s,t)) - F(s,t)),$$

where g is the generating function of the reproduction law.

Let us introduce the function

$$u(s) = a(g(s) - s), \quad s \in [0, 1].$$

The equation is written

$$\frac{\partial}{\partial t} F(s,t) = a\, (g(F(s,t)) - F(s,t)) = u(F(s,t)) \tag{5.4.2}$$

with $F(s, 0) = s$.

We will see how from this equation, we can deduce a non-explosion criterion, an extinction criterion, the dynamics of the expectation of the process and explicit computations in particular cases.

5.4.3 Non-Explosion Criterion

The process can have an infinite lifetime or only be defined over a time interval $[0, T_\infty[$, where T_∞ is finite with positive probability. We have the following theorem.

Theorem 5.4.5 *The continuous time branching process has an infinite lifetime almost surely if and only if $m = g'(1) < +\infty$, or if $m = +\infty$ and*

$$\int_{1-\varepsilon}^{1} \frac{\mathrm{d}s}{u(s)} = -\infty.$$

Before proving this theorem, let us note that the function u is positive if g has only the fixed point 1 (and in this case $m = g'(1) < +\infty$), and that it is positive before s_0 (since $g(0) = p_0$), and negative afterwards in the case where g admits a fixed point $s_0 < 1$.

Proof Let $h_i(t) = \mathbb{P}_i(T_\infty > t) = \sum_{j\geq 0} \mathbb{P}_i(Z_t = j)$. By the branching property, we know that $1 - h_i(t) = 1 - h(t)^i$, where $h(t) = F(1,t)$. The equation (5.4.2) for $s = 1$ gives that

$$h'(t) = u(h(t)), \quad t > 0. \tag{5.4.3}$$

A limit point of the function h as t tends to infinity is a stationary point of the equation (5.4.3) and therefore a fixed point for the function g. The function h is decreasing and $h(0) = 1$, so $h(t) \leq 1$ for all $t \geq 0$. In the case where g has only the fixed point 1, h is therefore constant equal to 1 and $\mathbb{P}_1(T_\infty = \infty) = \lim_{t\to\infty} h(t) = 1$.

Now suppose that g admits a fixed point $s_0 < 1$. Note that since $h(0) = 1$, the function h is positive and decreasing and $h(t) \in]s_0, 1]$. Suppose that the branching process has a finite lifetime T_∞ with positive probability. In this case h converges to s_0 and there exists a time t_0 such that $s_0 < h(t_0) < 1$. Let's fix $v \in]s_0, 1[$ and set $U(x) = \int_v^x \frac{\mathrm{d}s}{u(s)}$. Since the function $t - U(h(t))$ has zero derivative, it is constant, and $t - U(h(t)) = t_0 - U(h(t_0)) < +\infty$. By letting t tend to 0, we obtain that $U(1^-)$ has a finite value. We deduce that $\int_v^1 \frac{\mathrm{d}s}{u(s)}$ is finite, which implies that $m = +\infty$. (By an expansion of $g(s)$ in the neighborhood of 1, we see that if $m < +\infty$, the integral necessarily diverges.)

Conversely, suppose that $m = +\infty$ and that $\int_v^1 \frac{\mathrm{d}s}{u(s)}$ converges. We can then define the function

$$G(x) = \int_1^x \frac{\mathrm{d}s}{u(s)},$$

for $x \in]s_0, 1]$. Using $h' = u(h)$, we obtain that $t - G(h(t)) = 0$, which implies $h(t) < 1$ as soon as $t > 0$. □

5.4.4 The Moment Equation – Probability and Extinction Time

The Moment equation

Suppose that $m < +\infty$. Suppose that the expectation of X_t exists for all t and that the function F is sufficiently smooth. Recall that then $m(t) = \mathbb{E}(X_t) = \frac{\partial}{\partial s}F(s,t)_{|s=1}$. From (5.4.2), we can deduce an equation satisfied by the function m. Indeed

$$\frac{\partial}{\partial t}\mathbb{E}(X_t) = \frac{\partial}{\partial s}u(F(s,t))_{|s=1},$$

and $t \longrightarrow m(t)$ is a solution of the differential equation

$$m'(t) = u'(F(1,t))m(t) = u'(1)m(t),$$

with $u'(1) = a(g'(1)-1) = a(m-1)$. We deduce that

$$m(t) = \mathrm{e}^{a(m-1)t}. \tag{5.4.4}$$

The parameter $\rho = a(m-1)$ is the Malthusian parameter. In the supercritical case where $\rho > 0$, $m(t)$ tends to infinity; in the subcritical case where $\rho < 0$, $m(t)$ tends to 0. If $\rho = 0$, in the critical case, the population remains constant on average.

Extinction probability

Let us now focus on the eventual extinction of the population. We denote by T_0 the extinction time,

$$T_0 = \inf\{t \geq 0, X_t = 0\},$$

with the convention $\inf \emptyset = +\infty$, and introduce the extinction probability

$$q(t) = \mathbb{P}_1(T_0 \leq t), \quad t \in]0,+\infty].$$

Theorem 5.4.6 *The law of the extinction time is given implicitly by*

$$\int_0^{q(t)} \frac{\mathrm{d}s}{u(s)} = t\,, \quad t \geq 0.$$

Proof The arguments in the proof are close to those of Theorem 5.4.5. Observe that for all $t > 0$, $q(t) = \mathbb{P}_1(X_t = 0) = F(0,t)$, which is a bounded and increasing function over time because state 0 is absorbing. It is clear that $q(0^+) = 0$ and that $\lim_{t\to\infty} q(t) = s_0$, the smallest fixed point of g. Indeed, $\lim_{t\to\infty} q(t) = q_\infty$ is a stationary point for the dynamics defined by (5.4.2) and therefore satisfies $g(q_\infty) = q_\infty$. Thus $q_\infty = 1$ in the subcritical and critical cases, and $q_\infty = s_0 < 1$ in the supercritical case. From (5.4.2), we also deduce that

$$q'(t) = u(q(t)), \quad t > 0, \tag{5.4.5}$$

with $q(t) \in [0, s_0[$. The function $G(x) = \int_0^x \frac{ds}{u(s)}$ is well defined for $x < s_0$. Indeed, in this case, $g(s) > s$ and $u(s) \neq 0$. If we integrate (5.4.5), we finally obtain that $G(q(t)) = t$. □

5.4.5 The Binary Case

Case of the Yule process

A Yule process with rate $a > 0$ is a continuous time binary branching process for which each individual lives an exponential time of parameter a before producing two descendants ($p_2 = 1$). This model can represent bacteria dynamics with cell division in continuous time. It can also represent a reproduction mechanism where an individual reproduces according to a Poisson process of intensity a and produces at each birth event a unique descendant which will follow the same dynamics, independently of its ancestor and the past.

In this case, the generating function of the reproduction law is $g(s) = s^2$ and Equation (5.4.2) becomes for each $t > 0$,

$$\frac{\partial}{\partial t} F(s,t) = a \left(F(s,t)^2 - F(s,t) \right) \quad \text{with} \quad F(s,0) = s.$$

A simple computation then gives

$$F(s,t) = \frac{s\mathrm{e}^{-at}}{1 - s(1 - \mathrm{e}^{-at})}.$$

We recognize the generating function of a geometric law of parameter e^{-at}. We therefore have the following proposition.

Proposition 5.4.7 *The number of particles at time t of a Yule process with rate a, and starting from a single individual, has a geometric distribution of parameter* e^{-at}.

The general case

We now suppose that $p_0 + p_2 = 1$, with $p_0 > 0$ and $p_2 > 0$, that is, at a jump instant, either the individual dies, or it has a single descendant. This model will thus integrate the possible deaths of bacteria in addition to cell division.

Each individual has a death rate of parameter $d > 0$ and a reproduction rate of parameter $b > 0$. In this case, the total jump rate a per individual is $a = b + d$ and the jump law is defined as follows: with probability $p_2 = \frac{b}{b+d}$, there is a birth, and with probability $p_0 = \frac{d}{b+d}$, there is a death.

We then have $m = 2p_2 = \dfrac{2b}{b+d}$ and $r = b - d$ is the growth rate. The process will therefore be supercritical (resp. critical or subcritical) if and only if $r > 0$ (resp. $r = 0$ or $r < 0$).

The function u is then written $u(s) = d - (b+d)s + bs^2$ and $s_0 = \min(1, \frac{d}{b})$. In the critical binary case where $b = d$, $u(s) = b(1-s)^2$.

Note that, since $m < +\infty$, the process does not explode almost surely and $F(1,t) = 1$. In this binary case, we can obtain an explicit form of the generating function. Let us first compute it in the critical case. By (5.4.2), it satisfies

$$\frac{\partial}{\partial t} F(s,t) = b\,(1 - F(s,t))^2 \; ; \; F(s,0) = s,$$

from which, by elementary computation, we get

$$F(s,t) = 1 - \frac{1-s}{1 + bt\,(1-s)}.$$

In the case of non-critical binary reproduction where $b \neq d$, $u(s) = (1-s)(d-bs)$. We deduce from this by immediate integration that

$$F(s,t) = 1 - \frac{(1-s)\,(b-d)}{(bs-d)\,\mathrm{e}^{-(b-d)t} + b\,(1-s)}.$$

Recall that $r = b - d$. We know that the probability $q(t) = \mathbb{P}_1(T_0 \le t)$ satisfies $q(t) = F(0,t)$. We deduce from this that

$$q(t) = \begin{cases} d(\mathrm{e}^{rt} - 1)/(b\mathrm{e}^{rt} - d) & \text{if } b \neq d \\ bt/(1+bt) & \text{if } b = d. \end{cases}$$

We can deduce from this the extinction probability, equal to $\lim_{t\to\infty} F(0,t)$.

If $r \le 0$, this limit is 1 and the population size process goes almost surely to extinction.

If $r > 0$, the extinction probability is $s_0 = \dfrac{d}{b}$.

In Section 5.5 concerning birth and death processes (end of Section 5.5.2), we will see furthermore that the average extinction time is

$$\mathbb{E}_1(T_0) = \frac{1}{b} \log\left(\frac{1}{1 - b/d}\right) \quad \text{if } b < d,$$

and is infinite if $b \ge d$.

Remark 5.4.8 In the critical case and using (5.4.4), we observe that the average $m(t)$ is constant while the process almost surely goes to 0. This is a case where the hypotheses of the dominated convergence theorem are not satisfied.

5.4.6 Extensions

We can generalize this model and consider branching processes with *immigration* or with *logistic growth.*

Immigration

Let $\nu = (\nu_k)_{k\in\mathbb{N}}$ be a finite positive measure on $\mathbb{N}$, ($\forall k, \nu_k \geq 0$ and $\sum_k \nu_k < +\infty$), and let $\rho = \sum_{k\geq 0} \nu_k$. In this continuous time branching model with immigration,

- At rate ρ, groups of immigrants arrive in the population, independently of its state.
- A group is composed of k individuals with probability $\dfrac{\nu_k}{\rho}$.
- All individuals present in the population reproduce and die independently following the previous branching scheme.

The branching process with immigration has the transition rates:

$$\begin{cases} i \to i+k & \text{at rate } \; i\,p_{k+1} + \nu_k \\ i \to i-1 & \text{at rate } \quad\;\; i\,p_0. \end{cases}$$

Logistic Growth

In a population model with logistic growth,

- All individuals present in the population reproduce and die of natural causes independently, following the previous branching scheme. (p_0 is then called the *natural* or *intrinsic death rate*.)
- An individual experiences the pressure of other individuals on its survival. For example, individuals may be in competition for the sharing of resources. Thus the individual probability of survival will decrease as a function of the population size, which will result in an increase in the death rate.

If the impact of competition between two individuals is described by the parameter $c > 0$, the population process with logistic growth has the transition rates:

$$\begin{cases} i \to i+k & \text{at rate } \qquad\quad i\,p_{k+1} \\ i \to i-1 & \text{at rate } \; i p_0 + c\, i(i-1). \end{cases}$$

Due to the interaction, the branching property is no longer satisfied. In the next section we will develop the study of binary branching processes with logistic growth or, more generally, with individual birth and death rates depending on the size of the population.

5.5 Birth and Death Processes

5.5.1 Definition and Non-Explosion Criterion

Definition 5.5.1 A *birth and death process* is a pure jump $\mathbb{N}$-valued Markov process whose jump amplitudes are equal to ± 1. Its transition rates are given by

$$\begin{cases} i \to i+1 & \text{at rate } \lambda_i \\ i \to i-1 & \text{at rate } \mu_i, \end{cases}$$

with $(\lambda_i)_i$ and $(\mu_i)_i$ two sequences of non-negative real numbers, for $i \in \mathbb{N}$.

We assume that $\mu_0 = 0$. If the population is without immigration, then $\lambda_0 = 0$ and state 0 is absorbing.

Such a process extends the notion of binary branching process, with more general positive rate functions λ_i and μ_i defined for a population size i. With this dependence on the instantaneous state of the population, the independence and the branching property are lost. This model allows us to take into account some interaction between individuals.

The infinitesimal generator is given by

$$Q_{i,i+1} = \lambda_i, \;\; Q_{i,i-1} = \mu_i, \;\; Q_{i,i} = -(\lambda_i + \mu_i), \;\; Q_{i,j} = 0 \text{ otherwise.}$$

The overall jump rate for a population of size i is $\lambda_i + \mu_i$. Thus after an exponential time with parameter $\lambda_i + \mu_i$, the process increases by 1 with probability $\dfrac{\lambda_i}{\lambda_i + \mu_i}$ and decreases by -1 with probability $\dfrac{\mu_i}{\lambda_i + \mu_i}$. If $\lambda_i + \mu_i = 0$, the process is absorbed at i.

We can write the matrix of the generator $Q = (Q_{i,j})$.

$$\begin{pmatrix} \mu_1 & -(\lambda_1+\mu_1) & \lambda_1 & 0 & 0 \cdots \\ 0 & \mu_2 & -(\lambda_2+\mu_2) & \lambda_2 & 0 \cdots \\ 0 & 0 & \mu_3 & -(\lambda_3+\mu_3) & \lambda_3 \cdots \\ \cdots & & & & \\ \cdots & & & & \end{pmatrix}.$$

By Theorem 5.3.2, we have

$$P_{i,i+1}(h) = \lambda_i h + o(h); \;\; P_{i,i-1}(h) = \mu_i h + o(h); \;\; P_{i,i}(h) = 1 - (\lambda_i + \mu_i)h + o(h).$$

Examples:

1) The Yule process corresponds to $\lambda_i = i\lambda$, $\mu_i = 0$, with $\lambda > 0$.
2) The branching process or linear birth and death process corresponds to $\lambda_i = i\lambda$, $\mu_i = i\mu$, with $\lambda > 0$ and $\mu > 0$.
3) The birth and death process with immigration corresponds to $\lambda_i = i\lambda + \rho$, $\mu_i = i\mu$, with $\lambda > 0$, $\rho > 0$ and $\mu > 0$

4) The logistic birth and death process corresponds to $\lambda_i = i\lambda$, $\mu_i = i\mu + c\, i(i-1)$, with $\lambda > 0$, $c > 0$ and $\mu > 0$.

Birth rates can depend on i in different ways. For example, a cooperation phenomenon could be translated by a quadratic dependence of λ_i on i. The Allee effect (cf. [53]) describes a growth model when the population is small (quadratic birth rate in i) and a decrease when the population is large (cubic death rate) such that

$$\lambda_i - \mu_i = r\, i\left(1 - \frac{i}{K}\right)\left(\frac{i-A}{K}\right).$$

Note that if there exists an $I \in \mathbb{N}^*$ with $\lambda_I = 0$ then the population size cannot exceed I and the process remains bounded. On the other hand, if the birth rate grows too quickly compared to the death rate, the process can explode in finite time. The following theorem characterizes the non-explosion of the process in finite time, by a subtle balance between birth rates and death rates. If this is the case, we can define the process for all time $t \in \mathbb{R}_+$.

Theorem 5.5.2 *Suppose that $\lambda_i > 0$ for all $i \geq 1$. Then the birth and death process has an infinite lifetime almost surely if and only if*

$$R := \sum_{i\geq 1}\left(\frac{1}{\lambda_i} + \frac{\mu_i}{\lambda_i\lambda_{i-1}} + \cdots + \frac{\mu_i\cdots\mu_2}{\lambda_i\cdots\lambda_2\lambda_1}\right) \quad \textit{is infinite}.$$

The next corollary is then immediate.

Corollary 5.5.3 *If for all i, $\lambda_i \leq \lambda i$, with $\lambda > 0$, the process is well defined on $\mathbb{R}_+$.*

Remark 5.5.4 We deduce that the birth and death processes mentioned in the examples have an infinite lifetime almost surely.

Proof Let $(T_n)_n$ be the sequence of jump times of the process and $(S_n)_n$ the sequence of times between jumps,

$$S_n = T_n - T_{n-1}, \quad \forall n \geq 1; \quad T_0 = 0, \quad S_0 = 0.$$

We denote $\lim_n T_n$ by T_∞. The process does not explode almost surely and is well defined on $\mathbb{R}_+$ if and only if for all $i \in \mathbb{N}$, $\mathbb{P}_i(T_\infty < +\infty) = 0$.

We will show that the process does not explode almost surely if and only if the only positive and bounded solution $x = (x_i)_{i\in\mathbb{N}}$ of $Qx = x$ is the null solution, and we will see that this is equivalent to the non-explosion criterion for birth and death processes.

For all i, we set $h_i^{(0)} = 1$ and for $n \in \mathbb{N}^*$, $h_i^{(n)} = \mathbb{E}_i\big(\exp(-\sum_{k=1}^n S_k)\big)$. Let $q_i = \lambda_i + \mu_i$. The Markov property implies that

$$\mathbb{E}_i\Big(\exp\big(-\sum_{k=1}^{n+1} S_k\big)|S_1\Big) = \mathbb{E}_i\Big(\exp(-S_1)\,\exp\big(-\sum_{k=2}^{n+1} S_k\big)\,|S_1\Big)$$
$$= \exp(-S_1)\,\mathbb{E}_{X_{S_1}}\Big(\exp\big(-\sum_{k=1}^{n} S_k\big)\Big),$$

since for the process starting from X_{S_1}, the new jump times are the $T_n - S_1$.

By Theorem 5.3.2, we know that S_1 and X_{S_1} are independent given $X_0 = i$. Then

$$\mathbb{E}_i\Big(\exp\big(-\sum_{k=1}^{n+1} S_k\big)\Big) = \mathbb{E}_i\Big(\mathbb{E}_{X_{S_1}}\Big(\exp\big(-\sum_{k=1}^{n} S_k\big)\Big)\Big)\,\mathbb{E}_i\big(\exp(-S_1)\big).$$

We have

$$\mathbb{E}_i\Big(\mathbb{E}_{X_{S_1}}\Big(\exp\big(-\sum_{k=1}^{n} S_k\big)\Big)\Big) = \sum_{j\neq i} \mathbb{P}_i(X_{S_1} = j)\,\mathbb{E}_j\Big(\exp\big(-\sum_{k=1}^{n} S_k\big)\Big)$$
$$= \sum_{j\neq i} \frac{Q_{i,j}}{q_i}\,\mathbb{E}_j\Big(\exp\big(-\sum_{k=1}^{n} S_k\big)\Big)$$

and

$$\mathbb{E}_i\big(\exp(-S_1)\big) = \int_0^\infty q_i \mathrm{e}^{-q_i s}\mathrm{e}^{-s}\,\mathrm{d}s = \frac{q_i}{1+q_i}.$$

We finally deduce that

$$h_i^{(n+1)} = \sum_{j\neq i} \frac{Q_{i,j}}{1+q_i}\,h_j^{(n)}. \tag{5.5.1}$$

Let $(x_i)_i$ be a solution of $Qx = x$, positive and bounded by 1. We have $h_i^{(0)} = 1 \geq x_i$, and thanks to the previous formula, we easily deduce by recurrence that for all i and for all $n \in \mathbb{N}$, $h_i^{(n)} \geq x_i \geq 0$. Indeed, if $h_j^{(n)} \geq x_j$, we have $h_i^{(n+1)} \geq \sum_{j\neq i} \dfrac{Q_{i,j}}{1+q_i} x_j$. Since x is a solution of $Qx = x$, it yields $x_i = \sum_j Q_{i,j}\, x_j = Q_{i,i}x_i + \sum_{j\neq i} Q_{i,j}x_j = -q_i x_i + \sum_{j\neq i} Q_{i,j}x_j$ for any i, from which $\sum_{j\neq i} \frac{Q_{i,j}}{1+q_i} x_j = x_i$, and $h_i^{(n+1)} \geq x_i$.

If the process does not explode almost surely, we have $T_\infty = +\infty$ a.s., and $\lim_n h_i^{(n)} = 0$. By letting n tend to infinity in the previous inequality, we deduce that $x_i = 0$. Thus, in this case, the only positive bounded solution of $Qx = x$ is the null solution.

Now suppose that the process explodes with strictly positive probability. Let $z_i = \mathbb{E}_i(\mathrm{e}^{-T_\infty})$. There exists an i with $\mathbb{P}_i(T_\infty < +\infty) > 0$ and for this integer i, $z_i > 0$. As $T_\infty = \lim_n T_n$ and $T_n = \sum_{k=1}^n S_k$, we have $z_j = \lim_n h_j^{(n)}$. Formula (5.5.1) implies that for any integer i with $z_i > 0$, $z_i = \sum_{j \neq i} \frac{Q_{i,j}}{1+q_i} z_j$. We have obtained a solution z of the equation $Qz = z$, positive and bounded, with $z_i > 0$, and thus exhibited a non-trivial and bounded solution to the equation.

Let us apply this result to the birth and death process. Suppose that $\lambda_i > 0$ for $i \in \mathbb{N}^*$, and $\lambda_0 = \mu_0 = 0$. Let $(x_i)_{i \in \mathbb{N}}$ be a solution of the equation $Qx = x$. Introduce for $n \geq 1$, $\Delta_n = x_n - x_{n-1}$, and $r_n = \frac{1}{\lambda_n} + \sum_{k=1}^{n-1} \frac{\mu_{k+1} \cdots \mu_n}{\lambda_k \lambda_{k+1} \cdots \lambda_n} + \frac{\mu_1 \cdots \mu_n}{\lambda_1 \cdots \lambda_n}$.

The equation $Qx = x$ is then given by $x_0 = 0$ and for all $n \geq 1$ by

$$\lambda_n x_{n+1} - (\lambda_n + \mu_n) x_n + \mu_n x_{n-1} = x_n.$$

By setting $f_n = \frac{1}{\lambda_n}$ and $g_n = \frac{\mu_n}{\lambda_n}$, we obtain

$$\Delta_1 = x_1;\ \Delta_2 = x_2 - x_1;\ \Delta_{n+1} = \Delta_n\, g_n + f_n\, x_n.$$

Note that for all n, $\Delta_n \geq 0$, and therefore the sequence $(x_n)_n$ is increasing.

If $x_1 = 0$, the solution is clearly null. Otherwise, we deduce that

$$\Delta_{n+1} = \frac{1}{\lambda_n} x_n + \sum_{k=1}^{n-1} f_k\, g_{k+1} \cdots g_n x_k + g_1 \cdots g_n x_1.$$

Since $(x_k)_k$ is increasing, this implies that $r_n x_1 \leq \Delta_{n+1} \leq r_n x_n$, and by iteration

$$x_1 (1 + r_1 + \cdots + r_n) \leq x_{n+1} \leq x_1 \prod_{k=1}^{n} (1 + r_k).$$

We have therefore shown that the sequence $(x_n)_n$ is bounded if and only if the series with general term r_k converges, which completes the proof. □

5.5.2 Kolmogorov Equations and Invariant Measure

In this framework, we can write the two Kolmogorov equations.

Forward Kolmogorov Equation: for all $i, j \in \mathbb{N}$,

$$\begin{aligned} \frac{\mathrm{d}P_{i,j}}{\mathrm{d}t}(t) &= \sum_k P_{i,k}(t)\, Q_{k,j} = P_{i,j+1}(t) Q_{j+1,j} + P_{i,j-1}(t) Q_{j-1,j} + P_{i,j}(t) Q_{j,j} \\ &= \mu_{j+1} P_{i,j+1}(t) + \lambda_{j-1} P_{i,j-1}(t) - (\lambda_j + \mu_j) P_{i,j}(t). \end{aligned} \tag{5.5.2}$$

Backward Kolmogorov Equation: for all $i, j \in \mathbb{N}$,

$$\begin{aligned}\frac{\mathrm{d}P_{i,j}}{\mathrm{d}t}(t) &= \sum_k Q_{i,k}P_{k,j}(t) = Q_{i,i-1}P_{i-1,j}(t) + Q_{i,i+1}P_{i+1,j}(t) + Q_{i,i}P_{i,j}(t) \\ &= \mu_i P_{i-1,j}(t) + \lambda_i P_{i+1,j}(t) - (\lambda_i + \mu_i)P_{i,j}(t). \end{aligned} \tag{5.5.3}$$

Let us define for any $j \in \mathbb{N}$ the probability

$$p_j(t) = \mathbb{P}(X(t) = j) = \sum_i \mathbb{P}(X(t) = j \mid X_0 = i)\mathbb{P}(X(0) = i) = \sum_i \mathbb{P}(X(0) = i)P_{i,j}(t).$$

A simple computation shows that in this case, the forward Kolmogorov equation (5.5.2) is written as

$$\frac{\mathrm{d}p_j}{\mathrm{d}t}(t) = \lambda_{j-1}\,p_{j-1}(t) + \mu_{j+1}\,p_{j+1}(t) - (\lambda_j + \mu_j)\,p_j(t).$$

This equation describes the dynamics of the marginal distribution of the process at time t. It can in particular allow us to find a stationary solution when there is one. This amounts to finding a family $(\pi_j)_j$ of numbers between 0 and 1 such that $\sum_j \pi_j < +\infty$ and for all j,

$$\lambda_{j-1}\,\pi_{j-1} + \mu_{j+1}\,\pi_{j+1} - (\lambda_j + \mu_j)\,\pi_j = 0.$$

5.5.3 Extinction Criterion – Extinction Time

Here we assume that the process is defined over $\mathbb{R}_+$ and we are interested in the probability that the process reaches 0 starting from a strictly positive initial condition. Some of the computations in this section can be found in [49] or in [2] and they are rigorously developed in [8].

Let T_0 be the hitting time of 0 by the birth and death process,

$$T_0 = \inf\{t \geq 0, X_t = 0\},$$

with the convention $\{T_0 = +\infty\}$ if this state is never reached. Let us set $u_i = \mathbb{P}_i(T_0 < +\infty)$, the probability of extinction in finite time for a process starting from state i. We have $u_0 = 1$.

By conditioning on the first jump $X_{T_1} - X_0 \in \{-1, +1\}$, we obtain the following recurrence relation: for all $i \geq 1$,

$$\lambda_i u_{i+1} - (\lambda_i + \mu_i)u_i + \mu_i u_{i-1} = 0. \tag{5.5.4}$$

This equation can also be obtained from the backward Kolmogorov equation (5.5.3). Indeed, $u_i = \mathbb{P}_i(\exists t > 0, X_t = 0) = \mathbb{P}_i(\cup_t\{X_t = 0\}) = \lim_{t\to\infty} P_{i,0}(t)$, and

$$\frac{\mathrm{d}P_{i,0}}{\mathrm{d}t}(t) = \mu_i P_{i-1,0}(t) + \lambda_i P_{i+1,0}(t) - (\lambda_i + \mu_i)P_{i,0}(t).$$

Theorem 5.5.5

(i) *If* $\sum_{k=1}^{\infty} \frac{\mu_1 \cdots \mu_k}{\lambda_1 \cdots \lambda_k} = +\infty$, *then the extinction probabilities* u_i *are equal to* 1. *Thus the birth and death process almost surely goes to extinction in finite time for any non-zero initial condition.*

(ii) *If* $\sum_{k=1}^{\infty} \frac{\mu_1 \cdots \mu_k}{\lambda_1 \cdots \lambda_k} = U_\infty < +\infty$, *then for* $i \geq 1$,

$$u_i = (1 + U_\infty)^{-1} \sum_{k=i}^{\infty} \frac{\mu_1 \cdots \mu_k}{\lambda_1 \cdots \lambda_k}.$$

The process has a strictly positive probability of surviving, for any non-zero initial condition.

Note that the survival probability is strictly positive, but less than one, which means that the extinction probability is also positive.

Proof Let us solve (5.5.4). We know that $u_0 = 1$. First suppose that for a state N, $\lambda_N = 0$ and $\lambda_i > 0$ for $i < N$. Define $u_i^{(N)} = \mathbb{P}_i(T_0 < T_N)$, where T_N is the hitting time of N. Then $u_0^N = 1$ and $u_N^N = 0$. Let

$$U_N = \sum_{k=1}^{N-1} \frac{\mu_1 \cdots \mu_k}{\lambda_1 \cdots \lambda_k}.$$

Elementary computations using (5.5.4) show that for $i \in \{1, \ldots, N-1\}$

$$u_i^{(N)} = (1 + U_N)^{-1} \sum_{k=i}^{N-1} \frac{\mu_1 \cdots \mu_k}{\lambda_1 \cdots \lambda_k} \quad \text{and in particular} \quad u_1^{(N)} = \frac{U_N}{1 + U_N}.$$

To prove the general case, let N tend to infinity assuming that $\lambda_i > 0$ for all i. It is immediate that $u_i^{(N)}$ converges to u_i. Thus, extinction will take place almost surely in finite time or not, depending on whether the series $\sum_{k=1}^{\infty} \frac{\mu_1 \cdots \mu_k}{\lambda_1 \cdots \lambda_k}$ diverges or not. □

Application of Theorem 5.5.5 to the binary branching process (linear birth and death process)

Each individual is born at rate b ($\lambda_i = bi$) and dies at rate d ($\mu_i = di$). By applying the previous results, we obtain that when $b \leq d$, i.e. when the process is subcritical or critical, the sequence $(U_N)_N$ tends to infinity as $N \to +\infty$ and there is extinction with probability 1. If, on the other hand $b > d$, the sequence $(U_N)_N$ converges to $\frac{d}{b-d}$ and a simple computation shows that $u_i = (d/b)^i$. We thus find again the result obtained in Section 5.4.5.

Application of Theorem 5.5.5 to the logistic birth and death process

Here we assume that the birth and death rates are

$$\lambda_i = \lambda i; \quad \mu_i = \mu i + ci(i-1).$$

The parameter c models the competition pressure between two individuals. One can observe (by D'Alembert's Criterion) that in this case, the series $\sum_{k=1}^{\infty} \frac{\mu_1 \cdots \mu_k}{\lambda_1 \cdots \lambda_k}$ diverges, leading to almost sure extinction of the process. Thus, competition between individuals makes extinction unavoidable.

Let us now return to the general case and suppose that the series $\sum_{k=1}^{\infty} \frac{\mu_1 \cdots \mu_k}{\lambda_1 \cdots \lambda_k}$ diverges. The extinction time T_0 is well defined and we wish to compute its moments.

We use the classical notation (cf. [49])

$$\pi_1 = \frac{1}{\mu_1}; \quad \pi_n = \frac{\lambda_1 \ldots \lambda_{n-1}}{\mu_1 \ldots \mu_n} \qquad \forall n \geq 2.$$

Let us also introduce the sequence of hitting times T_n of state n. Since T_0 is almost surely finite, we can ensure the same for T_n, as soon as the process originates from an initial condition greater than n.

Proposition 5.5.6 *Suppose that*

$$\sum_{k=1}^{\infty} \frac{\mu_1 \cdots \mu_k}{\lambda_1 \cdots \lambda_k} = \sum_n \frac{1}{\lambda_n \pi_n} = +\infty. \tag{5.5.5}$$

Then

(i) *for all* $a > 0$ *and* $n \geq 1$,

$$G_n(a) = \mathbb{E}_{n+1}(\exp(-aT_n)) = 1 + \frac{\mu_n + a}{\lambda_n} - \frac{\mu_n}{\lambda_n} \frac{1}{G_{n-1}(a)}. \tag{5.5.6}$$

(ii) $\mathbb{E}_1(T_0) = \sum_{k\geq 1} \pi_k$ *and for all* $n \geq 2$,

$$\mathbb{E}_n(T_0) = \sum_{k\geq 1} \pi_k + \sum_{k=1}^{n-1} \frac{1}{\lambda_k \pi_k} \sum_{i\geq k+1} \pi_i = \sum_{k=1}^{n-1} \left(\sum_{i\geq k+1} \frac{\lambda_{k+1}\dots\lambda_{i-1}}{\mu_{k+1}\dots\mu_i} \right). \quad (5.5.7)$$

Proof (i) Let τ_n be a random variable with the same distribution as T_n under $\mathbb{P}_{n+1}$ and consider the Laplace transform of τ_n. Following [4, p. 264] and using the Markov property and Corollary 5.3.5, we can write

$$\tau_{n-1} \stackrel{(d)}{=} \mathbb{1}_{\{Y_n=-1\}}\mathcal{E}_n + \mathbb{1}_{\{Y_n=1\}}\left(\mathcal{E}_n + \tau_n + \tau'_{n-1}\right),$$

where Y_n, $\mathcal{E}_n$, τ'_{n-1} and τ_n are independent random variables, $\mathcal{E}_n$ follows an exponential law with parameter $\lambda_n + \mu_n$, τ'_{n-1} is distributed like τ_{n-1} and $\mathbb{P}(Y_n = 1) = 1 - \mathbb{P}(Y_n = -1) = \lambda_n/(\lambda_n + \mu_n)$. We deduce

$$G_{n-1}(a) = \frac{\lambda_n + \mu_n}{a + \lambda_n + \mu_n} \left(G_n(a) G_{n-1}(a) \frac{\lambda_n}{\lambda_n + \mu_n} + \frac{\mu_n}{\lambda_n + \mu_n} \right)$$

and (5.5.6) follows.

(ii) Differentiating (5.5.6) at $a = 0$, we obtain that

$$\mathbb{E}_n(T_{n-1}) = \frac{\lambda_n}{\mu_n}\, \mathbb{E}_{n+1}(T_n) + \frac{1}{\mu_n}, \qquad n \geq 1.$$

As in the proof of Theorem 5.5.5, we first study the particular case where $\lambda_N = 0$. We have $\mathbb{E}_N(T_{N-1}) = \frac{1}{\mu_N}$ and a simple induction gives

$$\mathbb{E}_n(T_{n-1}) = \frac{1}{\mu_n} + \sum_{i=n+1}^{N} \frac{\lambda_n \dots \lambda_{i-1}}{\mu_n \dots \mu_i}.$$

We deduce that $\mathbb{E}_1(T_0) = \sum_{k=1}^{N} \pi_k$ and by writing $\mathbb{E}_n(T_0) = \sum_{k=1}^{n} \mathbb{E}_k(T_{k-1})$, we obtain

$$\mathbb{E}_n(T_0) = \sum_{k=1}^{N} \pi_k + \sum_{k=1}^{n-1} \frac{1}{\lambda_k \pi_k} \sum_{i=k+1}^{N} \pi_i.$$

For the general case, let $N > n$. The process is right continuous and left limited and thanks to (5.5.5), the process does not explode in finite time and T_0 is finite almost surely, whatever the initial condition. Thus $\sup_{t\geq 0} X_t < +\infty$ $\mathbb{P}_n$-a.s. and $T_N = +\infty$ for N large enough. The monotone convergence theorem then gives

$$\mathbb{E}_n(T_0; T_0 \leq T_N) \longrightarrow_{N\to\infty} \mathbb{E}_n(T_0).$$

Let us consider a birth and death process X^N with birth and death rates $(\lambda_k^N, \mu_k^N; k \geq 0)$ such that $(\lambda_k^N, \mu_k^N) = (\lambda_k, \mu_k)$ for $k \neq N$ and $\lambda_N^N = 0$, $\mu_N^N = \mu_N$.

Since $(X_t, t \le T_N)$ and $(X_t^N, t \le T_N^N)$ have the same law under $\mathbb{P}_n$, we have

$$\mathbb{E}_n\left(T_0; T_0 \le T_N\right) = \mathbb{E}_n\left(T_0^N; T_0^N \le T_N^N\right),$$

which leads to

$$\mathbb{E}_n(T_0) = \lim_{N\to\infty} \mathbb{E}_n\left(T_0^N; T_0^N \le T_N^N\right) \le \lim_{N\to\infty} \mathbb{E}_n\left(T_0^N\right),$$

where the convergence of the last term is due to the stochastic monotonicity of T_0^N with respect to N under $\mathbb{P}_n$. Using now that T_0^N is stochastically less than T_0 under $\mathbb{P}_n$, we also have

$$\mathbb{E}_n(T_0) \ge \mathbb{E}_n(T_0^N).$$

Finally,

$$\mathbb{E}_n(T_0) = \lim_{N\to\infty} \mathbb{E}_n(T_0^N) = \lim_{N\to\infty} \sum_{k=1}^{N} \pi_k + \sum_{k=1}^{n-1} \frac{1}{\lambda_k \pi_k} \sum_{i=k+1}^{N} \pi_i,$$

which concludes the proof. □

Remark 5.5.7 Formula (5.5.6) allows us to compute all the moments of T_n. Exercise 5.7.4 focuses on moments of orders 2 and 3.

Application to binary branching processes

Let us return to the example of the binary branching process (see Section 5.4.5), with individual birth rate b and individual death rate d, in the case where $d > b$. Almost sure extinction takes place and we can compute $\mathbb{E}_1(T_0)$. Indeed

$$\mathbb{E}_1(T_0) = \sum_{k\ge 1} \pi_k = \sum_{k\ge 1} \frac{b^{k-1}}{k\, d^k} = \frac{1}{b} \sum_{k\ge 1} \frac{1}{k} \left(\frac{b}{d}\right)^k = \frac{1}{b} \log\left(\frac{1}{1-b/d}\right).$$

Note also that in this case, by applying (5.5.7) and for $n \in \mathbb{N}^*$,

$$\mathbb{E}_n(T_0) = \frac{1}{b} \sum_{j\ge 1} \left(\frac{b}{d}\right)^j \sum_{k=1}^{n-1} \frac{1}{k+j}.$$

A simple comparison shows that

$$\int_1^n \frac{1}{x+j}\,\mathrm{d}x \le \sum_{k=1}^{n-1} \frac{1}{k+j} \le \int_0^{n-1} \frac{1}{x+j}\,\mathrm{d}x.$$

Thus,

$$\frac{1}{b}\sum_{j\geq 1}\left(\frac{b}{d}\right)^j\Big(\log(n+j)-\log(1+j)\Big)\leq \mathbb{E}_n(T_0)\leq \frac{1}{b}\sum_{j\geq 1}\left(\frac{b}{d}\right)^j\Big(\log(n-1+j)-\log j\Big).$$

We deduce that

$$\mathbb{E}_n(T_0)\sim_{n\to\infty}\frac{\log n}{d-b}. \tag{5.5.8}$$

Indeed, it is easy to show, by the Dominated Convergence Theorem, that the series of general term $\left(\frac{b}{d}\right)^j\frac{\log(n+j)}{\log n}$ converges, as n tends to infinity, to $\sum_{j\geq 1}\left(\frac{b}{d}\right)^j=\frac{b}{b-d}$. This is because for j and n large enough, $\frac{\log(n+j)}{\log n}\leq\frac{\log(nj)}{\log n}\leq 1+\log(1+j)$ and the series $\sum_{j\geq 1}\left(\frac{b}{d}\right)^j(1+\log(1+j))$ converges.

Equation (5.5.8) tells us that even if the initial population size is very large, the average extinction time is short as soon as the process is subcritical. We have already observed this type of behavior for the Galton–Watson process (see Theorem 3.2.12). But what happens if n is extremely large, as is the case, for example, if we consider cohorts of bacteria (whose size quickly reaches around 10^{10})? To study this question, we will assume in the following subsection that the initial condition n is of the form Kx_0, where K tends to infinity.

5.5.4 Trajectorial Representation of Birth and Death Processes

The following can also be found in [8]. As before, we consider a birth and death process with birth rates $(\lambda_n)_n$ and death rates $(\mu_n)_n$. We write $\lambda_n=\lambda(n)$ and $\mu_n=\mu(n)$, where $\lambda(\cdot)$ and $\mu(\cdot)$ are two functions defined on $\mathbb{R}_+$. We assume further that there exist $\bar{\lambda}>0$ and $\bar{\mu}>0$ such that for any $x\geq 0$,

$$\lambda(x)\leq\bar{\lambda}x;\qquad \mu(x)\leq\bar{\mu}(1+x^2), \tag{5.5.9}$$

and $\mu_0=0$. This assumption is satisfied for the logistic case where $\lambda(x)=\lambda x$ and $\mu(x)=cx(x-1)+\mu x$.

Assumption (5.5.9) is a sufficient condition ensuring the existence of the process on $\mathbb{R}_+$, as observed in Corollary 5.5.3.

Proposition 5.5.8 *On the same probability space, we consider a Poisson point measure $Q(\mathrm{d}s,\mathrm{d}u)$ with intensity $\mathrm{d}s\mathrm{d}u$ on $\mathbb{R}_+\times\mathbb{R}_+$ (see Appendix B). We also consider an $\mathbb{N}$-valued random variable Z_0 independent of Q and introduce the filtration $(\mathcal{F}_t)_t$ given by $\mathcal{F}_t=\sigma(Z_0,Q((0,s]\times A),s\leq t,A\in\mathcal{B}(\mathbb{R}_+))$.*

The left-continuous and right-limited $\mathbb{N}$*-valued Markov process* $(Z_t)_{t\geq 0}$ *defined by*

$$Z_t = Z_0 + \int_0^t \int_{\mathbb{R}_+} \left(\mathbf{1}_{\{u\leq\lambda(Z_{s-})\}} - \mathbf{1}_{\{\lambda(Z_{s-})\leq u\leq\lambda(Z_{s-})+\mu(Z_{s-})\}}\right) Q(\mathrm{d}s,\mathrm{d}u) \quad (5.5.10)$$

is a birth and death process with birth (resp. death) rates $(\lambda_n)_n$ *(resp.* $(\mu_n)_n$*).*

If for $p \geq 1$, $\mathbb{E}(Z_0^p) < +\infty$, *then for any* $T > 0$,

$$\mathbb{E}\big(\sup_{t\leq T} Z_t^p\big) < +\infty. \quad (5.5.11)$$

Proof For $n \in \mathbb{N}$, let us introduce the stopping times

$$T_n = \inf\{t > 0, Z_t \geq n\}.$$

For $t \geq 0$, we have

$$\begin{aligned} Z^p_{t\wedge T_n} = Z_0^p &+ \int_0^{t\wedge T_n} \int_{\mathbb{R}_+} \left((Z_{s-}+1)^p - Z_{s-}^p\right) \mathbf{1}_{\{u\leq\lambda(Z_{s-})\}}\, Q(\mathrm{d}s,\mathrm{d}u) \\ &+ \int_0^{t\wedge T_n} \int_{\mathbb{R}_+} \left((Z_{s-}-1)^p - Z_{s-}^p\right) \mathbf{1}_{\{\lambda(Z_{s-})\leq u\leq\lambda(Z_{s-})+\mu(Z_{s-})\}}\, Q(\mathrm{d}s,\mathrm{d}u). \end{aligned}$$

The second part of the r.h.s. is non-positive and the first part is increasing in time, yielding the upper bound

$$\sup_{s\leq t} Z^p_{s\wedge T_n} \leq Z_0^p + \int_0^{t\wedge T_n} \int_{\mathbb{R}_+} \left((Z_{s-}+1)^p - Z_{s-}^p\right) \mathbf{1}_{\{u\leq\lambda(Z_{s-})\}}\, Q(\mathrm{d}s,\mathrm{d}u).$$

Since there exists $C > 0$ such that $(1+x)^p - x^p \leq C(1+x^{p-1})$ for any $x \geq 0$ and by (5.5.9), we get

$$\begin{aligned} \mathbb{E}(\sup_{s\leq t} Z^p_{s\wedge T_n}) &\leq \mathbb{E}(Z_0^p) + C\bar{\lambda}\mathbb{E}\left(\int_0^{t\wedge T_n} Z_s\,(1+Z_s^{p-1})\,\mathrm{d}s\right) \\ &\leq \bar{C}\left(1 + \int_0^t \mathbb{E}\big(\sup_{u\leq s\wedge T_n} Z_u^p\big)\,\mathrm{d}s\right), \end{aligned}$$

where $\bar{C}$ is a positive number independent of n. Since the process is bounded by n before T_n, Gronwall's Lemma implies the existence (for any $T > 0$) of a constant $C_{T,p}$, independent of n, such that

$$\mathbb{E}\big(\sup_{t\leq T\wedge T_n} Z_t^p\big) \leq C_{T,p}. \quad (5.5.12)$$

In particular, the sequence $(T_n)_n$ tends to infinity almost surely. Indeed, otherwise there would exist $T_0 > 0$ such that $\mathbb{P}(\sup_n T_n < T_0) > 0$. Hence $\mathbb{E}\big(\sup_{t\le T_0\wedge T_n} Z_t^p\big) \ge n^p\,\mathbb{P}(\sup_n T_n < T_0)$, which contradicts (5.5.12). Letting n tend to infinity in (5.5.12) and using Fatou's Lemma yields (5.5.11). □

Let us note that given Z_0 and Q, the process defined by (5.5.10) is unique. Indeed it can be inductively constructed. It is thus unique in law. We now recall its infinitesimal generator and give some martingale properties.

Theorem 5.5.9 *Let us assume that* $\mathbb{E}(Z_0^p) < \infty$, *for* $p \ge 2$.

(i) *The infinitesimal generator of the Markov process* Z *is defined for any bounded measurable function* ϕ *from* $\mathbb{R}_+$ *into* $\mathbb{R}$ *by*

$$L\phi(z) = \lambda(z)(\phi(z+1)-\phi(z)) + \mu(z)(\phi(z-1)-\phi(z)).$$

(ii) *For any measurable function* ϕ *such that* $|\phi(x)|+|L\phi(x)| \le C(1+x^p)$, *the process* M^ϕ *defined by*

$$M_t^\phi = \phi(Z_t) - \phi(Z_0) - \int_0^t L\phi(Z_s)\,\mathrm{d}s$$

is a left-limited and right-continuous (càdlàg) $(\mathcal{F}_t)_t$*-martingale.*

(iii) *The process* M *defined by*

$$M_t = Z_t - Z_0 - \int_0^t (\lambda(Z_s)-\mu(Z_s))\,\mathrm{d}s \tag{5.5.13}$$

is a square-integrable martingale with quadratic variation

$$\langle M\rangle_t = \int_0^t (\lambda(Z_s)+\mu(Z_s))\,\mathrm{d}s. \tag{5.5.14}$$

Note that the drift term of (5.5.13) involves the difference between the birth and death rates (i.e. the growth rate), while (5.5.14) involves the sum of both rates. Indeed the drift term describes the mean behavior whereas the quadratic variation reports the random fluctuations.

Proof (i) comes from (5.3.2).

(ii) The process $\phi(Z)$ is equal to the sum of its jumps : $\phi(Z_t) = \phi(Z_0) + \sum_{s\le t}\Delta\phi(Z_s) = \phi(Z_0) + \sum_{s\le t}(\phi(Z_s)-\phi(Z_{s-}))$. Then, using Theorem B.3, the assumption on ϕ and (5.5.11), we have

$$\phi(Z_t) - \phi(Z_0) =$$

$$\int_0^t \int_{\mathbb{R}_+} \Big(\mathbf{1}_{\{u \le \lambda(Z_{s-})\}} (\phi(Z_{s-}+1) - \phi(Z_{s-}))$$

$$- \mathbf{1}_{\{\lambda(Z_{s-}) \le u \le \lambda(Z_{s-}) + \mu(Z_{s-})\}} (\phi(Z_{s-}-1) - \phi(Z_{s-})) \Big) Q(\mathrm{d}s, \mathrm{d}u)$$

$$= M_t^\phi + \int_0^t \big(\lambda(Z_s)(\phi(Z_s+1) - \phi(Z_s)) - \mu(Z_s)(\phi(Z_s-1) - \phi(Z_s)) \big) \, \mathrm{d}s, \quad (5.5.15)$$

where

$$M_t^\phi = \int_0^t \int_{\mathbb{R}_+} \Big(\mathbf{1}_{\{u \le \lambda(Z_{s-})\}} (\phi(Z_{s-}+1) - \phi(Z_{s-}))$$

$$- \mathbf{1}_{\{\lambda(Z_{s-}) \le u \le \lambda(Z_{s-}) + \mu(Z_{s-})\}} (\phi(Z_{s-}-1) - \phi(Z_{s-})) \Big) \widetilde{Q}(\mathrm{d}s, \mathrm{d}u).$$

Indeed, all the terms of the r.h.s. of (5.5.15) are integrable. Therefore M^ϕ is a martingale.

(iii) We first assume that $\mathbb{E}(Z_0^3) < \infty$. By (5.5.9) and (5.5.11), we can again apply Theorem B.3 in the case with $\phi = \textit{identity}$ to prove that

$$\langle M \rangle_t = \int_0^t \int_{\mathbb{R}_+} \big(\mathbf{1}_{\{u \le \lambda(Z_{s-})\}} (Z_{s-} + 1 - Z_{s-})$$

$$- \mathbf{1}_{\{\lambda(Z_{s-}) \le u \le \lambda(Z_{s-}) + \mu(Z_{s-})\}} (Z_{s-} - 1 - Z_{s-}) \big)^2 \, \mathrm{d}u \mathrm{d}s$$

$$= \int_0^t (\lambda(Z_s) + \mu(Z_s)) \, \mathrm{d}s.$$

The general case $\mathbb{E}(Z_0^2) < +\infty$ follows by a standard localization argument. □

5.6 Continuous Approximations: Deterministic and Stochastic Models

When the population size is very large, the jump rates increase dramatically, so that the times between jumps are infinitesimal. It is therefore very difficult to take into account all the jump events and the dynamics of population size will be close to that of a continuous time process. In this section, we obtain approximations valid for large populations, which will be either deterministic or stochastic depending on the scaling choice between demographic parameters and the order of magnitude of resources (or population size). We thus justify classical models of population dynamics. These different approximations give qualitatively different results for the long-term behavior of the population and thus allow us to reflect on the relevance of the choice of a model.

5.6.1 Deterministic Approximations – Malthusian and Logistic Equations

We assume that the birth and death process depends on a parameter $K \in \mathbb{N}$ that scales the system size. The main assumption is that the initial condition Z_0^K is of order K, for K tending to infinity, in the sense that

$$\lim_{K\to+\infty} \frac{1}{K} Z_0^K = x_0 \in \mathbb{R}_+^*, \tag{5.6.1}$$

the limit being in distribution (or in probability since the limit x_0 is deterministic). The birth rates λ_i^K and death rates μ_i^K can also depend on K. Let $(Z_t^K, t \geq 0)$ be the birth and death process thus defined.

We study the asymptotic behavior of the process $(Z_t^K, t \geq 0)$ as $K \to +\infty$. To obtain a non-trivial approximation of the process, we normalize it by $\frac{1}{K}$ (this amounts to giving the weight $\frac{1}{K}$ to each individual), and consider the process $(X_t^K, t \geq 0)$ defined by $X_t^K = \frac{1}{K} Z_t^K$. The states taken by the process X^K are then of the form $\frac{i}{K}$, $i \in \mathbb{N}$.

We therefore study the limit (in law) of the sequence of processes $(X_t^K, t \geq 0)$, as K tends to infinity. Let us first observe that, as the jumps of the birth and death process $(Z_t^K, t \geq 0)$ are of amplitude ± 1, the jumps of the process $(X_t^K, t \geq 0)$ are of amplitude $\pm\frac{1}{K}$ and tend to 0 as $K \to \infty$. Thus, if the processes $(X_t^K, t \geq 0)$ converge to a limit process $(X_t, t \geq 0)$, we expect the process $(X_t, t \geq 0)$ to be continuous in time. Also note that the process X can take any positive real value.

We will see that the asymptotic behavior of the sequence $(X^K)_K$ is related to the dependence on K of the birth and death rates. Two cases are particularly interesting:

– the linear birth and death processes for which

$$\lambda_i^K = \lambda i = \lambda \frac{i}{K} K; \qquad \mu_i^K = \mu i = \mu \frac{i}{K} K.$$

Then, setting $r = \lambda - \mu$, we have for all $x \in \mathbb{R}_+$,

$$\lim_{K\to\infty, \frac{i}{K}\to x} \frac{1}{K} (\lambda_i^K - \mu_i^K) = rx; \quad \lim_{K\to\infty, \frac{i}{K}\to x} \frac{1}{K^2} (\lambda_i^K + \mu_i^K) = 0. \tag{5.6.2}$$

– the logistic birth and death processes where

$$\lambda_i^K = \lambda i = \lambda \frac{i}{K} K; \qquad \mu_i^K = \mu i + \frac{c}{K} i(i-1) = \mu \frac{i}{K} K + c \frac{i}{K} K (\frac{i}{K} - \frac{1}{K}),$$

with $c > 0$. Thus, for all $x \in \mathbb{R}_+$,

$$\lim_{K\to\infty, \frac{i}{K}\to x} \frac{1}{K} (\lambda_i^K - \mu_i^K) = rx - cx^2; \qquad \lim_{K\to\infty, \frac{i}{K}\to x} \frac{1}{K^2} (\lambda_i^K + \mu_i^K) = 0. \tag{5.6.3}$$

The linear birth and death process does not assume *any interaction* between individuals, which can be considered unrealistic. Under a fixed global resource assumption, there is generally *competition between individuals* for resource sharing, which increases the death rate in large populations. The logistic birth and death process allows for the regulation of population size in cases where the individual growth rate $\lambda - \mu$ is positive. Each individual is in competition with the $i-1$ other individuals in the population and its biomass is the energy it can devote to competition. We can assume that it is proportional to its individual resources and therefore proportional to $\frac{1}{K}$ since the population size is of order K. This explains the competition coefficient of the form $\frac{c}{K}$.

In the results presented below, we generalize the above assumptions, but always focusing on the asymptotic behavior of $\frac{1}{K}(\lambda_i^K - \mu_i^K)$ and $\frac{1}{K^2}(\lambda_i^K + \mu_i^K)$. We will more generally assume in the rest of this subsection that the birth and death rates satisfy

$$\lim_{K\to\infty,\frac{i}{K}\to x} \frac{1}{K}(\lambda_i^K - \mu_i^K) = H(x); \qquad \lim_{K\to\infty,\frac{i}{K}\to x} \frac{1}{K^2}(\lambda_i^K + \mu_i^K) = 0, \quad (5.6.4)$$

where the function H is continuous.

Theorem 5.6.1 *Let $T > 0$. Assume that the sequence $(X_0^K)_K$ converges in law, as K tends to infinity, to a deterministic value x_0 and let us place ourselves under Assumptions* (5.6.4). *Assume further that the differential equation*

$$\frac{\mathrm{d}x(t)}{\mathrm{d}t} = H(x(t)); \quad x(0) = x_0. \qquad (5.6.5)$$

admits a unique solution.

Then, the process $(X_t^K, t \in [0,T])$ converges in law (and therefore in probability) to the deterministic solution $(x(t), t \in [0,T])$ of (5.6.5).

Proof (sketch) This theorem is proved by a compactness-uniqueness result on the set of trajectories, which is the set of right-continuous and left-limited functions from $[0,T]$ to $\mathbb{R}_+$. This set is called the *Skorohod space* and denoted by $\mathbb{D}([0,T],\mathbb{R}_+)$. It is equipped with a topology that makes it a complete separable metric space, which allows the application of general results characterizing the convergence in law (see Billingsley [11]). By Prokhorov's Theorem, the relative compactness of the sequence of laws of the processes X^K is characterized by the uniform tightness of these laws:

$$\forall \varepsilon > 0, \exists \text{ a compact } K_\varepsilon \subset \mathbb{D}([0,T],\mathbb{R}_+), \text{ such that } \sup_K \mathbb{P}(X^K \in K_\varepsilon^c) \le \varepsilon.$$

A characterization of the compact sets of $\mathbb{D}([0,T],\mathbb{R}_+)$ is then possible by a criterion of equi-continuity and equi-boundedness adapted from the Ascoli criterion (given usually for continuous functions). This makes it possible to give criteria of tightness (see [11] or Aldous [1]), which we will not develop in more detail here, but which are satisfied in our case (see for example Bansaye–Méléard [8].)

The sequence of laws of the processes X^K is therefore relatively compact and there exists a subsequence of $(X^K)_K$ that converges in law to a process X. We want to show that X is a solution of the differential equation (5.6.5), which will ensure its uniqueness. For simplicity, let's still denote the subsequence by $(X^K)_K$.

A first, elementary approach is to study the increment of the process $(X^K_t, t \geq 0)$ between times t and $t+h$. By Theorem 5.3.2, for all $i \geq 1$ and $K \in \mathbb{N}$,

$$\begin{aligned}\mathbb{E}(Z^K_{t+h} - Z^K_t \mid Z^K_t = i) &= ((i+1-i)\lambda^K_i + (i-1-i)\mu^K_i)\,h + o(h)\\ &= (\lambda^K_i - \mu^K_i)h + o(h).\\ \mathrm{Var}(Z^K_{t+h} - Z^K_t \mid Z^K_t = i) &= (\lambda^K_i + \mu^K_i)h - (\lambda^K_i - \mu^K_i)^2 h^2 + o(h).\end{aligned}$$

Thus, we deduce that

$$\begin{aligned}\mathbb{E}(X^K_{t+h} - X^K_t \mid Z^K_t = i) &= \frac{1}{K}(\lambda^K_i - \mu^K_i)h + \frac{1}{K}o(h),\\ \mathrm{Var}(X^K_{t+h} - X^K_t \mid Z^K_t = i) &= \frac{1}{K^2}(\lambda^K_i + \mu^K_i)h + \frac{1}{K^2}o(h).\end{aligned}$$

Let us assume that the infinitesimal quantities $o(h)$ appearing in the formulas above are uniform in K. Thus, the limit process X satisfies

$$\begin{aligned}X_0 &= x_0,\\ \lim_{h\to 0}\frac{1}{h}\mathbb{E}(X_{t+h} - X_t \mid X_t) &= H(X_t),\\ \lim_{h\to 0}\frac{1}{h}\mathrm{Var}(X_{t+h} - X_t \mid X_t) &= 0.\end{aligned}$$

The variances of the increments tend to 0 faster than h and the stochasticity thus disappears at the limit. We deduce that X is deterministic, continuous in time and a solution of the differential equation

$$\frac{\mathrm{d}X_t}{\mathrm{d}t} = H(X_t);\quad X_0 = x_0.$$

The uniqueness of the solution of this equation then implies that the sequence $(X^K)_K$ converges to the unique solution of the equation. The size Z^K_t of the population is then close to $Kx(t)$, as K tends to infinity.

A second, less elementary proof of this identification consists in studying the convergence of the generators. The generator of X^K is defined by (5.3.2). Consider a bounded function $f \in C^2_b$. We can write

$$\begin{aligned}Q^K f\left(\frac{i}{K}\right) &= \sum_{y\in\frac{\mathbb{N}}{K}} Q^K\left(\frac{i}{K}, y\right)\left(f(y) - f\left(\frac{i}{K}\right)\right)\\ &= \lambda^K_i\left(f\left(\frac{i}{K}+\frac{1}{K}\right) - f\left(\frac{i}{K}\right)\right) + \mu^K_i\left(f\left(\frac{i}{K}-\frac{1}{K}\right) - f\left(\frac{i}{K}\right)\right)\end{aligned}$$

$$= \frac{(\lambda_i^K - \mu_i^K)}{K} f'(\frac{i}{K}) + o(\frac{\|f''\|_\infty}{K}),$$

which converges uniformly to $Qf(x) = H(x)f'(x)$, as $K \to \infty$ and $\frac{i}{K} \to x$. We recognize the generator of the deterministic process, the unique solution of (5.6.5).□

The Malthusian equation

Assume that Hypothesis (5.6.2) is satisfied, namely that $H(x) = rx$. Then, the limit in "large population" is described by the solution of the differential equation

$$\dot{x}(t) = r\, x(t); \;\; x(0) = x_0.$$

This is the *Malthusian equation*. The long-term behavior of the solution $x(t)$ depends on the sign of the growth rate r. If $r > 0$, then $x(t) \to +\infty$ and the population size explodes at infinity. If $r < 0$, then $x(t) \to 0$, interpreted as population extinction. If $r = 0$, the solution stays constant.

The Logistic equation

Let us now assume Hypothesis (5.6.3). Then, the limit in "large population" is described by the solution of the differential equation

$$\dot{x}(t) = x(t)(r - c\,x(t)); \;\; x(0) = x_0, \tag{5.6.6}$$

called the *logistic equation*. This equation is famous in population dynamics (introduced by Verhulst in 1838, cf. [6]) in the case where the growth rate r is positive but where competition leads to a regulation of the population and prevents it from exploding.

So let us assume $r > 0$. The equation has two equilibria: the unstable equilibrium 0 and the stable one $\frac{r}{c} \neq 0$. It is easy to find the explicit solution of Equation (5.6.6):

$$x(t) = \frac{r x_0 e^{rt}}{r - C x_0 + C x_0 e^{rt}}.$$

This solution converges as $t \to \infty$, as soon as $x_0 \neq 0$, to the quantity r/c, called the *carrying capacity*. We underline the difference of long-term behavior between the logistic equation and the Malthusian equation, and also between the stochastic logistic birth and death process going almost surely to extinction (cf. Section 5.5.3) and the deterministic logistic equation with positive limit. Thus, *the limits as $K \to \infty$ and $t \to \infty$ do not commute!*

The use of such a deterministic model only makes sense for large populations. It does not take into account the stochastic variations due to small numbers. Thus, if the population size falls below a certain number, the modeling by a deterministic model loses its meaning and it is relevant to consider the stochastic model. Recent research seeks to define the threshold size from which the deterministic model is no longer suitable, cf. Coron et al. [29]. See also Chazottes et al. [22].

5.6.2 Stochastic Approximation – Demographic Stochasticity, Feller's Equation

Still considering a linear or logistic birth and death process, we now assume that the individual birth and death rates are of the order of magnitude K. This assumption makes sense if we consider a population of very small individuals. Indeed, the metabolic theory that links the mass of living beings with their characteristics (mortality, fecundity, etc.) predicts that individual characteristic times (age at maturity, lifespan, etc.) increase with the mass of individuals (see [14], [78]). The corresponding individual rates will then be all the greater as the individuals are small. More specifically, we assume that the individual birth and death rates are respectively of the form

$$\widehat{\lambda}_i^K = \gamma i K + \lambda_i^K \quad \text{and} \quad \widehat{\mu}_i^K = \gamma i K + \mu_i^K,$$

where λ_i^K and μ_i^K are chosen as in (5.6.4). Thus,

$$\lim_{K\to\infty, \frac{i}{K}\to x} \frac{1}{K}(\widehat{\lambda}_i^K - \widehat{\mu}_i^K) = H(x) \tag{5.6.7}$$

$$\lim_{K\to\infty, \frac{i}{K}\to x} \frac{1}{K^2}(\widehat{\lambda}_i^K + \widehat{\mu}_i^K) = 2\gamma x. \tag{5.6.8}$$

Compared to the previous study, only the variances computation changes. Here the limit of variances is not zero. More specifically, we obtain that

$$\mathrm{Var}(X_{t+h} - X_t \mid X_t) = 2\gamma X_t \, h + o(h).$$

Let us also recall that $\mathbb{E}(X_{t+h} - X_t \mid X_t) = H(X_t)\, h + o(h)$. We still assume (5.6.1), allowing the limit X_0 to be possibly random. The sequence of processes $(X_t^K, t \geq 0)$ then converges towards a stochastic process.

Theorem 5.6.2 *Let $T > 0$. Assume that the sequence $(X_0^K)_K$ converges in law, as K tends to infinity, to a random square integrable variable X_0. Assume further that (5.6.7) and (5.6.8) are satisfied. Then, the sequence of processes $(X_t^K, t \in [0,T])$ converges in law to the solution $(X_t, t \in [0,T])$ of the stochastic differential equation*

$$\mathrm{d}X_t = H(X_t)\,\mathrm{d}t + \sqrt{2\gamma X_t}\,\mathrm{d}B_t; \; X_0, \tag{5.6.9}$$

as long as a solution to the equation (5.6.9) exists and is unique, where $(B_t, t \geq 0)$ is a standard Brownian motion.

In the case where the function H is Lipschitzian, Theorem 4.6.11 ensures the existence and pathwise uniqueness of the solution to (5.6.9). This is particularly true for the Malthusian case, where the sequence of processes $(X_t^K, t \in [0,T])$ converges to the solution $(X_t, t \in [0,T])$ of the stochastic differential equation

$$\mathrm{d}X_t = r\, X_t\, \mathrm{d}t + \sqrt{2\gamma X_t}\, \mathrm{d}B_t;\ \ X_0. \tag{5.6.10}$$

This is the Feller equation studied in Section 4.7.1 and in Exercise 4.8.6.

In the logistic case, the limit equation becomes

$$\mathrm{d}X_t = (r - c\, X_t) X_t\, \mathrm{d}t + \sqrt{2\gamma X_t}\, \mathrm{d}B_t;\ \ X_0.$$

This is the logistic Feller equation studied in Section 4.7.2 and in Exercise 4.8.7. By general arguments (see [42]), we can show that the solution exists and is unique until time $T_0 \wedge T_\infty$ and that 0 is an absorbing point. However, we have shown in Section 4.7.2 that $T_\infty = \infty$ a.s. and that $\mathbb{P}(T_0 < +\infty) = 1$. Thus the solution is well defined and is unique on $\mathbb{R}_+$.

Proof The proof of Theorem 5.6.2 proceeds in a similar way to that of Theorem 5.6.1, particularly the proof of uniform tightness. However, the identification of the limit law is more subtle. For a complete proof, we refer to Lipow [58] or to [8]. Let us give some indications here.

Compared to the previous analysis, the term that changes in the computation of the generator is the order 2 term, which corresponds to the infinitesimal variance. Here this term is not null. More specifically, the generator of X^K is written for any function $f \in C_b^3$ as

$$\begin{aligned} Q^K f(\frac{i}{K}) &= \sum_{y \in \frac{\mathbb{N}}{K}} Q^K(\frac{i}{K}, y)\Big(f(y) - f(\frac{i}{K})\Big) \\ &= \widehat{\lambda}_i^K \left(f(\frac{i}{K} + \frac{1}{K}) - f(\frac{i}{K})\right) + \widehat{\mu}_i^K \left(f(\frac{i}{K} - \frac{1}{K}) - f(\frac{i}{K})\right) \\ &= \frac{(\lambda_i^K - \mu_i^K)}{K} f'(\frac{i}{K}) + \frac{1}{2K^2}(\widehat{\lambda}_i^K + \widehat{\mu}_i^K) f''(\frac{i}{K}) + o(\frac{\|f'''\|_\infty}{K^2}), \end{aligned}$$

which converges uniformly to $Qf(x) = H(x) f'(x) + \gamma x f''(x)$, as K tends to infinity and $\frac{i}{K}$ tends to x. We recognize the generator of the diffusion process

$$\mathrm{d}X_t = H(X_t)\, \mathrm{d}t + \sqrt{2\gamma X_t}\, \mathrm{d}B_t.$$

□

Remark 5.6.3 We have introduced and justified different Markovian, probabilistic or deterministic, continuous in time or jump models, which describe the dynamics of the same population size in various scalings. These models have different asymptotic behaviors.

Let us summarize the different logistic models:

- *The birth and death logistic process.* This describes the dynamics of a small population size subject to random fluctuations. The process tends to 0 as time tends to infinity. In this case, we obtained the average extinction time as the sum of a series.
- *The (deterministic) logistic equation.* This is a large population approximation of the birth and death process. The solution of the logistic equation converges as time tends to infinity to a non-zero limit, called the carrying capacity, which describes the equilibrium state of the population. This model makes sense for studying the dynamics of a large population size, which, although random, is approximated by this deterministic approximation. The probability that the birth and death process deviates from the carrying capacity is small but not zero and can be quantified by a large deviations theorem. The average extinction time is then of the order of e^{CK}, where $C > 0$ depends on the birth and death parameters (cf. [22]).
- *The Feller logistic stochastic differential equation.* This model is also a large population approximation, but for very high birth and death rates. It is relevant for populations with numerous and very small individuals reproducing and dying very quickly (populations of insects or bacteria, for example). The "noise" created by the numerous jumps due to birth and death events is so important that it persists in the limit. This explains the appearance of a new demographic stochasticity, through the Brownian stochastic term. In this case, we have shown by stochastic computation arguments that the process almost surely converges to 0 and we can compute the extinction time distribution.

5.6.3 The Lotka–Volterra and Predator-Prey Models

One can also define multi-type branching processes or birth and death processes in continuous time. We only develop here two-dimensional models leading to competitive Lotka–Volterra or predator-prey systems. For an initial population of size K, we define for any population state $i = (i_1, i_2)$ the following parameters:

- $r_i^{1,K}$: growth rate of the sub-population of type 1 in state i,
- $r_i^{2,K}$: growth rate of the sub-population of type 2 in state i,
- $\frac{c_{11}}{K} > 0$: competition rate between two individuals of type 1,
- $\frac{c_{12}}{K} > 0$: competition rate of an individual of type 2 on an individual of type 1,
- $\frac{c_{21}}{K} > 0$: competition rate of an individual of type 1 on an individual of type 2,
- $\frac{c_{22}}{K} > 0$: competition rate between two individuals of type 2.

We assume that for every state $x = (x_1, x_2)$ such that $\frac{i}{K} \to x$ we have

$$\lim_{K\to\infty, \frac{i}{K}\to x} \frac{1}{K} r_i^{1,K} = r_1\, x_1; \quad \lim_{K\to\infty, \frac{i}{K}\to x} \frac{1}{K} r_i^{2,K} = r_2\, x_2\,, \tag{5.6.11}$$

where r_1 and r_2 are two strictly positive real numbers. This means that in the absence of any competition, the two populations would tend to develop at an exponential speed.

We can adapt the arguments of Section 5.6.1 and show that the birth and death process $(\frac{1}{K}Z_t^K, t \in [0,T]) = \big((\frac{1}{K}Z_t^{1,K}, \frac{1}{K}Z_t^{2,K}), t \in [0,T]\big)$, composed of the two sub-processes describing the sizes of the sub-populations of type 1 and type 2 normalized by $\frac{1}{K}$, converges as K tends to infinity to the deterministic solution $(x(t), t \in [0,T])) = ((x_1(t), x_2(t)), t \in [0,T]))$ of the following differential system:

$$\begin{aligned} \dot{x}_1(t) &= r_1 x_1(t) - c_{11}(x_1(t))^2 - c_{12} x_1(t) x_2(t), \\ \dot{x}_2(t) &= r_2 x_2(t) - c_{21} x_1(t) x_2(t) - c_{22}(x_2(t))^2. \end{aligned} \tag{5.6.12}$$

Such systems have been extensively studied by theoretical biologists and specialists in dynamical systems. They are known as *competitive Lotka–Volterra systems*. It can be easily shown that the system has an unstable equilibrium $(0,0)$ and three stable equilibria: $(\frac{r_1}{c_{11}}, 0)$, $(0, \frac{r_2}{c_{22}})$, which represent the fixation of one population and the disappearance of the other, and a non-trivial co-existence equilibrium (x_1^*, x_2^*) that can be easily computed.

It is possible to generalize Assumption (5.6.11) as in Section 5.6.1 or to consider parameters $c_{ij} < 0$ modeling cooperation, leading to a more complicated dynamical system. A very famous case of such a dynamical system is the predator-prey model. The historical model of predation is due to Volterra (1926) and, almost contemporaneously, to Lotka (see [6]). Let us think of the two types of populations as, respectively, prey and predators of this prey. We assume that:

- In the absence of predators, the prey population size grows exponentially.
- In the absence of prey, the predator population size decreases exponentially.
- The rates of death of prey and growth of predators are proportional to the number of encounters between a prey and a predator.

We then obtain the following model:

$$\begin{aligned} \dot{x}_1(t) &= \alpha_1 x_1(t) - \beta_1 x_1(t) x_2(t), \\ \dot{x}_2(t) &= -\alpha_2 x_2(t) + \beta_2 x_1(t) x_2(t), \end{aligned} \tag{5.6.13}$$

where the parameters α_1, α_2, β_1 and β_2 are positive. For the study of Systems (5.6.12) and (5.6.13), we particularly refer to the book by Jacques Istas [43]. See also Renshaw [70] and Kot [53].

There is also a stochastic version of the model (5.6.12) where we can add to each equation a term of the form $\sqrt{\gamma_i X_i(t)}\, \mathrm{d}B_i(t)$, where B_i is a Brownian motion. This stochastic Lotka–Volterra system is obtained as an approximation of the multi-type birth and death process in the case of very rapid births and deaths, as for the Feller equation (see Cattiaux–Méléard [17]).

5.7 Exercises

Exercise 5.7.1 (Effect of fishing on a sardine shoal)
We model the size dynamics of a sardine shoal by a process $(X_t, t \geq 0)$. The sardines reproduce and their number grows at a rate $g > 0$. At random times $(T_i : i \geq 1)$, fishing causes the disappearance of a part of the sardine population.

The process counting the fishing events, $N_t := \text{Card}\{i \geq 1 : T_i \leq t\}$, is modeled by a Poisson process with intensity $r > 0$. Moreover, successive fishing events have independent and identically distributed (reducing) effects whose law is that of a random variable $F \in [0, 1]$, independent of $(N_t : t \geq 0)$. We assume that $X_0 = 1$.

1. Show that the size of the sardine population at time $t \geq 0$ is given by the process

$$X_t = \mathrm{e}^{gt} \prod_{i=1}^{N_t} F_i,$$

where $(F_i : i \geq 1)$ are independent and identically distributed random variables with the same law as F.

2. Assume that $\mathbb{E}(\log(\frac{1}{F})) < +\infty$. Let us recall that $\frac{N_t}{t} \longrightarrow r$ a.s. as $t \longrightarrow +\infty$ (see Proposition 5.2.15).

(a) Show that

$$\frac{\log(X_t)}{t} \overset{t\to\infty}{\longrightarrow} g - r\,\mathbb{E}(\log(\frac{1}{F})) \qquad \text{a.s.}$$

(b) Deduce the almost sure limit of X_t, when $g \neq r\,\mathbb{E}(\log(\frac{1}{F}))$.

3. Assume now that $\mathbb{E}(F) < +\infty$.

(a) Show that

$$Y_t = \mathrm{e}^{-gt+rt(1-\mathbb{E}(F))} X_t$$

is a martingale for the filtration $(\mathcal{F}_t = \sigma(X_s, s \leq t), t \geq 0)$.

(b) Deduce that Y_t converges a.s. to a random variable Y_∞, finite almost surely.

(c) Assume that $\mathbb{P}(F = 1) < 1$. Show that $Y_\infty = 0$ a.s.

4. Legislation for the preservation of sardines could consist in fixing the average effect of fishing, that is to say, fixing $c = \mathbb{E}(F) \in]0, 1[$.
Is this constraint wise for the preservation of sardines?
For a value c fixed by the legislation, is there a growth rate g that ensures the survival of the sardines?

Exercise 5.7.2
Assume that the population of individuals infected with the HIV virus grows according to a Poisson process of unknown intensity λ. Denote by N_t the number of infected individuals at time t. Deaths are not taken into account.

Each infected individual undergoes an incubation period between the moment he is infected with HIV and the moment the symptoms of AIDS appear. The duration of this incubation period is random. The incubation periods for the different individuals are independent and identically distributed, with a common distribution on $\mathbb{R}_+$ characterized by the distribution function G. Define for $t \in \mathbb{R}_+$,

$$\tilde{G}(t) = 1 - G(t).$$

Denote by N_t^1 the number of individuals who, at time t, show symptoms of AIDS, and by N_t^2 the number of individuals who, at time t, are infected with HIV but do not yet show symptoms of AIDS.

1. *Preliminaries*: Let $n \in \mathbb{N}^*$. Denote by S_n the set of permutations of $\{1,\ldots,n\}$. Let $t > 0$ and $X_1,\ldots,X_n$ be independent random variables, with uniform law on $[0,t]$. Let $(Y_1,\ldots,Y_n)$ be the same sequence ordered in ascending order.
   Show that the random vector $(Y_1,\ldots,Y_n)$ follows the uniform law on the set

   $$\{(y_1,\ldots,y_n);\ y_1 \le \cdots \le y_n \le t\},$$

   and give its density. (It may be helpful to note that for any bounded continuous f,

   $$\mathbb{E}(f(Y_1,\ldots,Y_n)) = \sum_{\sigma\in S_n} \mathbb{E}(f(X_{\sigma(1)},\cdots,X_{\sigma(n)})\mathbf{1}_{\{X_{\sigma(1)}\le\cdots\le X_{\sigma(n)}\}}).)$$

2. Let $T_1,,T_2,\ldots,T_n$ be the n first jump times of the process $(N_t, t \ge 0)$.
   Let $0 < t_1 < \cdots < t_n < t$. Compute

   $$\mathbb{P}(T_1 < t_1 < T_2 < t_2 < T_3 < \cdots < T_n < t_n | N_t = n).$$

   Deduce that the conditional law of the random vector $(T_1,\ldots,T_n)$ given $N_t = n$ is the law found in (1).
3. Let $n \in \mathbb{N}^*$ and $(T_1,\ldots,T_n)$ be defined as above. Let $(Z_1,\ldots,Z_n)$ be the vector $(T_{\sigma(1)},\ldots,T_{\sigma(n)})$, where σ is a random permutation independent of the process $(N_t, t \ge 0)$ and drawn uniformly in S_n.
   Give the law of $(Z_1,\ldots,Z_n)$ conditionally on $N_t = n$.
4. What is the probability that an individual infected with HIV at time s will see symptoms appear at time t? From this, deduce the probability p_t that the individual infected at time Z_i will be sick at time t, and the probability q_t that he will not be. (Note that they do not depend on i.)
5. Compute, for $k,l \in \mathbb{N}$, the probability $\mathbb{P}(N_t^1 = k;\ N_t^2 = l)$.
   Deduce that for all $t > 0$, N_t^1 and N_t^2 are two independent random variables with Poisson distributions, the intensity of which should be specified.

Exercise 5.7.3
Let $(N_t^1, t \ge 0)$ and $(N_t^2, t \ge 0)$ be two independent Poisson processes with respective intensities λ_1 and λ_2. These two processes are defined by their sequences of jump times: $(T_n^1)_n$ and $(T_m^2)_m$, as follows:

$$N_t^1 = \sum_{j\ge1} \mathbb{1}_{\{T_j^1 \le t\}};\quad N_t^2 = \sum_{j\ge1} \mathbb{1}_{\{T_j^2 \le t\}}.$$

1. What is the law of the pair of random variables (T_1^1, T_1^2)?
2. From this, deduce the value of $\mathbb{P}(T_1^1 < T_1^2)$.
3. What is the law of the random variable $\tau = \inf(T_1^1, T_1^2)$.
4. Let $0 \leq s \leq t$. Give the generating function of the random variable $N_t^1 - N_s^1$.
5. Define the process $(N_t, t \geq 0)$ by $N_t = N_t^1 + N_t^2$.
   Give the generating function of the random variable $N_t - N_s$.
   From this, deduce that $(N_t, t \geq 0)$ is a Poisson process with intensity $\lambda_1 + \lambda_2$.
6. Denote by $(R_k)_k$ the sequence of jump times of $(N_t)_t$:

$$N_t = \sum_{j \geq 1} \mathbb{1}_{\{R_j \leq t\}}.$$

   Introduce a sequence $(X_j)_j$ of independent random variables, with the same Bernoulli law of parameter $\frac{\lambda_1}{\lambda_1+\lambda_2}$, and independent of the Poisson process $(N_t)_t$. Define the processes $(M_t^1)_t$ and $(M_t^2)_t$ by

$$M_t^1 = \sum_{j \geq 1} \mathbb{1}_{\{R_j \leq t\}} \mathbb{1}_{\{X_j = 1\}};$$
$$M_t^2 = \sum_{j \geq 1} \mathbb{1}_{\{R_j \leq t\}} \mathbb{1}_{\{X_j = 0\}}.$$

   Compute, by conditioning on the value of N_t, the probability

$$\mathbb{P}(M_t^1 = k_1; M_t^2 = k_2),$$

   for $k_1, k_2 \in \mathbb{N}$.
   Deduce that M_t^1 and M_t^2 are two independent Poisson random variables with respective parameters λ_1 and λ_2.

It will also be assumed that the processes $(M_t^1)_t$ and $(M_t^2)_t$ are independent Poisson processes with respective intensities λ_1 and λ_2.

7. Recall that R_{n+m-1} is the $(n+m-1)$-th jump time of the process (N_t). Show that for any integer $k \leq n+m-1$, we have

$$\mathbb{P}(M^1_{R_{n+m-1}} = k) = \binom{n+m-1}{k} \left(\frac{\lambda_1}{\lambda_1+\lambda_2}\right)^k \left(\frac{\lambda_2}{\lambda_1+\lambda_2}\right)^{n+m-1-k}.$$

8. Show that for $n, m \in \mathbb{N}^*$, $\{T_n^1 < T_m^2\}$ has the same probability as $\{M^1_{R_{n+m-1}} \geq n\}$, where T_n^1 and T_m^2 were defined at the beginning of the exercise.
9. Infer that

$$\mathbb{P}(T_n^1 < T_m^2) = \sum_{k=n}^{n+m-1} \binom{n+m-1}{k} \left(\frac{\lambda_1}{\lambda_1+\lambda_2}\right)^k \left(\frac{\lambda_2}{\lambda_1+\lambda_2}\right)^{n+m-1-k}.$$

Exercise 5.7.4
Consider a birth and death process with the notations of Section 5.5.2 and assume that (5.5.5) is satisfied. Show that for all $n \geq 0$,

$$\mathbb{E}_{n+1}(T_n^2) = \frac{2}{\lambda_n \pi_n} \sum_{i \geq n} \lambda_i \pi_i \left(\mathbb{E}_{i+1}(T_i)\right)^2;$$

$$\mathbb{E}_{n+1}(T_n^3) = \frac{6}{\lambda_n \pi_n} \sum_{i \geq n} \lambda_i \pi_i \, \mathbb{E}_{i+1}(T_i) \, \mathrm{Var}_{i+1}(T_i).$$

Exercise 5.7.5
Consider a population consisting of N individuals of two types, denoted by A and a. The dynamics of this population are as follows: each individual A gives birth to another individual A at rate 1 and each individual a gives birth to another individual a at rate $1+s$, where $s > -1$ represents the selective advantage of a. Simultaneously with each birth, an individual is randomly chosen among the N individuals in the population, and dies to make room for the newborn. Thus, the population remains of constant size equal to N.

Denote by N_t the number of individuals of type a present at time t.

1. Assume that there are k individuals of type a in the population. What is the birth rate of an individual of type a? What is the probability that an individual of type A is replaced by an individual of type a?
   Deduce the infinitesimal generator of the Markov process $(N_t,\ t \geq 0)$. What can be said about states 0 and N for the process $(N_t,\ t \geq 0)$?
2. Let $T_N = \inf\{t \geq 0 :\ N_t = N\}$ and for $k \in \{0, \dots, N\}$ set $u(k) = \mathbb{P}_k(T_N < +\infty) = \mathbb{P}(T_N < +\infty | N_0 = k)$.
   For $k \in \{1, \dots, N-1\}$, give a relation linking $u(k+1)$, $u(k)$ and $u(k-1)$.
3. Deduce an expression for u. (The case $s = 0$ can be treated separately.)
4. For $k \in \{0, \dots, N\}$ let $w(k) = \mathbb{E}_k(T_N \mathbb{1}_{T_N < +\infty}) = \mathbb{E}(T_N \mathbb{1}_{T_N < +\infty} | N_0 = k)$. Show that for all $k \in \{1, \dots, N-1\}$ we have the relation

$$\frac{k(N-k)}{N}(1+s)(w(k+1) - w(k)) + \frac{k(N-k)}{N}(w(k-1) - w(k)) = -u(k).$$

5. In the case where $s = 0$, deduce an expression for $\mathbb{E}_1(T_N | T_N < +\infty)$.

Exercise 5.7.6
Let $(Z_t, t \geq 0)$ be a binary branching process with birth and death rates b and d, respectively. The growth rate, denoted by $r = b - d$, is assumed to be positive. The law of the process starting from $Z_0 = k$ individuals is denoted by $\mathbb{P}_k$.

1. Show that for any integer k and $t \in \mathbb{R}_+$, $\mathbb{E}_k(Z_t) = k\, u(t)$, where u is a function only depending on t.
   Deduce that for any integer k and $t \in \mathbb{R}_+$, $\mathbb{E}_k(Z_t) = k \exp(rt)$. (You may use the backward Kolmogorov equation.)
2. Using the Markov property, show that $(Z_t \exp(-rt); t \geq 0)$ is a martingale (for the filtration generated by Z).

3. Denote by T_n the first time the integer n is reached by the process $(Z_t, t \geq 0)$.

   (a) Show that
   $$T_n > T_0 \Rightarrow T_n = +\infty.$$
   (b) Deduce rigorously that $\mathbb{E}_k(\exp(-rT_n)) = k/n,\ \forall\, n \geq k$.

4. Let ρ be a random variable with an exponential distribution of parameter $r > 0$, independent of the process $(Z_t, t \geq 0)$.

   (a) Show that $\mathbb{P}_k(T_n \leq \rho) = \frac{k}{n}$.
   (b) Let $S_t := \sup_{s \leq t} Z_s$. Find the law of S_ρ.
   What is $\mathbb{E}(S_\rho)$?

Exercise 5.7.7
Let $(X_t, t \geq 0)$ be a continuous time birth and death process, starting from $X_0 > 0$ and with generator

$$Q_{i,i+1} = \lambda\, i, \quad \forall i \geq 0,$$
$$Q_{i,i-1} = \mu\, i(i-1), \quad \forall i \geq 1.$$

1. (a) Can the process $(X_t, t \geq 0)$ reach 0?
   (b) Give the transition probabilities of the embedded Markov chain.
   (c) Show that the embedded Markov chain is irreducible on $\mathbb{N}^*$. Is it irreducible on $\mathbb{N}$?
2. Let $\pi = (\pi_j, j \in \mathbb{N}^*)$ be a stationary probability measure on $\mathbb{N}^*$, that is, such that $\pi Q = 0$.

   (a) Determine the equation satisfied by this stationary probability measure π.
   (b) Show that, for each $i \in \mathbb{N}^*$, one has $\pi_i = \dfrac{\lambda^{i-1}}{i!\,\mu^{i-1}}\, \pi_1$.
   (c) Show that $\pi_1 = \dfrac{\lambda}{\mu(\mathrm{e}^{\lambda/\mu} - 1)}$.

3. Let Y be a Poisson random variable with intensity $a > 0$.

   (a) Compute $\mathbb{P}(Y > 0)$.
   Deduce the value of $\mathbb{P}(Y = i | Y > 0)$ for all $i \in \mathbb{N}^*$.
   (b) Show that π is the law of a Poisson random variable conditioned to remain strictly positive, whose intensity will be given.

Exercise 5.7.8 (Modeling population dynamics with immigration)
Let $(X_t, t \geq 0)$ be a birth and death process, with values in $\mathbb{N}$, of generator Q given for all $k \geq 1$ (k integer) by:

$$Q_{k,k+1} = \lambda k + a,$$
$$Q_{k,k-1} = \mu k,$$
$$Q_{k,k} = -(\lambda k + \mu k + a),$$

where a, λ and μ are three (strictly) positive real numbers.

1. Describe the embedded Markov chain.
2. Let $P_{ij}(t) = \mathbb{P}(X_t = i|X_0 = j)$. Write the partial differential equation satisfied by P.
3. Let $\mathbb{E}_i(X_t) = \mathbb{E}(X_t|X_0 = i)$. Show that

$$\partial_t \mathbb{E}_i(X_t) = (\lambda - \mu)\mathbb{E}(X_t|X_0 = i) + a.$$

4. Compute $\mathbb{E}_i(X_t)$.
5. Study the limit of $\mathbb{E}_i(X_t)$, as $t \to +\infty$, distinguishing cases according to the respective values of λ and μ.
6. Let $\pi = (\pi_j, j \in \mathbb{N})$ be a probability measure on $\mathbb{N}$ such that $\pi Q = 0$.
   Assuming that $\lambda < \mu$, show that for all $k \geq 1$

$$\pi_k = \frac{a(\lambda+a)(2\lambda+a)\cdots((k-1)\lambda+a)}{\mu.2\mu\cdots(k-1)\mu.k\mu}\pi_0.$$

7. Show that

$$\pi_0 = \left(1 - \frac{\lambda}{\mu}\right)^{\frac{a}{\lambda}}.$$

   One can use the following formula: for all $x \in [0,1[$, for all $N \in \mathbb{R}_+$,

$$\frac{1}{(1-x)^N} = \sum_{k=0}^{\infty}\binom{N+k-1}{k}x^k,$$

   where for all $N \in \mathbb{R}_+$,

$$\binom{N+k-1}{k} = \frac{(N+k-1)(N+k-2)\cdots N}{k!}.$$

8. Deduce π.

Exercise 5.7.9 (Binary branching process and Feller diffusion – Population division model)

Part I: Binary branching process

Consider a birth and death process $(X_t : t \geq 0)$ where, independently, individuals reproduce at rate b and die at rate d, with $b \neq d$.

0. What is the lifetime distribution of a given individual?
   What biological assumption(s) do these mathematical characteristics correspond to?
   What is the law of the number of times an individual will reproduce during his lifetime?
1. What is the generator of the process $(X_t : t \geq 0)$?

2. Let $u_\lambda(t,x) = \mathbb{E}(\exp(-\lambda X_t) \mid X_0 = x)$ for $\lambda \geq 0$ and $x \in \mathbb{N}$.

   (a) Show that for all $t \geq 0$ and $x \in \mathbb{N}$,

$$u_\lambda(t,x) = u_\lambda(t,1)^x.$$

   (b) Show that u_λ satisfies, for all $t \geq 0$,

$$\frac{\partial}{\partial t} u_\lambda(t,1) = (u_\lambda(t,1) - 1)(b\, u_\lambda(t,1) - d), \qquad u_\lambda(0,1) = \exp(-\lambda).$$

   Solving this equation by separation of variables we get

$$u_\lambda(t,1) = 1 - \frac{(b-d)}{b - [b - d\exp(\lambda)][1 - \exp(\lambda)]^{-1}\exp(-(b-d)t)}.$$

3. Show that $\mathbb{P}_1(X_t = 0)$ can be obtained from the function $\lambda \to u_\lambda(t,1)$. Conclude that for all $t \geq 0$,

$$\mathbb{P}_1(X_t > 0) = \frac{(b-d)}{b - d\exp(-(b-d)t)}.$$

   Note: We thus find the extinction rate of the binary branching process obtained in Section 5.4.5.

Part II: Feller diffusion and asymptotic behavior

The Feller diffusion $(Y_t : t \geq 0)$ with parameter $\sigma > 0$ is the process that satisfies the stochastic differential equation:

$$Y_t = y_0 + \int_0^t Y_s \,\mathrm{d}s + \int_0^t \sigma \sqrt{2Y_s}\,\mathrm{d}B_s,$$

with $y_0 \geq 0$.

0. Consider the solution Z of the SDE

$$Z_t = z_0 + \int_0^t h(Z_s)\,\mathrm{d}s + \int_0^t g(s, Z_s)\,\mathrm{d}B_s.$$

   Show that for a function $f \in C^{1,2}(\mathbb{R}_+ \times \mathbb{R}, \mathbb{R})$,

$$\begin{aligned} f(t, Z_t) &= f(0, z_0) \\ &\quad + \int_0^t \left[\frac{\partial f}{\partial s}(s, Z_s) + \frac{\partial f}{\partial y}(s, Z_s) h(Z_s) + \frac{1}{2}\frac{\partial^2 f}{\partial y^2}(s, Z_s) g(s, Z_s)^2\right] \mathrm{d}s \\ &\quad + \int_0^t \frac{\partial f}{\partial y}(s, Z_s) g(s, Z_s)\,\mathrm{d}B_s. \end{aligned}$$

1. Show that $\bar{Y}_t = Y_t e^{-t}$ satisfies the stochastic differential equation

$$\bar{Y}_t = y_0 + \int_0^t e^{-s/2}\sigma\sqrt{2\bar{Y}_s}\,dB_s.$$

2. Let $t_0 \geq 0$. Set for $t \in [0, t_0]$, and $\lambda, y \geq 0$,

$$v_\lambda(t,y) = \exp\left(-\frac{\lambda y}{\sigma^2\lambda(\exp(-t)-\exp(-t_0))+1}\right).$$

(a) Show that this function satisfies for all $s, y \geq 0$,

$$\frac{\partial v_\lambda}{\partial s}(s,y) + \frac{\partial^2 v_\lambda}{\partial y^2}(s,y)\,\sigma^2 y e^{-s} = 0.$$

(b) Show that $(v_\lambda(t,\bar{Y}_t) : 0 \leq t \leq t_0)$ satisfies the stochastic differential equation

$$v_\lambda(t,\bar{Y}_t) = v_\lambda(0,y_0) - \int_0^t \lambda \frac{v_\lambda(s,\bar{Y}_s)}{\sigma^2\lambda(\exp(-s)-\exp(-t_0))+1} e^{-s/2}\sigma\sqrt{2\bar{Y}_s}\,dB_s.$$

Thus $(v_\lambda(t,\bar{Y}_t) : 0 \leq t \leq t_0)$ is a martingale with respect to the Brownian filtration.

(c) Deduce the value of $\mathbb{E}_{y_0}(v_\lambda(t_0,\bar{Y}_{t_0}))$ and then that

$$\mathbb{E}_{y_0}(\exp(-\lambda Y_t)) = \exp\left(-\frac{y_0}{\sigma^2(1-\exp(-t))+(\lambda\exp(t))^{-1}}\right).$$

(d) Briefly show that the law of Y_t starting from $y_0 + y_0'$ is the same as that of $Y_t^1 + Y_t^2$, where Y^1 and Y^2 are two independent Feller diffusions with the same law as Y starting respectively from y_0 and y_0'.

3. Show that

$$\mathbb{P}_{y_0}(Y_t > 0) = 1 - \exp\left(-\frac{y_0}{\sigma^2(1-\exp(-t))}\right),$$

then that $\mathbb{P}_{y_0}(\forall t \geq 0 : Y_t > 0) = 1 - \exp(-y_0/\sigma^2)$.

4. Consider for $N \geq 1$ the binary branching process $(X_t^N : t \geq 0)$ with individual birth rate $b_N = b + \sigma^2 N$ and individual death rate $d_N = b - 1 + \sigma^2 N$, where $b > 1$.

(a) In your opinion, what do the birth and death rates of this form model? What is the average population growth rate?

(b) Check that $u_\lambda^{(N)}(t) = \mathbb{E}_1(\exp(-\lambda X_t^N))$ satisfies

$$1 - u_{\lambda/N}^{(N)}(t) \sim \frac{1}{N}\frac{\lambda}{\lambda\sigma^2 + (1-\lambda\sigma^2)\exp(-t)} \qquad \text{when } N \to \infty.$$

(c) Deduce that for all $t, \lambda \geq 0$,

$$\mathbb{E}(\exp(-\lambda X_t^N/N) \mid X_0 = N) \overset{N\to\infty}{\longrightarrow} \mathbb{E}_1(\exp(-\lambda Y_t)).$$

(d) Explain how to simulate a Feller diffusion with parameter σ from independent exponential random variables with parameter 1.

Part III: Branching Feller Diffusion

An insect population develops with rapid births and deaths. Its size Y_t at time t is modeled by the Feller diffusion with parameter $\sigma > 0$ defined in Part II. We assume that the initial size of the population is $Y_0 = 1$ a.s.

1. Let $T_x = \inf\{t \geq 0 : Y_t = x\}$ for $x \geq 0$.

(a) Show that $\alpha = \sup\{\mathbb{P}_x(Y_1 > 0) : x \in [0,2]\} < 1$ and that for all $n \in \mathbb{N}$,

$$\mathbb{P}_1(Y_{n+1} > 0,\ Y_n < 2, Y_{n-1} < 2, \ldots, Y_1 < 2) \leq \alpha^{n+1}.$$

Deduce that $\mathbb{P}_1(T_0 = \infty, T_2 = \infty) = 0$.

(b) Show using the previous part that

$$\mathbb{P}_1(T_2 < T_0) \geq 1 - \exp(-\frac{1}{\sigma^2}) > 0.$$

If the insect population reaches size 2, it splits into two subpopulations of size 1 due to resource constraints. The two populations then evolve independently according to Feller diffusions with parameter σ starting from 1. These, similarly, either go extinct or reach level 2 and again split into two subpopulations of size 1, etc.

2. Consider the Galton–Watson process $(Z_n)_n$ where each individual dies with probability $1 - p$ and has two children with probability p, where $p = \mathbb{P}_1(T_2 < T_0)$. Show that the probability that the number of alive insect subpopulations reaches zero in finite time is the extinction probability of the process $(Z_n)_n$.
3. What is the probability that the total size of the insect population (obtained by summing over the sizes of each subpopulation) tends to infinity in infinite time? What then can be said about the asymptotic behavior of the number of present populations?

Exercise 5.7.10 (Quasi-stationary behavior of a pure jump Markov process with finite state space)

We wish to understand the long-term behavior of pure jump Markov processes with finite state space. It was seen earlier that a subcritical Galton–Watson process attains 0 in finite time with probability 1, but conditioning it to survive for a long time allows us to define a *quasi-stationary* distribution describing a transitory equilibrium before extinction. We will generalize this result to pure jump Markov processes with finite state space.

We are therefore interested in a Markov process $(Z_t)_{t\geq 0}$ taking its values in the space $\{0, 1, \ldots, N\}$ (where $N \geq 1$ is a fixed integer) and whose only absorbing state is 0. We assume that the hitting time T_0 of 0 is finite a.s.

We recall that if ν is a distribution on $\{1, \ldots, N\}$, $\mathbb{P}_\nu$ denotes the probability measure given by

$$\mathbb{P}_\nu(Z_t = j) = \sum_{i=1}^{N} \nu(i)\, \mathbb{P}_i(Z_t = j), \qquad \forall t \geq 0,\ j \in \{0, \ldots, N\},$$

under which the initial value Z_0 has law ν.

1. Give an example of a biological situation in which it seems relevant to represent the population by a Markov process with finite state space and absorbed at 0.
2. Recall that a distribution α is said to be *quasi-stationary* for Z if for all $t \geq 0$ and all $j \in \{1, \ldots, N\}$,

$$\sum_{i=1}^{N} \alpha(i)\, \mathbb{P}_i(Z_t = j \,|\, Z_t > 0) = \alpha(j).$$

   (a) Suppose that Z admits a quasi-stationary distribution α. Show that for all $s, t \geq 0$ and $j \in \{1, \ldots, N\}$, we have

$$\mathbb{P}_\alpha(T_0 > t+s,\, Z_t = j) = \mathbb{P}_\alpha(T_0 > t+s \,|\, Z_t = j)\; \mathbb{P}_\alpha(Z_t = j \,|\, T_0 > t)\; \mathbb{P}_\alpha(T_0 > t).$$

   Deduce that

$$\mathbb{P}_\alpha(T_0 > t+s) = \mathbb{P}_\alpha(T_0 > t)\; \mathbb{P}_\alpha(T_0 > s).$$

   (b) Deduce that if Z_0 has law α, then the extinction time of Z follows an exponential law whose parameter will be denoted by $\theta(\alpha)$.

In the following, we denote by $\bar{P}_t$ the matrix of coefficients $\bar{P}_t(i, j) = \mathbb{P}_i(X_t = j)$ for all $i, j \in \{1, \ldots, N\}$ (observe that the matrix $\bar{P}_t$ of size $N \times N$ is obtained from P_t by forgetting state 0). We assume that $\bar{P}_1$ has strictly positive entries and that $\mathbb{P}_i(Z_1 = 0) > 0$ for all i.

3. (a) Show that for all $i \in \{1, \ldots, N\}$,

$$\sum_{j=1}^{N} \bar{P}_1(i, j) < 1.$$

   (b) Show that the Perron–Frobenius Theorem (Theorem 3.7.7) applies and that there exist a unique real eigenvalue $\lambda_0 > 0$, greater in modulus than all the others (real or complex), and a unique left eigenvector u and a unique right eigenvector v such that $u_i > 0$, $v_i > 0$, for all i, and such that $\sum_{i=1}^{N} u_i = 1$ and $\sum_{i=1}^{N} u_i v_i = 1$, satisfying

$$\sum_{i=1}^{N} u_i \bar{P}_1(j,i) = \lambda_0 u_j \quad \forall j \in \{1,\ldots,N\} \qquad \text{and}$$
$$\sum_{j=1}^{N} \bar{P}_1(i,j) v_j = \lambda_0 v_i \quad \forall i \in \{1,\ldots,N\}.$$

(c) Show that $\lambda_0 < 1$. Therefore, it can be written as $\lambda_0 = \mathrm{e}^{-\theta'}$, for a real number $\theta' > 0$.

We will assume that for all $t \geq 0$, $\bar{P}_t$ is written in the form

$$\bar{P}_t = \mathrm{e}^{-\theta' t} A + O(\mathrm{e}^{-\chi t}), \tag{5.7.1}$$

where A is the matrix of coefficients $A_{i,j} = v_i u_j$, $\chi > \theta'$ and $O(\mathrm{e}^{-\chi t})$ is a matrix whose entries do not exceed $C\mathrm{e}^{-\chi t}$ for a given constant C. We will use this result to obtain the behavior before extinction of Z.

4. (a) Let $i \in \{1,\ldots,N\}$. Using (5.7.1), provide an equivalent of $\mathbb{P}_i(T_0 > t)$ as $t \to \infty$. *(Hint: what are the values that Z_t can take when $T_0 > t$?)*
   (b) Deduce that for all $i, j \in \{1,\ldots,N\}$,
   $$\lim_{t\to\infty} \mathbb{P}_i(Z_t = j \,|\, T_0 > t) = u_j.$$

5. In this question, we will show that $u = (u_1,\ldots,u_N)$ is a quasi-stationary distribution for Z.

   (a) Show that for all $s,t \geq 0$ and any $j \neq 0$, we have
   $$\mathbb{P}_u(Z_{t+s} = j \,|\, T_0 > t+s) = \sum_{i=1}^{N} \mathbb{P}_u(Z_t = i \,|\, T_0 > t)\, \mathbb{P}_i(Z_s = j)\, \frac{\mathbb{P}_u(T_0 > t)}{\mathbb{P}_u(T_0 > t+s)}. \tag{5.7.2}$$
   (b) Show by cutting according to the possible values of Z_t that
   $$\lim_{t\to\infty} \mathbb{P}_u(T_0 > t+s \,|\, T_0 > t) = \mathbb{P}_u(T_0 > s).$$
   (c) By letting t tend to infinity in (5.7.2), show that u is indeed a quasi-stationary distribution for Z.
   (d) Recalling the notation from question 2(b), provide an expression for $\theta(u)$.

Chapter 6
Genetic Evolution Processes

Those we called brutes had their revenge when Darwin proved to us that they were our cousins.

George Bernard Shaw (1856–1950)

6.1 An Idealized Model of Infinite Population: The Hardy–Weinberg Model

In this chapter, we will focus on the reproduction of individuals through the transmission of their alleles (for given loci) over time. In a haploid population, the proportions of individuals carrying the different alleles related to a given gene provide all the information on the genotypic frequency in the population. However, if the population is diploid, the genes are paired in individuals. It is then necessary to distinguish two types of frequencies to describe the genetic composition of the population at a considered locus: the genotypic frequencies, which are the frequencies of the different genotypes at a considered locus, and the allelic frequencies, which are the frequencies of the different alleles at the considered locus. In general, genotypic frequencies cannot be deduced from allelic frequencies. For example, suppose we have two alleles A and a. A population in which the allelic composition is 50% A and 50% a may consist entirely of homozygotes (AA and aa), entirely of heterozygotes (Aa), or any mixture of these three genotypes. However, in the ideal Hardy–Weinberg model that we develop in this first section, strong simplifying assumptions allow us to directly link genotypic frequencies and allelic frequencies.

The Hardy–Weinberg model is the first published model concerning the genotypic structure of a population, introduced simultaneously by Hardy and Weinberg in 1908 (see [6]). The assumptions that are made greatly simplify the computations but do not allow the integration of Darwinian ideas of adaptation and evolution. These assumptions are as follows.

S. Méléard, *Random Models in Biology, Ecology and Evolution*, Texts in Applied Mathematics 87, https://doi.org/10.1007/978-3-662-73483-4_6

- Gametes associate randomly regardless of the genes considered (the panmixia hypothesis). This assumption amounts to saying that there is an infinite reservoir of gametes that are paired randomly, without taking into account the sex of the individual.
- The population has an infinite size. By the law of large numbers, we will replace the frequency of each allele by its probability.
- The frequency of genes is not modified from one generation to the next.

Under these assumptions, suppose that at one locus, the probabilities of alleles A and a are p and $q = 1 - p$. In the second generation, after pairing a male gamete and a female gamete, we obtain the genotype AA with probability p^2, the genotype aa with probability q^2 and the genotype Aa with probability $2pq$. This genotypic structure is known as the Hardy–Weinberg structure. But then, since each individual has two copies of each gene, the probability of appearance of the allele A in the population, in the second generation, will be $\frac{2p^2+2pq}{2} = p^2 + pq = p$. Similarly, the frequency of the allele a will be q.

Thus, under the previous assumptions, we can state the law of Hardy–Weinberg: *In an isolated population of unlimited size, not subject to selection and in which there is no mutation, allelic frequencies remain constant.* If the mating is panmictic, the genotypic frequencies are directly deduced from the allelic frequencies and therefore remain constant, and equal to $p^2, 2pq, q^2$. The model does not present random variation. It is reduced to a probability computation on allelic frequencies and we will not be able to observe the fixation or the disappearance of an allele after a certain time.

6.2 Finite Population: The Wright–Fisher Model

6.2.1 Wright–Fisher Process

While in a population of infinite size the allelic frequencies are stable over generations in the absence of selection and mutation (the law of large numbers), the allelic frequencies vary randomly in populations of finite size. (This is due to the variability in the distribution of genes from one generation to the next.) To allow a not too complicated mathematical treatment, the Wright–Fisher process models the transmission of genes from one generation to the next in a very schematic way. It assumes separate generations, which is a considerable simplification of the reproduction cycle. Here, a population of M individuals is represented by a vector of $N = 2M$ alleles.

We will denote by $n \in \mathbb{N}$ the generation indices. In the Wright–Fisher model, the parent of each individual of the generation $n+1$ is distributed uniformly in the n-th generation. It is also assumed that the population of individuals is of finite and constant size M. Each individual is characterized by one of the two alleles A and a that it transmits by heredity. We therefore have $N = 2M$ alleles. The individuals reproduce independently of each other and the new generation is formed of N

genes chosen uniformly in an infinite reservoir of gametes, and in which the allelic distribution is that of the generation n. The Wright–Fisher model is a neutral model. This means that there is no selective advantage associated with either of the two types (which would favor the reproduction of one of the alleles). Due to the neutrality of the model, all individuals are interchangeable, and only the distribution of alleles matters.

Let us define the quantity

$$X_n^N = \text{number of type } A \text{ alleles at generation } n.$$

We study the dynamics of $(X_n^N, n \in \mathbb{N})$. Since M individuals are chosen (and therefore $N = 2M$ alleles) at generation $n+1$, the distribution of the number of alleles A among these N alleles should follow a hypergeometric law. But the assumption of an infinite reservoir of gametes allows us to approximate this hypergeometric law (simultaneous drawing) by a binomial law that does not take into account the number of individuals but only the proportions of the two alleles in the population of gametes (drawing with replacement, see the study of urn models, Méléard [59, Section 2.2.3]).

We have the following result.

Proposition 6.2.1 *For all $n \in \mathbb{N}$, for all i and j in $\{0, \ldots, N\}$,*

$$\mathbb{P}(X_{n+1}^N = j | X_n^N = i) = \binom{N}{j} \left(\frac{i}{N}\right)^j \left(1 - \frac{i}{N}\right)^{N-j}. \tag{6.2.1}$$

Proof By definition, for all $n \in \mathbb{N}$, the variable X_n^N takes its values in $\{0, \ldots, N\}$. It is clear that

$$\mathbb{P}(X_{n+1}^N = 0 | X_n^N = 0) = 1$$

by definition, and likewise,

$$\mathbb{P}(X_{n+1}^N = N | X_n^N = N) = 1.$$

More generally, let $i \in \{1, \ldots, N-1\}$. Conditionally on the fact that $X_n^N = i$, the frequency of the allele A at generation n is $\frac{i}{N}$ and that of the allele a is $\frac{N-i}{N} = 1 - \frac{i}{N}$. At the next generation, each individual randomly chooses its parent, without considering the other individuals. This therefore corresponds to a drawing with replacement and the number of individuals with allele A at the $(n+1)$-th generation, knowing that $X_n^N = i$, then follows a binomial law $\mathcal{B}(N, \frac{i}{N})$, which yields (6.2.1). □

Remark 6.2.2 In the non-neutral models that we will see later (with mutation or selection), individuals will no longer be interchangeable, and the probability of "drawing" an individual with allele A knowing $X_n^N = i$ will no longer be uniform.

Remark 6.2.3 According to Proposition 6.2.1 and denoting $\frac{i}{N}$ by p_i, we have that

$$\mathbb{E}(X_{n+1}^N | X_n^N = i) = Np_i = i \text{ and therefore } \mathbb{E}(X_{n+1}^N - X_n^N | X_n^N = i) = 0;$$
$$\mathrm{Var}(X_{n+1}^N | X_n^N = i) = \mathrm{Var}(X_{n+1}^N - X_n^N | X_n^N = i) = Np_i(1-p_i) = i(1-\frac{i}{N}). \quad (6.2.2)$$

We are interested in the transmission of the allele A. We can ask ourselves the following questions: will the allele A invade the entire population? Or the allele a? In how much time? Will the two alleles coexist?

Theorem 6.2.4 *The sequence $(X_n^N)_{n\in\mathbb{N}}$ is a Markov chain with finite state space and is a bounded martingale. The transition matrix P of the chain is given for all i and j in $\{0,\dots,N\}$ by*

$$P_{i,j} = \binom{N}{j}\left(\frac{i}{N}\right)^j\left(1-\frac{i}{N}\right)^{N-j}. \quad (6.2.3)$$

The states 0 and N are two absorbing states for the chain.

Proof By definition, the process $(X_n^N, n \geq 0)$ takes its values in the finite set $\{0,\dots,N\}$. It is clear that this process defines a Markov chain since the law of X_{n+1}^N conditioned on the past, up to time n, only depends on X_n^N. The form of the transition matrix was computed in Proposition 6.2.1.

Moreover, by repeating the computations of Remark 6.2.3 and if $\mathcal{F}_n$ denotes the σ-field generated by $X_0^N,\dots,X_n^N$, we have

$$\begin{aligned}
\mathbb{E}(X_{n+1}^N | \mathcal{F}_n) &= \sum_{j=0}^N j\,\mathbb{P}(X_{n+1}^N = j | \mathcal{F}_n) \\
&= \sum_{j=0}^N j\,\mathbb{P}(X_{n+1}^N = j | X_n^N) = \sum_{j=0}^N j\binom{N}{j}\left(\frac{X_n^N}{N}\right)^j\left(1-\frac{X_n^N}{N}\right)^{N-j} \\
&= N\frac{X_n^N}{N} = X_n^N.
\end{aligned}$$

Thus, the process $(X_n^N)_n$ is a martingale for the filtration $(\mathcal{F}_n)_n$, bounded by N.

It is obvious that the states 0 and N are absorbing (see (6.2.1)). □

Remark 6.2.5 The process $(X_n^N)_n$ is a martingale and therefore, in this neutral model, the size of the allele A population is constant on average: there is no selective advantage for an allele. We will see in Section 6.2.4 how selection influences this dynamics.

We now study how alleles become fixed in a population of size N, as the number of generations n tends to infinity.

Theorem 6.2.6 *As n tends to infinity, the sequence of random variables $(X_n^N)_n$ converges almost surely to a random variable X_∞^N and*

$$X_\infty^N \in \{0, N\}.$$

Moreover,

$$\mathbb{P}_i(X_\infty^N = N) = \frac{i}{N} = \frac{1}{N}\mathbb{E}(X_\infty^N \,|\, X_0^N = i).$$

Note that $\mathbb{P}_i(X_\infty^N = N)$ is the fixation probability of the allele A in the population and that $\mathbb{P}_i(X_\infty^N = 0)$ is its extinction probability and therefore the fixation probability of the allele a. The fixation of an allele can therefore occur without selection, due solely to random fluctuations. This is what biologists call the *genetic drift* effect.

Example 6.2.7 Example of fixation of the allele a. There is a single individual of allele a at time 0 among 10 individuals. The probability of having fixation of a is $1/10$.

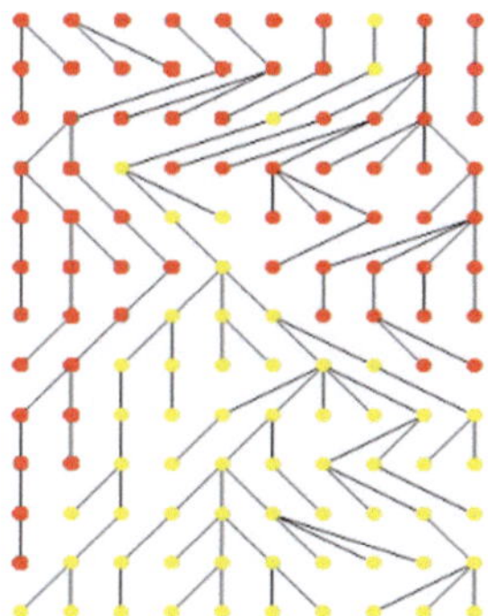

Proof Since $(X_n^N)_n$ is a bounded martingale, it converges almost surely, as $n \to \infty$, to a random variable X_∞^N (see Theorem 2.5.5). The process $(X_n^N)_n$ is a Markov chain with finite state space. Then the two absorbing points are positive recurrent states and all other states are transient. Indeed, if $i \in \{1, \dots, N-1\}$ and according to (6.2.1),

$$p_{i0} = \mathbb{P}(X_1^N = 0 | X_0^N = i) = \left(1 - \frac{i}{N}\right)^N > 0,$$

so i leads to 0 but 0 does not lead to i since 0 is absorbing. We therefore have two recurrence classes $\{0\}$ and $\{N\}$ and a transient class $\{1, \dots, N-1\}$ from which the chain almost surely exits after a certain time. Thus, X_∞^N takes its values in $\{0, N\}$. Let us introduce the fixation time

$$\tau = \inf\{n \geq 0;\, X_n^N = 0 \text{ or } X_n^N = N\} = T_0 \wedge T_N,$$

where for $m \in \{0, \dots, N\}$, $T_m = \inf\{n, X_n^N = m\}$. The stopping time τ is therefore finite almost surely. Let $\mathbb{P}_i$ be the law of the chain starting from i. As $(X_n^N)_n$ is a bounded martingale, the stopping theorem applies (see Theorem 2.5.4) and therefore

$$\mathbb{E}_i(X_\tau^N) = \mathbb{E}_i(X_0^N) = i.$$

This implies in particular that

$$P_i(X_\infty^N = N) = \frac{i}{N}. \qquad \square$$

To study the time it takes for fixation, let us study the probability that two copies of the same locus chosen at random (without replacement) carry different alleles (heterozygosity). Let $h(n)$ be this probability, in the case where these copies are chosen in the n-th generation. First, let us compute the probability of having heterozygosity, conditional on $X_n^N = i$. It is given by

$$H_n(i) = \frac{\binom{i}{1}\binom{N-i}{1}}{\binom{N}{2}} = \frac{2i(N-i)}{N(N-1)}.$$

More generally, H_n is the random variable

$$H_n = \frac{\binom{X_n^N}{1}\binom{N-X_n^N}{1}}{\binom{N}{2}} = \frac{2X_n^N(N-X_n^N)}{N(N-1)}$$

and by (6.2.2),

$$H_n = \frac{2}{(N-1)}\mathrm{Var}(X_{n+1}^N \,|\, X_n^N).$$

We can then show that

Proposition 6.2.8

$$h(n) = \mathbb{E}(H_n) = \left(1 - \frac{1}{N}\right)^n \mathbb{E}(H_0). \tag{6.2.4}$$

This proposition gives the speed of (exponential) decay of heterozygosity over time.

Proof Let us write $\mathbb{E}(H_n)$ in terms of $\mathbb{E}(H_{n-1})$. We know that $\mathbb{E}(X_n^N) = \mathbb{E}(X_{n-1}^N)$. Moreover,

$$\mathbb{E}((X_n^N)^2 \,|\, X_{n-1}^N) = \mathrm{Var}(X_n^N \,|\, X_{n-1}^N) + (\mathbb{E}(X_n^N \,|\, X_{n-1}^N))^2 = \frac{(N-1)H_{n-1}}{2} + (X_{n-1}^N)^2.$$

We easily deduce that

$$\mathbb{E}(H_n) = \frac{2}{N(N-1)}\left(\mathbb{E}(NX_{n-1}^N - (X_{n-1}^N)^2) - \frac{(N-1)\mathbb{E}(H_{n-1})}{2}\right) = \left(1 - \frac{1}{N}\right)\mathbb{E}(H_{n-1}).$$

$\square$

Note that if we start with a single copy of allele A, then $X_0^N = 1$, and $H_0 = \frac{2}{N}$. Thus, $h(0) = \frac{2}{N}$, and in this case, thanks to (6.2.8), the probability that, at generation n, two copies chosen at random have two different alleles is $\frac{2}{N}\left(1 - \frac{1}{N}\right)^n$.

Remark 6.2.9 As we have just seen, the proof of Proposition 6.2.8 follows from a very simple computation. However, we now propose another proof, introducing a dual approach based on the genealogy of individuals. This approach was developed by Durrett [33] and it will be useful in Section 6.5. We call *individuals* the N copies of the locus considered and we will study their genealogy going back in time. Indeed, two individuals at time n will be heterozygous if they do not have a common ancestor and if at time 0, their parents were also heterozygous.

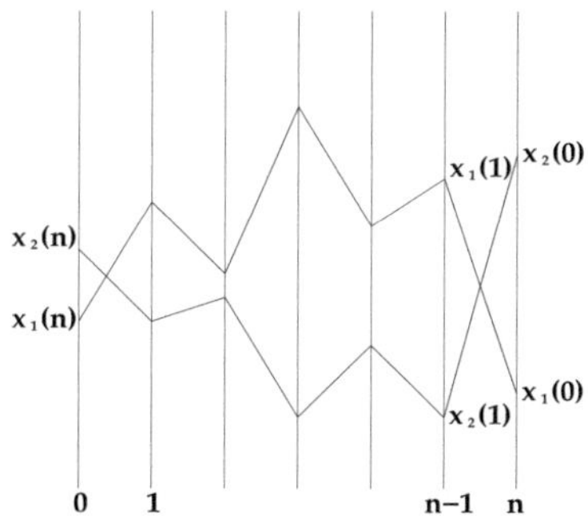

Fig. 6.1 The genealogy of two alleles.

Let us choose at time n two distinct alleles numbered $x_1(0)$ and $x_2(0)$. The individuals $i = 1,2$ are each descendants of individuals marked $x_i(1)$ at time $n-1$, who are themselves descendants of individuals marked $x_i(2)$ at time $n-2$... Thus the sequence $x_i(m)$, $0 \le m \le n$, describes the genealogy of $x_i(0)$, that is, the sequence of its ancestors if the direction of time is reversed. See Figure 6.1.

Note that if $x_1(m) = x_2(m)$, the two individuals had the same ancestor m generations earlier and $x_1(l) = x_2(l)$ for $m \le l \le n$. If $x_1(m) \ne x_2(m)$, since the choices of the two parents of the individuals are made independently of each other, the probability that these individuals have the same common ancestor at the $(m+1)$-th generation is $\frac{1}{N}$ and the probability that $x_1(m+1) \ne x_2(m+1)$ is $1 - \frac{1}{N}$. A proof by induction then leads to the probability of having $x_1(n) \ne x_2(n)$, which is equal to $\left(1 - \frac{1}{N}\right)^n$. In this case, the two lineages are disjoint up to this past generation n. Conditionally on this event, the two alleles $x_1(n)$ and $x_2(n)$ are therefore chosen at random in the population at real time 0 and the probability that they are different is the frequency H_0 with expectation $h(0)$, which concludes the proof.

There is no exact formula for the average fixation time, but we will be able to give an approximation when the size N of the population tends to infinity.

Proposition 6.2.10 *We assume that N tends to infinity. Let us fix $x_0 \in]0,1[$ and assume that the initial frequency of the allele A is $\frac{[Nx_0]}{N}$, converging to $x_0 \in]0,1[$. Then, the average fixation time behaves approximately like*

$$\mathbb{E}_{[Nx_0]}(T_N \wedge T_0) \sim_{N\to\infty} -2N\,(x_0 \ln x_0 + (1-x_0)\,\ln(1-x_0)).$$

Proof For this proof, we assume that there exists a function T defined on $[0,1]$, three times differentiable and with a third derivative bounded by C, such that $T\left(\frac{k}{N}\right) = \mathbb{E}_k(T_N \wedge T_0)$ for all $k \in \{0,\ldots,N\}$.

Recall that if the number of individuals with allele A is $[Nx_0]$, then this number in the next generation follows a binomial distribution $\mathcal{B}(N, \frac{[Nx_0]}{N})$, from which $\mathbb{E}(\frac{X_1^N-[Nx_0]}{N}) = 0$ and $\mathrm{Var}(\frac{X_1^N-[Nx_0]}{N}) = \frac{1}{N}\frac{[Nx_0]}{N}(1-\frac{[Nx_0]}{N})$. We use a conditioning with respect to the first jump of the chain X^N and an expansion around the point $\frac{[Nx_0]}{N}$. Then

$$\begin{aligned}
\mathbb{E}_{[Nx_0]}(T_N \wedge T_0) &= \sum_{k=0}^{N} \mathbb{E}_{[Nx_0]}(T_N \wedge T_0 \,|\, X_1^N = k)\,\mathbb{P}(X_1^N = k)\\
&= \sum_{k=0}^{N} \mathbb{P}(X_1^N = k)\left(T\left(\frac{k}{N}\right)+1\right)\\
&= 1+\sum_{k=0}^{N} \mathbb{P}(X_1^N = k)\left(T\left(\frac{[Nx_0]}{N}\right)+\left(\frac{k-[Nx_0]}{N}\right)T'\left(\frac{[Nx_0]}{N}\right)\right.\\
&\quad +\frac{1}{2}\left(\frac{k-[Nx_0]}{N}\right)^2 T''\left(\frac{[Nx_0]}{N}\right)\\
&\quad \left.+\frac{1}{6}\left(\frac{k-[Nx_0]}{N}\right)^3 T^{(3)}\left(\frac{[Nx_0]}{N}+\theta\,\frac{k-[Nx_0]}{N}\right)\right), \qquad \theta \in]0,1[,\\
&= 1+T\left(\frac{[Nx_0]}{N}\right)+\frac{1}{2N}\frac{[Nx_0]}{N}\left(1-\frac{[Nx_0]}{N}\right)T''\left(\frac{[Nx_0]}{N}\right)\\
&\quad +\frac{1}{6}\mathbb{E}\left(\left(\frac{X_1^N-[Nx_0]}{N}\right)^3 T^{(3)}\left(\frac{[Nx_0]}{N}+\theta\,\frac{X_1^N-[Nx_0]}{N}\right)\right).
\end{aligned}$$

A simple computation (using the moments up to order 3 of the binomial law) shows that

$$\begin{aligned}
&\mathbb{E}\left(\left(\frac{X_1^N-[Nx_0]}{N}\right)^3 T^{(3)}\left(\frac{[Nx_0]}{N}+\theta\frac{X_1^N-[Nx_0]}{N}\right)\right)\\
&\qquad\qquad \le \frac{C}{N^3}\left(N\frac{[Nx_0]}{N}\left(1-\frac{[Nx_0]}{N}\right)\left(1-2\frac{[Nx_0]}{N}\right)\right).
\end{aligned}$$

Thus, this term behaves like $\frac{1}{N^2}$ and is therefore negligible compared the first terms for large N. It follows that $T(\frac{[Nx_0]}{N})$ is approximately equal to $1+T(\frac{[Nx_0]}{N})+\frac{1}{2N}\frac{[Nx_0]}{N}(1-\frac{[Nx_0]}{N})T''(\frac{[Nx_0]}{N})$, from which

$$\frac{[Nx_0]}{N}\left(1-\frac{[Nx_0]}{N}\right)T''\left(\frac{[Nx_0]}{N}\right) \sim_{N\to\infty} -2N.$$

As $T(0) = T(1) = 0$, we deduce the result by solving the equation $x_0\,(1-x_0)\,T''(x_0) = -2N$. □

6.2.2 Quasi-Stationary Distribution for a Wright–Fisher Process

We have seen that the points 0 and N are absorbing for the Markov chain $(X_n^N)_n$. As in the case of the Galton–Watson process, we want to study the quasi-stationary distributions by conditioning on the non-attainment of the absorbing set $\{0, N\}$ (see Section 4.4). As before, we denote by $\tau_N = T_0 \wedge T_N$ the absorbing time.

Quasi-stationary distributions for a Markov chain with finite state space have been studied by Darroch and Seneta [30]. We also refer to [61] and to [56], where the Wright–Fisher process is considered as a particular case of the Cannings model.

Theorem 6.2.11 *The Yaglom limit $u = (u_i)_{i=1}^{N-1}$ of the Wright–Fisher chain exists and is the unique quasi-stationary distribution of the process $(X_n^N)_n$. It is a probability on $\{1, \dots, N-1\}$, i.e. a vector of positive numbers $(u_1, \dots, u_{N-1})$ such that $\sum_{i=1}^{N-1} u_i = 1$.*

Moreover, the extinction rate $\rho(u)$ associated with u (see Proposition 3.4.6) is equal to $1 - \frac{1}{N}$ and there exists a unique vector v, such that $(u_1 v_1, \dots, u_{N-1} v_{N-1})$ is a probability measure on $\{1, \dots, N-1\}$ and for all $i, j \in \{1, \dots, N-1\}$,

$$\lim_{n\to\infty} \Big(1 - \frac{1}{N}\Big)^n\, \mathbb{P}_i(X_n^N = j) = v_i\, u_j \tag{6.2.5}$$

and

$$\lim_{n\to\infty} \frac{\mathbb{P}_i(\tau_N > n+m)}{\mathbb{P}_j(\tau_N > n)} = \frac{v_i}{v_j}\Big(1 - \frac{1}{N}\Big)^m. \tag{6.2.6}$$

Note that $\mathbb{P}(\tau_N > n)$ can decrease very slowly for large N.

Proof The proof is a consequence of the Perron–Frobenius theorem (see Theorem 3.7.7), from which we take the notation. We consider the restriction of the matrix $P = (P_{i,j})$ defined in (6.2.3) to $i, j \in \{1, \dots, N-1\}$. It has strictly positive entries. The Perron–Frobenius theorem then tells us that there exist $\lambda_0 > 0$ and a probability u on $\{1, \dots, N-1\}$ such that for all $i, j \in \{1, \dots, N-1\}$,

$$\lambda_0^{-n}\, \mathbb{P}_i(X_n^N = j) = \lambda_0^{-n}\, P_{i,j}^n = v_i u_j + \lambda_0^{-n} B_{i,j}^n,$$

and none of the elements of the matrix B^n exceeds ρ^n, where $0 < \rho < \lambda_0$. Moreover, $\sum_{j=1}^{N-1} P_{i,j} < 1$ (for all i) implies that $\lambda_0 < 1$. Indeed, let i_0 be such that $v_{i_0} = \sup_i v_i$. Since v is a right eigenvector of P associated with λ_0, we have

$$\lambda_0\, v_{i_0} = \sum_{j=1}^{N-1} P_{i_0,j} v_j \le v_{i_0} \sum_{j=1}^{N-1} P_{i_0,j} < v_{i_0}.$$

Summing over $j \in \{1, \dots, N-1\}$, we deduce that

$$\lambda_0^{-n}\,\mathbb{P}_i(\tau_N > n) = v_i + \frac{O(\rho^n)}{\lambda_0^n}.$$

It follows that for all $i, j \in \{1, \ldots, N-1\}$,

$$\mathbb{P}_i(X_n^N = j \,|\, \tau_N > n) = \frac{\mathbb{P}_i(X_n^N = j)}{\mathbb{P}_i(\tau_N > n)} \xrightarrow[n\to\infty]{} u_j.$$

Thus the Yaglom limit exists and is equal to u. It is a quasi-stationary distribution, as proved in Proposition 3.4.4. For any initial distribution ν on $\{1, \ldots, N-1\}$,

$$\begin{aligned}\lim_{n\to\infty} \mathbb{P}_\nu(X_n^N = j \,|\, \tau_N > n) &= \sum_{i\in\{1,\cdots,N-1\}} \nu_i \lim_{n\to\infty} \mathbb{P}_i(X_n^N = j \,|\, \tau_N > n)\\ &\xrightarrow[n\to\infty]{} \sum_{i\in\{1,\cdots,N-1\}} \nu_i u_j = u_j.\end{aligned}$$

In particular, this is true for any quasi-stationary distribution μ. We therefore deduce that $\mu_j = u_j$ for all j and that the Yaglom limit u is the unique quasi-stationary distribution of X^N.

Note that by Proposition 3.4.6, we also have that $\sum_{i\in\{1,\cdots,N-1\}} u_i\,\mathbb{P}_i(\tau_N > n) = \rho(u)^n$. This quantity is also equal to λ_0^n, and therefore $\rho(u) = \lambda_0$. Lemma 6.2.12 below gives us this extinction rate for the Wright–Fisher chain, which is $\lambda_0 = 1 - \frac{1}{N}$. We immediately deduce (6.2.5) and (6.2.6). □

Lemma 6.2.12 *The eigenvalues of the transition matrix P of X^N, arranged in descending order, are* 1, 1 *and $a_j = \mathbb{E}(\eta_1\eta_2\cdots\eta_j)$ for $2 \le j \le N$, where for all $k \in \{1, \ldots, N\}$, the random variable η_k denotes the number of descendants of individual k. In particular,*

$$a_2 = \mathbb{E}(\eta_1\eta_2) = 1 - \frac{1}{N}.$$

Proof The proof is adapted from Lambert [56]. It is clear that 1 is a double eigenvalue, corresponding to the two eigenvectors $(1, 0, \ldots, 0)$ and $(0, \ldots, 0, 1)$. Let us introduce the invertible matrix Z of size $(N+1) \times (N+1)$ defined by $z_{ij} = i^j$ for $0 \le i, j \le N$. There exists a triangular matrix A such that $a_{00} = 1$ and $a_{jj} = \mathbb{E}(\eta_1\eta_2\cdots\eta_j)$ for $1 \le j \le N$ and $P = ZAZ^{-1}$. Indeed, it is enough to write $PZ = ZA$, leading to

$$\sum_{k=0}^{N} P_{i,k}\, k^j = \sum_{k=0}^{j} i^k\, a_{kj} \iff \mathbb{E}_i((X_1^N)^j) = R_j(i),$$

where R_j is a polynomial of degree j. However, $\mathbb{E}_i((X_1^N)^j) = \mathbb{E}((\eta_1 + \eta_2 + \cdots + \eta_i)^j)$.

It is easy to see that the random variables η_k are exchangeable, that $\sum_{k=1}^N \eta_k = N$ and that $(\eta_1, \ldots, \eta_N)$ follows a multinomial distribution Multinom$(N, \frac{1}{N}, \ldots, \frac{1}{N})$. We deduce from this exchangeability that

$$R_j(i) = \sum_{\sigma \in S_i, n_1+\cdots+n_i=j} \mathbb{E}(\eta_{\sigma(1)}^{n_1} \cdots \eta_{\sigma(i)}^{n_i})$$
$$= \sum_{k=1}^{j} i(i-1)\cdots(i-k+1) \sum_{n_1+\cdots+n_k=j} \mathbb{E}(\eta_1^{n_1} \cdots \eta_k^{n_k}),$$

where we have grouped permutations of subsets of size $k \le i$. The dominant term of $R_j(X)$ is then $a_{jj} = \mathbb{E}(\eta_1\eta_2\cdots\eta_j)$. The eigenvalues of P other than 1 are therefore given by these values a_{jj}, $j = 2, \ldots, N$.

Let us compute a_2. Since $\eta_1 + \cdots + \eta_N = N$, we have by exchangeability

$$N^2 = \sum_{i=1}^{N} \mathbb{E}(\eta_i^2) + \sum_{i,j=1}^{N} \mathbb{E}(\eta_i\eta_j) = N\mathbb{E}(\eta_1^2) + N(N-1)\mathbb{E}(\eta_1\eta_2).$$

However, η_1 follows a binomial distribution $\mathcal{B}(N, \frac{1}{N})$, from which $\mathbb{E}(\eta_1^2) = \mathrm{Var}(\eta_1) + 1 = 2 - \frac{1}{N}$. Thus,

$$a_2 = \mathbb{E}(\eta_1\eta_2) = 1 - \frac{1}{N}.$$ □

Remark 6.2.13 Unfortunately, we do not know how to specify an eigenvector corresponding to the eigenvalue a_2 and therefore we do not know how to specify the quasi-stationary distribution of the Wright–Fisher chain, which limits the interest of the Perron–Frobenius theorem in this particular case. We refer to Exercise 6.5.2 for another model of population genetics, called the Moran model, for which the quasi-stationary distribution can be specified.

6.2.3 The Wright–Fisher Process With Mutation

Let us now assume that during reproduction, a mutation from allele A to allele a can occur with probability α_1 and from allele a to allele A with probability α_2. The model is always assumed to be neutral; both alleles therefore have the same selective advantage. The parent of an individual is always chosen uniformly at random from the previous generation, but if this parent is of allele A, the individual inherits allele A with probability $1-\alpha_1$ or undergoes a mutation and carries allele a with probability α_1. If this parent is of allele a, the individual inherits allele a with probability $1-\alpha_2$ or undergoes a mutation and carries allele A with probability α_2.

If $(X_n^N)_n$ is the process describing the number of alleles A at generation n, the law of X_{n+1}^N is a binomial law with parameters that depend, as before, on the distribution of allele A at generation n. If there are i copies of allele A at generation n, this distribution is given by

$$p_i = \frac{i(1-\alpha_1) + (N-i)\alpha_2}{N}. \tag{6.2.7}$$

In this case, X_{n+1}^N follows a binomial law $\mathcal{B}(N, p_i)$, and we have

$$\mathbb{P}(X_{n+1}^N = j | X_n^N = i) = \binom{N}{j} (p_i)^j (1-p_i)^{N-j}. \tag{6.2.8}$$

The Markov chain has a very different long-term behavior from that of the neutral case. Indeed, if $\alpha_1 > 0$ and $\alpha_2 > 0$, there is no possibility of fixation. As the chain has finite support, it has a unique non-trivial invariant probability π, the solution of $\sum_{i=0}^N \pi_i P_{i,j} = \pi_j$, where $P_{i,j}$ is given by (6.2.8).

6.2.4 The Wright–Fisher Process With Selection

We assume here that allele A has a selective advantage over allele a, in the sense that a copy carrying allele A has a higher chance of replicating (its fertility is greater). Let $s > 0$ be the parameter describing this advantage. If at generation n, there are i individuals carrying allele A, the selective advantage is modeled by the fact that the probability of obtaining allele A is given by

$$\widehat{p}_i = \frac{(1+s)\,i}{(1+s)\,i + N - i} \geq \frac{i}{N}. \tag{6.2.9}$$

We then have

$$\mathbb{P}(X_{n+1}^N = j | X_n^N = i) = \binom{N}{j} (\widehat{p}_i)^j (1-\widehat{p}_i)^{N-j}.$$

We can likewise model an unfavorable or deleterious allele by a selective disadvantage: in this case $s < 0$.

More generally, we can also define a Wright–Fisher model with selection and mutation. In this case, the choice of the parent is made by giving a selective advantage to a carrier of allele A and any individual can undergo a mutation at birth, as described in Section 6.2.3. We therefore assume that

$$\bar{p}_i = \frac{[(1+s)\,i\,(1-\alpha_1)] + (N-i)\,\alpha_2}{(1+s)\,i + N - i}. \tag{6.2.10}$$

Note that the difference $\bar{p}_i - \frac{i}{N}$ of the fraction of alleles A at the end of the cycle of the n-th generation expresses the difference due to selection and mutation. The statistical fluctuations related to the finite size of the population appear through the random behavior of the Markov chain, described by the transition probabilities $P_{i,j}$. It is difficult, for such a chain, to compute the probabilities of interest, such as the probabilities of extinction or fixation of allele A or the average times to reach these states. We will see in the next section that, depending on whether the population size N is large and the individual mutation and selection effects are weak, the process $(X_n^N)_n$ can be approximated by a diffusion process, for which it will be easier to obtain quantitative results.

6.3 Demographic Diffusion Models

6.3.1 The Wright–Fisher Diffusion Process

In this subsection, we obtain an approximation of the neutral Wright–Fisher model, in the case where the population size N tends towards infinity. We assume that $\frac{1}{N}X_0^N$ has a limit z as $N \to \infty$. In this case, the number of individuals of allele A is of the order of the population size and the expectation and variance of the chain computed in Remark 6.2.3 are of order N. To hope for an interesting limit, we will consider the process $(\frac{1}{N}X_n^N, n \geq 0)$. Thus, if $Y_n^N = \frac{1}{N}X_n^N$, we will have

$$\begin{aligned}
&\mathbb{E}(Y_{n+1}^N | Y_n^N = x) = \frac{1}{N}\mathbb{E}(X_{n+1}^N | X_n^N = Nx) = x \text{ and therefore} \\
&\mathbb{E}(Y_{n+1}^N - Y_n^N | Y_n^N = x) = 0, \\
&\mathrm{Var}(Y_{n+1}^N - Y_n^N | Y_n^N = x) = \frac{1}{N^2}\mathrm{Var}(X_{n+1}^N - X_n^N | X_n^N = Nx) = \frac{1}{N}x(1-x).
\end{aligned}$$

We see that if $N \to \infty$, the expectation and variance of increments are asymptotically zero and the process seems to remain constant. The time scale is therefore not the right one to observe a non-trivial limit: the reproduction events are not on the same scale as the population size (their fluctuations are much too small on this size scale). We will consider a unit of time that depends on the population size and study the process on a very long time scale, which corresponds to an acceleration of the reproduction process. For $t \in [0,T]$, we set $n = \lfloor Nt \rfloor$, $\lfloor x \rfloor$ denoting the integer part of x.

Let's redo the previous computations by setting

$$Z_t^N = \frac{1}{N}X_{\lfloor Nt \rfloor}^N = Y_{\lfloor Nt \rfloor}^N.$$

Since $n = \lfloor Nt \rfloor$ and setting $\Delta t = \frac{1}{N}$, we get

$$\begin{aligned}
&\mathbb{E}(Z_{t+\Delta t}^N | Z_t^n = x) = x \text{ and therefore} \\
&\mathbb{E}(Z_{t+\Delta t}^N - Z_t^n | Z_t^N = x) = 0, \\
&\mathrm{Var}(Z_{t+\Delta t}^N - Z_t^n | Z_t^n = x) = \frac{1}{N}x(1-x) = x(1-x)\,\Delta t.
\end{aligned}$$

In this time scale, $(Z_t^N, t \geq 0)$ is a martingale and the variance of the process does not explode as N tends towards infinity. We can also see that the jumps of the process $(Z_t^N, t \geq 0)$ are of amplitude $\frac{1}{N}$ and tend to 0 as N tends to infinity. Thus, if the process converges, its limit will be continuous in time. We actually have the following theorem.

Theorem 6.3.1 *If the sequence $(\frac{1}{N}X_0^N)_N$ converges in law to $Z_0 \in [0,1]$ as $N \to \infty$, the process $(\frac{1}{N}X_{\lfloor Nt \rfloor}^N, t \geq 0)$ converges in law to the solution of the stochastic* *differential equation*

$$Z_t = Z_0 + \int_0^t \sqrt{Z_s(1-Z_s)}\,\mathrm{d}B_s, \tag{6.3.1}$$

where $(B_t, t \geq 0)$ is a standard Brownian motion. Uniqueness is obtained thanks to Theorem 4.6.11 and for all t, $Z_t \in [0,1]$.

Proof The rigorous proof of this theorem is beyond the scope of this book. We only give the main ideas.

The process $(Z_t^N - Z_0^N, t \geq 0)$ is a martingale with values in $[0,1]$ and for each t, the variance of $Z_t^N - Z_0^N$ equals $\int_0^t \mathbb{E}(Z_s^N(1-Z_s^N))\,\mathrm{d}s$. We can show the convergence in law of the process $(Z_t^N, t \in [0,T])$, as a process with right-continuous and left-limited trajectories on any time interval $[0,T]$, to the solution $(Z_t, t \in [0,T])$ of (6.3.1). The proof is done in two steps:

– Show that the sequence Z^N admits a limit point; this requires relative compactness arguments for process laws, which are probability measures on the functional space of right-continuous and left-limited functions (cf. Billingsley [11] for theoretical results and [8] for proofs in a close framework).

– Show that this limit point is unique. One has to identify this process as a solution of the equation (6.3.1) and show the uniqueness in law of this solution. This follows from Theorem 4.6.11. Uniqueness implies in particular that if Z reaches the edges, it stays there, i.e., the points 0 and 1 are absorbing. □

The Wright–Fisher diffusion process defined as the solution of (6.3.1) has already been introduced as an example in Section 4.7.5. There, the absorption time was studied, the average of which was given in (4.7.5):

$$\mathbb{E}_z(T_0 \wedge T_1) = -2(z \log z + (1-z)\log(1-z)).$$

This formula can be compared with the approximate formula obtained for the Wright–Fisher Markov chain in Proposition 6.2.10.

6.3.2 The Wright–Fisher Diffusion Process With Mutation or Selection

Let us repeat the previous computations, in the case of a Wright–Fisher model with mutation, with the parameters introduced in Section 6.2.3. In this case, and if $X_n^N = i$, the probability p_i is defined by (6.2.7). Suppose that at least one of the parameters α_1, α_2 is non-zero. We will, as before, study the process

$$Z_t^N = \frac{1}{N} X_{\lfloor Nt \rfloor}^N.$$

Computations give $\mathbb{E}(Z_{t+\Delta t}^N | Z_t^N = x) = p_{Nx}$, from which

$$\mathbb{E}(Z^N_{t+\Delta t} - Z^N_t | Z^N_t = x) = N\,(-\alpha_1 x + (1-x)\alpha_2)\,\Delta t$$

and

$$\begin{aligned}\mathbb{E}((Z^N_{t+\Delta t} - Z^N_t)^2 | Z^N_t = x) &= \mathbb{E}((Z^N_{t+\Delta t})^2 | Z^N_t = x) - 2x\mathbb{E}(Z^N_{t+\Delta t} | Z^N_t = x) + x^2 \\ &= \frac{1}{N} p_{Nx}(1-p_{Nx}) + (p_{Nx} - x)^2 \\ &= \Delta t \Big(x(1-x) + Q(x,\alpha_1,\alpha_2) + N(\alpha_1 x - (1-x)\alpha_2)^2 \Big),\end{aligned}$$

where $Q(x,\alpha_1,\alpha_2)$ is a second-degree polynomial in x, all of whose coefficients are factors of α_1 or α_2.

For the conditional expectation of the difference $Z^N_{t+\Delta t} - Z^n_t$ and its square to have a chance of converging, we will place ourselves under a hypothesis of rare mutations

$$\alpha_1 = \frac{\beta_1}{N}; \quad \alpha_2 = \frac{\beta_2}{N}, \tag{6.3.2}$$

where the probabilities of mutation tend to 0 with N. In this case,

$$\mathbb{E}(Z^N_{t+\Delta t} - Z^N_t | Z^N_t = x) = (-\beta_1 x + (1-x)\beta_2)\,\Delta t$$

and

$$\mathbb{E}((Z^N_{t+\Delta t} - Z^N_t)^2 | Z^N_t = x) = \Big(x(1-x) + o(\frac{1}{N})\Big)\,\Delta t.$$

It is then possible to prove the following proposition

Proposition 6.3.2 *Under the hypothesis of rare mutations* (6.3.2), *and if* $(\frac{1}{N}X^N_0)_N$ *converges in law to* $Z_0 \in [0,1]$ *as* $N \to \infty$, *the process* $(\frac{1}{N}X^N_{\lfloor Nt \rfloor}, t \geq 0)$ *converges in law to the solution of the stochastic differential equation*

$$Z_t = Z_0 + \int_0^t \sqrt{Z_s\,(1-Z_s)}\,\mathrm{d}B_s + \int_0^t (-\beta_1 Z_s + (1-Z_s)\,\beta_2)\,\mathrm{d}s.$$

The uniqueness is obtained as above, thanks to Theorem 4.6.11. Again, $Z_t \in [0,1]$.

Proof Under the assumption (6.3.2), $\mathbb{E}(Z^N_{t+\Delta t} - Z^n_t | Z^N_t = x)$ is equal to $(-\beta_1 x + (1-x)\,\beta_2)\,\Delta t$ and $\mathbb{E}((Z^N_{t+\Delta t} - Z^N_t)^2 | Z^N_t = x)$ converges to $x\,(1-x)\,\Delta t$ as $N \to \infty$. By arguments similar to those of Theorem 6.3.1, we deduce the result. □

We will now study the case of the model with selection (see Section 6.2.4) under a rare selection assumption: we assume that

$$s = \frac{r}{N}. \tag{6.3.3}$$

The process $Z^N_t = \frac{1}{N}X^N_{\lfloor Nt \rfloor}$ is then defined from the Wright–Fisher model with selection, with transition probabilities derived from the probabilities $\widehat{p_i}$ defined in (6.2.9). We then have the following proposition.

Proposition 6.3.3 *Under the rare selection assumption* (6.3.3), *and if* $(\frac{1}{N}X_0^N)_N$ *converges in law to* $Z_0 \in [0,1]$ *as* $N \to \infty$, *the process* $(\frac{1}{N}X_{\lfloor Nt \rfloor}^N, t \geq 0)$ *converges in law to the solution of the stochastic differential equation*

$$Z_t = Z_0 + \int_0^t \sqrt{Z_s(1-Z_s)}\,\mathrm{d}B_s + \int_0^t rZ_s(1-Z_s)\,\mathrm{d}s.$$

Proof We have

$$\mathbb{E}(Z_{t+\Delta t}^N - Z_t^N | Z_t^N = x) = N\left(\frac{-x^2 s + sx}{sx+1}\right)\Delta t;$$

$$\mathbb{E}((Z_{t+\Delta t}^N - Z_t^N)^2 | Z_t^N = x) = \left(x(1-x) + N(\frac{s^2x^2(1+2x+x^2)}{(xs+1)^2})\right)\Delta t.$$

Thus, assuming (6.3.3), we find that $\mathbb{E}(Z_{t+\Delta t}^N - Z_t^N | Z_t^N = x)$ converges to $rx(1-x)\Delta t$ as $N \to \infty$ and that $\mathbb{E}((Z_{t+\Delta t}^N - Z_t^N)^2 | Z_t^N = x)$ converges to $x(1-x)\Delta t$. By arguments similar to those of Theorem 6.3.1, we deduce the result. □

Proposition 6.3.4 *Let us now consider the most general model with mutation and selection, where the transition probabilities of* $(X_n^N, n \in \mathbb{N})$ *are obtained by* (6.2.10). *We place ourselves under the rare mutation and selection assumptions* (6.3.2) *and* (6.3.3). *Then, if* $(\frac{1}{N}X_0^N)_N$ *converges in law to* $Z_0 \in [0,1]$, *the process* $(Z_t^N, t \geq 0)$ *defined by* $Z_t^N = \frac{1}{N}X_{\lfloor Nt \rfloor}^N$ *converges in law, as N tends to infinity, to the solution of the stochastic differential equation*

$$Z_t = Z_0 + \int_0^t \sqrt{Z_s(1-Z_s)}\,\mathrm{d}B_s + \int_0^t \Big(-\beta_1 Z_s + (1-Z_s)\beta_2 + rZ_s(1-Z_s)\Big)\,\mathrm{d}s.$$

Remark 6.3.5 Note that computations of probabilities of fixation or extinction of an allele are almost impossible because they are too complicated on discrete models. They become accessible for diffusive approximations with the tools of stochastic calculus, as we have seen in Chapter 4 and recalled at the end of Section 6.3.1.

6.3.3 Another Time Scale Change

Let us take up the previous model by accelerating time a little less. We consider the process

$$R_t^N = \frac{1}{N}X_{\lfloor N^\gamma t \rfloor}^N,$$

where $0 < \gamma < 1$. By taking up the computations developed in the mutation model in Section 6.2.3 and now assuming that

$$\alpha_1 = \frac{\beta_1}{N^\gamma};\ \alpha_2 = \frac{\beta_2}{N^\gamma},$$

we find that $\mathbb{E}((Z^N_{t+\Delta t} - Z^N_t)^2 | Z^N_t = x)$ tends to 0 as N tends to infinity and that $\mathbb{E}(Z^N_{t+\Delta t} - Z^N_t | Z^N_t = x)$ converges to $(-\beta_1 x + (1-x)\beta_2)\Delta t$. We can then show that if the sequence $(\frac{1}{N} X^N_0)_N$ converges in law to a deterministic limit y_0 as $N \to \infty$, the process $(R^N_t, t \geq 0)$ converges in law to the (deterministic) solution (a continuous function) $(y(t), t \geq 0)$ of the differential equation

$$dy(t) = (-\beta_1 y(t) + \beta_2(1 - y(t)))\, dt; \;\; y(0) = y_0.$$

Similarly, in the selection model with $s = \dfrac{r}{N^\gamma}$, we can show that the process $(R^N_t, t \geq 0)$ converges in law to the (deterministic) solution $(y(t), t \geq 0)$ of the differential equation

$$dy(t) = r\, y(t)\, (1 - y(t)) dt; \;\; y(0) = y_0. \tag{6.3.4}$$

It is easy to prove that the solution of the deterministic approximation (6.3.4) converges to 1 or 0 depending on whether $r > 0$ or $r < 0$. Then one has fixation or extinction in this slow time scale depending on whether $r > 0$ or $r < 0$.

Remark 6.3.6 We have seen that if $0 < \gamma < 1$, the limit is deterministic, while for $\gamma = 1$, the limit is stochastic. This stochasticity comes from the extreme variability due to a very large number of zero expectation reproduction events (we are in a "Central Limit Theorem" type limit). Of course, if $\gamma > 1$, the limit explodes, and the time scale is no longer adapted to the size of the population.

It is important to keep in mind these relationships between size and time scales and this dichotomy between a deterministic behavior that takes into account only the macroscopic aspect of the phenomenon and a stochastic behavior that will also take into account the large individual variability of the system.

6.4 Coalescence: Description of Genealogies

Imagine a genealogy of diploid individuals, where we consider not only the individuals, but all their offspring. Each individual will have transmitted to each of his descendants one of the two copies of his genes at a given locus. From one generation to the next, some alleles will not be transmitted, but others may be transmitted in several copies. It is interesting to study the genealogy of a sample of genes (or individuals) observed in a certain generation. Our goal is to reconstruct the genealogical history of the genes up to their most recent common ancestor, taking into account the demographic constraints of the population and considering possible mutations. We will not take into account the entire population, but we will focus on the sample of interest. This is a retrospective approach. As before, we will assimilate the diploid population of size M to a haploid population of size $2M = N$ by focusing on the different alleles.

A *lineage* refers to the ancestry of a gene. When two lineages converge at an ancestral gene, it is said that they *coalesce* or that a *coalescence event* has occurred. The theory of coalescence therefore simply describes the process of coalescence of the ancestral lineages of individuals from a sample from the present generation to their most recent common ancestor. This is what we will model at the end of this chapter.

6.4.1 Asymptotic as N Tends to Infinity: The Kingman Coalescent

Let us first return to the neutral Wright–Fisher model (without mutation or selection). We consider a sample of k individuals from a population of size N. We assume that k is small but that N is very large ($N \longrightarrow +\infty$). Recall that the word 'individual' here refers to an allele.

We want to reconstruct the genealogy of the k individuals in the sample. We will first study the first common ancestor to (at least) two of these individuals. We have seen that the probability $h(n)$ of having heterozygosity for two individuals at time n is given, for a population of size N, by the formula (6.2.4):

$$h(n) = \left(1 - \frac{1}{N}\right)^n h(0).$$

Now suppose that N is very large. If time n is small, this formula provides little information, other than that $h(n)$ is of the order of $h(0)$. It is then much more interesting to study what happens in long time (n large). We will therefore assume that our observations are obtained at a time n of the order of N. Indeed, since $(1-x) \sim \mathrm{e}^{-x}$ when x is close to 0, we deduce that

$$h(n) \sim \mathrm{e}^{-\frac{n}{N}} h(0),$$

and $\frac{n}{N}$ is not negligible. Thus, if n and N are of the same order, the probability of heterozygosity (called the coefficient of heterozygosity in population genetics) decreases exponentially fast in the variable $\frac{n}{N}$.

We can reason similarly to compute the probability that two individuals (at least) have the same parent. This event will occur if one of the following three events is satisfied:

- exactly two individuals among the k individuals have the same parent,
- at least three individuals have the same parent,
- at least two pairs of individuals have a common parent.

The probability that exactly two individuals, among the k individuals, have the same parent is equal to

$$\frac{k(k-1)}{2} \frac{1}{N}.$$

Indeed the probability of having the same parent for two fixed individuals is $\frac{1}{N}$ and there are $\frac{k(k-1)}{2}$ possible pairs of individuals. The probability of the other two events is of the order of $\frac{1}{N^2}$. Indeed, the probability that exactly three individuals have the same parent will be $\binom{k}{3}N\frac{1}{N^3}$ and the probability that exactly two pairs of individuals have the same parent is $\binom{k}{2}\binom{k-2}{2}\frac{1}{N(N-1)}$. These events therefore have negligible probabilities compared to the probability of the first event. Thus, as N tends to infinity, the probability that two individuals have the same parent is of the order of $\frac{k(k-1)}{2}\frac{1}{N}$.

By reasoning as in the proof of (6.2.4) given in Remark 6.2.9, we then obtain that the probability that there have been no common parents for these k individuals in the first n generations is proportional to

$$\left(1-\frac{k(k-1)}{2}\frac{1}{N}\right)^n \sim_{N\to\infty} \exp\left(-\frac{k(k-1)}{2}\frac{n}{N}\right). \tag{6.4.1}$$

Let us now place ourselves at a given time. We are going to go back in time from the k individuals in the sample. Let us call T_1 the first coalescence time, that is, the first moment when the k individuals have a common ancestor in their genealogy. We deduce from (6.4.1) that

$$\mathbb{P}(T_1 > n) \sim_{N\to\infty} \exp\left(-\frac{k(k-1)}{2}\frac{n}{N}\right).$$

We now change the time scale and take N generations as the new time unit by replacing n with $\lfloor Nt \rfloor$ (which denotes the integer part of Nt). Thus, in this new time scale, and as N tends to infinity, the first coalescence time of k given individuals, which we will still denote by T_1, will satisfy, for all $t > 0$,

$$\mathbb{P}(T_1 > t) = \exp\left(-\frac{k(k-1)}{2}t\right).$$

The law of T_1 is therefore that of an exponential random variable with parameter $\frac{k(k-1)}{2}$ and therefore with mean $\frac{2}{k(k-1)}$.

Remark 6.4.1 The coalescence time of two lineages is an exponential random variable with parameter 1 and then, given k individuals, the first time at which a pair has a common ancestor is therefore the infimum of $\frac{k(k-1)}{2}$ independent exponential random variables with parameter 1.

We are thus able to define the k-coalescent as a limit process, as N tends to infinity, of the process that describes the genealogy of k individuals.

Definition 6.4.2 Let $k \in \mathbb{N}^*$. The *k-coalescent* is the Markov chain $(\Pi_t)_t$ with values in the set $\mathcal{P}_k$ of partitions of $\{1,\dots,k\}$ defined as follows:

- $\Pi_0 = \{\{1\},\dots,\{k\}\}$.

- Let T_i be the i-th coalescence time. We set $T_0 = 0$. Then the time intervals $T_i - T_{i-1}$ are independent and follow exponential laws with parameter $\dfrac{(k-i)(k-i+1)}{2}$.
- At each jump time, two blocks of the partition are chosen uniformly among the existing block pairs and coalesce, in the sense that the two sub-blocks are grouped into one.

Note that the definition of the k-coalescent allows for an easy simulation algorithm:

- We are given k individuals. We set $\tilde{\Pi}_0 = \{\{1\}, \dots, \{k\}\}$.
- We simulate an exponential random variable with parameter $\frac{k(k-1)}{2}$. To do this, we consider a random variable U with a uniform distribution over $[0,1]$ and we set $T_1 = \frac{2}{k(k-1)} \log(1/U)$ (cf. [59]).
- We randomly choose two distinct blocks of the partition (thus with probability $\frac{2}{k(k-1)}$). We group these two blocks into a single block. We then call this new partition $\tilde{\Pi}_1$.
- We repeat this procedure. After the $i-1$ step, we simulate an exponential random variable with parameter $\frac{(k-i+1)(k-i)}{2}$. We randomly choose two distinct blocks of the partition composed of $k-i+1$ blocks (thus with probability $\frac{2}{(k-i)(k-i+1)}$). We group these blocks into a single set. We thus obtain $\tilde{\Pi}_i$.
- We then set

$$\Pi_t = \sum_i \tilde{\Pi}_i \mathbf{1}_{\{T_i \le t < T_{i+1}\}}.$$

At each coalescence time, the number of elements in the partition decreases by 1, and therefore it will be reduced to one element after $k-1$ coalescence events. This element is the most recent common ancestor of the k individuals (MRCA). The time T when this ancestor is found is $T = T_{k-1}$.

A representation of a coalescent is given in Figure 6.2 (Section 6.4.2), for a sample of 8 individuals.

Theorem 6.4.3 *Let T be the time of the most recent common ancestor of the k individuals in the sample. Then*

$$\mathbb{E}(T) = 2\left(1 - \frac{1}{k}\right).$$

Proof We write

$$T = T_{k-1} = T_{k-1} - T_{k-2} + \dots + T_1.$$

T is therefore the sum of $k-1$ independent exponential random variables with parameters $\frac{(k-i+1)(k-i)}{2}$, for i ranging from 1 to $k-1$. Thus,

$$\mathbb{E}(T) = \sum_{i=1}^{k-1} \frac{2}{(k-i)(k-i+1)} = 2\sum_{i=1}^{k-1}\left(\frac{1}{k-i} - \frac{1}{k-i+1}\right) = 2\left(1-\frac{1}{k}\right). \qquad \square$$

Theorem 6.4.4 *(cf. Durrett [33]). If π is a partition of $\{1,\ldots,k\}$ into ℓ blocks, let $\pi = (B_1, B_2, \ldots, B_\ell)$. Then this partition can be realized at time $T_{k-\ell}$ and*

$$\mathbb{P}(\Pi_{T_{k-\ell}} = \pi) = \frac{\ell!}{k!}\frac{(k-\ell)!(\ell-1)!}{(k-1)!}\prod_{i=1}^{\ell}(\mathrm{Card}(B_i))!. \tag{6.4.2}$$

We thus have a description of the law of the k-coalescent.

In the following, we define

$$c_{k,\ell} = \frac{\ell!}{k!}\frac{(k-\ell)!(\ell-1)!}{(k-1)!}$$

and if $B_1, \ldots, B_\ell$ are the blocks of the partition π, we define

$$w(\pi) = \prod_{i=1}^{\ell}(\mathrm{Card}(B_i))!.$$

Thus the probability that, at the $(k-\ell)$-th coalescence time, the k-coalescent with ℓ blocks is equal to π is the product of the term $c_{k,\ell}$, which depends only on k and ℓ, and the weight $\prod_{i=1}^{\ell}(\mathrm{Card}(B_i))!$, which favors unequal partitions. (The reader can check as an example that a partition of 5 individuals composed of two blocks of 2 and 3 individuals will have a weight equal to 12 while a partition composed of two blocks of 4 and 1 individuals will have a weight of 24.)

Proof The proof consists of a downward induction on ℓ. When $\ell = k$, the only possible partition π is the partition composed of singleton blocks, and the probability of realization of π is 1, which is also given by the right-hand side of (6.4.2) for $\ell = k$.

Suppose that (6.4.2) is true for any partition of size ℓ and consider a partition η of size $\ell-1$. We write $\xi < \eta$ if $\mathrm{Card}(\xi) = \mathrm{Card}(\eta)+1$ and if the partition η is obtained after merging two blocks of ξ. In this case, there is exactly one coalescence event that transitions from ξ to η. We then have

$$\mathbb{P}(\Pi_{T_{k-\ell+1}} = \eta | \Pi_{T_{k-\ell}} = \xi) = \begin{cases} \frac{2}{\ell(\ell-1)} & \text{if } \xi < \eta \\ 0 & \text{otherwise.} \end{cases}$$

We therefore have

$$\mathbb{P}(\Pi_{T_{k-\ell+1}} = \eta) = \frac{2}{\ell(\ell-1)}\sum_{\xi<\eta}\mathbb{P}(\Pi_{T_{k-\ell}} = \xi).$$

Given a partition ξ of $\{1,\dots,k\}$ into ℓ blocks, we order its blocks into $\xi_1,\xi_2,\dots,\xi_\ell$, where ξ_1 is the block containing 1, ξ_2 is the one containing the smallest number not in ξ_1, and so on.

If the integers $\lambda_1,\dots,\lambda_{\ell-1}$ are the sizes of the blocks of the partition η, then for j such that $1\le j\le \ell-1$, and m such that $1\le m<\lambda_j$, we associate a partition $\xi<\eta$, where the blocks of ξ have the sizes $\lambda_1,\dots,\lambda_{j-1},m,\lambda_j-m,\lambda_{j+1},\dots,\lambda_{\ell-1}$. Note that there are $\frac{1}{2}\binom{\lambda_j}{m}$ choices of such a partition $\xi<\eta$, where the j-th block has been cut into λ_j-m and m individuals.

Using the recurrence hypothesis, we then obtain that

$$\begin{aligned}\mathbb{P}(\Pi_{T_{k-\ell+1}}=\eta)&=\frac{2}{\ell(\ell-1)}\sum_{j=1}^{\ell-1}\sum_{m=1}^{\lambda_j-1}\frac{1}{2}\binom{\lambda_j}{m}\,c_{k,\ell}\,\lambda_1!\cdots\lambda_{j-1}!m!(\lambda_j-m)!\lambda_{j+1}!\cdots\lambda_\ell!\\&=w(\eta)\frac{c_{k,\ell}}{\ell(\ell-1)}\sum_{j=1}^{\ell-1}\sum_{m=1}^{\lambda_j-1}1.\end{aligned}$$

The double sum is

$$\sum_{j=1}^{\ell-1}(\lambda_j-1)=k-(\ell-1).$$

We then easily check that

$$\frac{c_{k,\ell}}{\ell(\ell-1)}(k-\ell+1)=c_{k,\ell-1}.$$

The result is therefore proven by induction. □

Theorem 6.4.5 *(cf. Durrett [33]). Let σ be a permutation chosen at random among the permutations of $\{1,\dots,\ell\}$. Let $\Lambda_j=\mathrm{Card}(\xi_{\sigma(j)})$ be the size of the j-th block of the partition $\Pi_{T_{k-\ell}}$, when the blocks are ordered according to σ. Then $(\Lambda_1,\dots,\Lambda_\ell)$ follows a uniform law on the vectors of $(\mathbb{N}^*)^\ell$ whose coordinates sum to k.*

For example, for $\ell=2$, the law of the size of one of the two blocks of $\Pi_{T_{k-2}}$ drawn at random is uniformly distributed over $\{1,\dots,k-1\}$.

Proof Each ordered rearrangement of the ℓ blocks of the partition $\Pi_{T_{k-\ell}}=\pi$ (consisting of k individuals) has the probability $\frac{c_{k,\ell}w(\pi)}{\ell!}$ of being chosen. By conditioning on the fact that the k individuals are distributed into the ℓ blocks of the partition π, we obtain

$$\mathbb{P}((\Lambda_1,\dots,\Lambda_\ell)=(\lambda_1,\dots,\lambda_\ell))=\frac{c_{k,\ell}w(\pi)}{\ell!}\frac{k!}{\lambda_1!\lambda_2!\dots\lambda_\ell!}=\frac{(k-\ell)!(\ell-1)!}{(k-1)!}=\frac{1}{\binom{k-1}{\ell-1}}.$$

The final quantity depends only on k and ℓ, and not on the vector $(\lambda_1,\dots,\lambda_\ell)$. Thus, the distribution is uniform. Let us check that the denominator of the last fraction gives the number of vectors composed of positive integers whose coordinates sum to k. Imagine k balls lined up, separated into ℓ groups by $\ell-1$ pieces of cardboard. The

$\ell - 1$ pieces of cardboard can placed in the $k-1$ spaces between the balls. Therefore, there are $\binom{k-1}{\ell-1}$ possible choices of vectors of size ℓ of positive integers whose sum of coordinates is equal to k. □

As a consequence of this theorem, we have the following surprising result.

Theorem 6.4.6 *The probability that the most recent common ancestor of a group of k individuals is the same as that of the total population converges to $\frac{k-1}{k+1}$ as the size N of the population tends to infinity.*

For example, when $k = 2$, this probability is $\frac{1}{3}$ and when $k = 10$, it is $\frac{9}{11}$.

Proof Let N be the size of the population. The k individuals do not have the same MRCA (most recent common ancestor) as the total population of size N if and only if at the first branching (in the direction of physical time) in the coalescence tree of the population, there is a partition into two sub-trees, and the k individuals in the sample must necessarily belong to one or the other of these sub-blocks. Let $\mathrm{MRCA}_N \neq \mathrm{MRCA}_k$ denote the event "the MRCA of the total population of size N differs from that of the group of individuals of size k". Let $\mathcal{P}(i, N-i)$ be the event "At the first branching, there is a partition into 2 blocks of size $(i, N-i)$". We can then write

$$\mathbb{P}(\mathrm{MRCA}_N \neq \mathrm{MRCA}_k) = \sum_{i=1}^{N} \mathbb{P}(\mathrm{MRCA}_N \neq \mathrm{MRCA}_k | \mathcal{P}(i, N-i)) \times \mathbb{P}(\mathcal{P}(i, N-i)).$$

The probability of existence of the partition $(i, N-i)$ is uniform, given by Theorem 6.4.5 for $\ell = 2$, and is therefore $\frac{1}{N-1}$. Knowing $\mathcal{P}(i, N-i)$, the event $\mathrm{MRCA}_N \neq \mathrm{MRCA}_k$ will be realized if the k individuals are either in the block of size i, or in the block of size $N-i$. As N is very large, we can assume that the sampling of the k individuals among the N is done following a draw with replacement. Thus we obtain

$$\mathbb{P}(\mathrm{MRCA}_N \neq \mathrm{MRCA}_k | \mathcal{P}(i, N-i)) \sim_{N\to\infty} \left(\frac{i}{N}\right)^k + \left(\frac{N-i}{N}\right)^k,$$

then

$$\mathbb{P}(\mathrm{MRCA}_N \neq \mathrm{MRCA}_k) \sim_{N\to\infty} \frac{1}{N-1} \sum_{i=1}^{k} \left(\left(\frac{i}{N}\right)^k + \left(\frac{N-i}{N}\right)^k \right).$$

Using the convergence of Riemann sums, we deduce that

$$\lim_{N\to\infty} \mathbb{P}(\mathrm{MRCA}_N \neq \mathrm{MRCA}_k) = \int_0^1 \left(x^k + (1-x)^k\right) \mathrm{d}x = \frac{2}{k+1}.$$

Finally $\lim_{N\to\infty} \mathbb{P}(\mathrm{MRCA}_N = \mathrm{MRCA}_k) = 1 - \lim_{N\to\infty} \mathbb{P}(\mathrm{MRCA}_N \neq \mathrm{MRCA}_k) = 1 - \frac{2}{k+1}$, which concludes the proof. □

6.4.2 The Coalescent With Mutation

The previous model is very simple because it does not include mutation, however this can be justified on certain time scales. In fact many mutations occur with each reproduction. We assume that mutations occur on the tree and that there are so many possible alleles that each mutation is always of a new type. This assumption is biologically reasonable. Indeed, if for example a gene consists of 500 nucleotides (the bases A, T, G or C), the number of possible DNA sequences is 4^{500}, and the probability of having two mutations at the same nucleotide in a finite time is negligible.

There are two ways to account for these mutations: either we count the total number of mutations (the infinite sites mutation model), or we count the number of different alleles that we obtain in a sample of size k (the infinite alleles mutation model), see for example Figure 6.2 to understand the difference. The first approach is simpler from a mathematical point of view, but the second one corresponds to biological observations. It is this latter perspective that we will choose. As before, k denotes the size of the sample we are considering and A_k will denote the number of different alleles among these k individuals. In the rest of this chapter, we will be interested in the law of the random variable A_k and the distribution of alleles in the sample (also called the *frequency spectrum*). Indeed, biological observations will only tell us the number of alleles carried by an individual, by two individuals, ..., by j individuals, for j ranging from 1 to k.

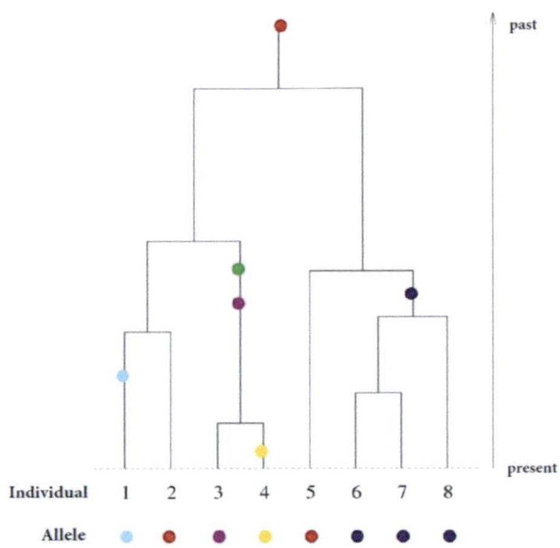

Fig. 6.2 Mutations on the coalescent. $k = 8$, there are 6 mutations but $A_8 = 5$. *(Thanks to Amandine Veber for the picture.)*

Construction of the coalescent with mutation

We first see how to construct a coalescent with mutations. Let us go back to the Wright–Fisher model for a population of size N, in which we consider a sample of size k. As before, the individuals of generation n choose their parent in generation $n-1$, uniformly among the N individuals, independently of each other. We have seen that the probability of having a coalescence event is of the order of $\frac{k(k-1)}{2}\frac{1}{N}$ (as $N \to +\infty$) and that the probability that there is no common parent for the k individuals of the chosen sample is of the order of $\exp\left(-\frac{k(k-1)}{2}\frac{n}{N}\right)$. We also assume that the allele of the descendant is not automatically inheritable from that of the parent, but may undergo a mutation with probability μ. In this case, a new type appears. We will be interested in the limit model, in the asymptotics of Kingman's coalescent.

If we assume that $\mu \sim_{N\to\infty} \frac{\theta}{2N}$, the probability that there was a mutation in the previous generation is of the order of $k\frac{\theta}{2N}$. Indeed, $k\mu$ is the probability that there was exactly one mutation and the probability that there were at least two mutations is of the order $\frac{1}{N^2}$. Thus, we obtain that the probability that there was no mutation in the first n generations for the k individuals is of the order of

$$\left(1-k\frac{\theta}{2N}\right)^n \sim_{N\to\infty} \exp\left(-k\frac{\theta}{2}\frac{n}{N}\right).$$

So, if S denotes the first time when at least one individual among the k individuals in the sample comes from a mutation, we have

$$\mathbb{P}(S > n) \sim_{N\to\infty} \exp\left(-k\frac{\theta}{2}\frac{n}{N}\right).$$

Thus, the change of time scale $n = \lfloor Nt \rfloor$, which allows us to obtain Kingman's coalescent, also implies that the time of appearance of a mutation (at this scale) follows an exponential law of parameter $k\frac{\theta}{2}$, and therefore of average $\frac{2}{k\theta}$.

We thus obtain a coalescent model with infinite size and mutation rate of $\frac{\theta}{2}$. We have on each branch a rate of mutation $\frac{\theta}{2}$ and on each pair of branches a rate of coalescence 1. The algorithmic construction of the coalescent with mutation is done exactly like that of Definition 6.4.2. However, there is on each branch of the coalescence tree an exponential clock of parameter $\frac{\theta}{2}$. If the clock rings, the parent's allele is replaced by a new type, which will then be transmitted. Between the coalescence times T_{i-1} and T_i, there are i individuals and the mutation time will follow an exponential law of parameter $i\frac{\theta}{2}$.

If we go back in time in the Kingman coalescent, starting from a sample of size $k+1$, we know that a coalescence event will occur at the rate $\frac{k(k+1)}{2}$ and a mutation at the rate $(k+1)\frac{\theta}{2}$. A mutation will occur before a coalescence event with probability

$$\int_0^\infty (k+1)\frac{\theta}{2}\mathrm{e}^{-(k+1)\frac{\theta}{2}u}\int_u^\infty \frac{k(k+1)}{2}\mathrm{e}^{-\frac{k(k+1)}{2}v}\,\mathrm{d}v\mathrm{d}u$$
$$=\int_0^\infty (k+1)\frac{\theta}{2}\mathrm{e}^{-(k+1)\frac{\theta}{2}u}\mathrm{e}^{-\frac{k(k+1)}{2}u}\,\mathrm{d}u=\frac{\theta}{\theta+k}.$$

Similarly, a coalescence event will occur before a mutation with probability

$$\frac{k}{\theta+k}.$$

Three equivalent models

We present here three models equivalent to our process of coalescence with mutation.

- **Hoppe's urn** (cf. Hoppe [40]). Hoppe's urn contains a black ball of mass θ and colored balls of mass 1. The process of filling the urn is iterative. Initially, there is a black ball. At the k-th draw, a ball is drawn (draw with replacement) in proportion to its mass. If a colored ball is chosen, it is put back in the urn and another ball of the same color is added. If the drawn ball is black, it is returned to the urn and a ball of a new color is added. It is assumed that the number of colors is infinite. The choice of the black ball corresponds to a new mutation and the choice of a colored ball to a coalescence event. Before the $(k+1)$-th draw, there are therefore $k+1$ balls in the urn. A mutation will therefore be chosen with probability $\frac{\theta}{\theta+k}$ and a coalescence event with probability $\frac{k}{\theta+k}$. Here, A_k the number of different colors after k draws.
  The genealogy of k particles in a Kingman coalescent model with mutation to an infinity of alleles can be obtained (from the point of view of the law) by simulating $k-1$ steps of the process of drawing and replacing colors in a Hoppe urn.
  This construction can be practical for the simulation of such coalescents.
- **The Chinese restaurant model.** We consider an infinity of tables and k guests sitting at certain tables. A new guest arrives. With probability $\frac{\theta}{\theta+k}$, he sits at a new table, and with probability $\frac{1}{\theta+k}$, he chooses a guest at random and sits next to him. The number of distinct alleles (species) has the same law as the number of occupied tables.
- **Hubbell's ecological model** (cf. Hubbell [41]). The approximation leading to the coalescent with mutation is also of interest in ecology, through Hubbell's ecological model. In this model of quantitative ecology, we consider a forest comprising N sites occupied by N trees of various species. Each tree that dies gives way to a new tree whose species is either re-sampled from the remaining $k-1$ trees with probability $1-\mu$, or replaced by a new species with probability μ. We assume that the migrant species come from a metacommunity of trees composed of an infinite number of species. The probability μ is called the *speciation rate*.

6.4.3 Law of the Number of Distinct Alleles, Ewens' Formula

Let us study the law of A_k, which, in the Chinese restaurant model, is the number of tables occupied by the first k guests, or the number of colors after k draws in Hoppe's model.

Theorem 6.4.7

1) *The random variable A_k can be written as $A_k = \sum_{i=1}^k \varepsilon_i$, where the random variables ε_i are independent and follow a Bernoulli distribution with*

$$\mathbb{P}(\varepsilon_i = 1) = \frac{\theta}{\theta + i - 1}.$$

2) *It follows that*

$$\mathbb{E}(A_k) \sim_{k\to\infty} \theta \log k; \qquad \mathrm{Var}(A_k) \sim_{k\to\infty} \theta \log k.$$

3) *If Φ denotes the distribution function of the standard normal distribution, then for any real number x, we have*

$$\mathbb{P}\left(\frac{A_k - \mathbb{E}(A_k)}{\sqrt{\mathrm{Var}(A_k)}} \le x\right) \longrightarrow_{k\to+\infty} \Phi(x).$$

Proof In the different models, we set $\varepsilon_i = 1$ if the i-th allele comes from a mutation, or if the i-th guest chooses a new table, or if the i-th draw gives a new color. We set $\varepsilon_i = 0$ otherwise. The ε_i are independent Bernoulli random variables with parameter $\frac{\theta}{\theta+i-1}$. We therefore have

$$\mathbb{E}(A_k) = \sum_{i=1}^{k} \mathbb{E}(\varepsilon_i) = \sum_{i=1}^{k} \frac{\theta}{\theta + i - 1}; \qquad \mathrm{Var}(A_k) = \sum_{i=1}^{k} \frac{\theta(i-1)}{(\theta + i - 1)^2}.$$

Using the decrease of $x \longrightarrow \frac{\theta}{\theta+x}$, it is easy to show that

$$\int_0^k \frac{\theta}{\theta + x}\,\mathrm{d}x \le \mathbb{E}(A_k) \le 1 + \int_0^{k-1} \frac{\theta}{\theta + x}\,\mathrm{d}x.$$

We deduce from this that

$$\mathbb{E}(A_k) \sim_{k\to\infty} \theta \log(k).$$

We also have that

$$\mathrm{Var}(A_k) - \mathbb{E}(A_k) = -\sum_{i=0}^{k-1} \left(\frac{\theta}{\theta + i}\right)^2,$$

which is the partial sum of a convergent series. This implies that $\mathrm{Var}(A_k)$ behaves like $\mathbb{E}(A_k)$, as k tends to infinity. The third assertion follows from a central limit theorem that generalizes the classic central limit theorem: here, the random variables are independent but not identically distributed. We therefore apply the following theorem.

Theorem 6.4.8 *(see Billingsley [10]). Let $(X_i)_i$ be a sequence of independent centered square integrable variables. Introduce $S_n = X_1 + \cdots + X_n$ and $s_n^2 = \mathbb{E}(S_n^2)$. Then the Lindenberg condition*

$$\forall \varepsilon > 0, \quad \frac{1}{s_n^2} \sum_{i=1}^{n} \mathbb{E}\left(X_i^2 \mathbb{1}_{|X_i| > \varepsilon s_n}\right) \longrightarrow_{n\to\infty} 0$$

is fulfilled if and only if $\lim_n \left(s_n^{-2} \max_{1\le i\le n} \mathbb{E}(X_i^2)\right) = 0$ and in this case, the sequence $\left(\frac{S_n}{s_n}\right)_n$ converges in distribution to a centered reduced normal random variable.

In our context, we have

$$\max_{1\le i\le n} \mathbb{E}((\varepsilon_i - \mathbb{E}(\varepsilon_i))^2) = \max_{1\le i\le n} \frac{\theta(i-1)}{(\theta+i-1)^2} = \frac{\theta}{(\theta+1)^2},$$

and we know that $s_n^2 \sim_{n\to+\infty} \theta \log n$.

Thus the condition $\lim_n \left(s_n^{-2} \max_{1\le i\le n} \mathbb{E}((\varepsilon_i - \mathbb{E}(\varepsilon_i))^2)\right) = 0$ is fulfilled and we can apply the theorem: the sequence $\left(\frac{A_k - \mathbb{E}(A_k)}{\sqrt{\mathrm{Var}(A_k)}}\right)_k$ converges in distribution to a centered reduced normal law. □

An immediate consequence of Theorem 6.4.7 is that the random variable $\frac{A_k}{\log k}$ is an asymptotically normal estimator of the unknown parameter θ. We also know that its variance tends slowly to 0, at the speed of $\frac{1}{\log(k)}$. This gives a very poor convergence speed of the estimator and therefore requires large sample sizes. For example, if we want to estimate θ with an error of 0.01, we must use a sample size of approximately $k = \mathrm{e}^{100}$ (cf. Durrett [33]).

The last result of this chapter is due to Ewens (cf. Ewens [36]). It gives the complete distribution of alleles in the sample. For $j \in \{1, \dots, k\}$, let us define N_j as the number of alleles carried by j individuals. For example, N_1 is the number of singleton alleles. The N_j's satisfy in particular

$$\sum_{j=1}^{k} j N_j = k.$$

We then have the following theorem.

Theorem 6.4.9 (Ewens' Formula) *Consider k integers $n_1, \dots, n_k$ less than or equal to k with $\sum_{j=1}^k j n_j = k$. Then we have*

$$\mathbb{P}(N_1 = n_1, \dots, N_k = n_k) = \prod_{j=1}^{k} \frac{j}{j-1+\theta} \frac{\left(\frac{\theta}{j}\right)^{n_j}}{n_j!} = \frac{k!}{\theta_{(k)}} \prod_{j=1}^{k} \frac{\left(\frac{\theta}{j}\right)^{n_j}}{n_j!}, \quad (6.4.3)$$

where $\theta_{(k)} = \theta(\theta+1)\cdots(\theta+k-1)$.

The formula (6.4.3) is written as $C_{k,\theta} \prod_{j=1}^k \mathrm{e}^{-\frac{\theta}{j}} \frac{\left(\frac{\theta}{j}\right)^{n_j}}{n_j!}$, where $C_{k,\theta}$ is a constant that depends on θ and k, such that the sum of the probabilities equals 1. The allelic partition therefore has the same law as the distribution of $(Y_1, \dots, Y_k)$ conditionally on $\sum_j jY_j = k$, where the random variables $Y_1, \dots, Y_j, \dots Y_k$ are independent and have a Poisson distribution with parameters $\theta, \dots, \frac{\theta}{j}, \dots, \frac{\theta}{k}$. This result will be justified later.

Proof It suffices to show that the distribution of colors in Hoppe's urn at time k is given by Ewens' formula. We show this by induction on k. When $k = 1$, the partition $N_1 = 1$ has probability 1 and the result is proven. Now suppose that the property holds for all times less than $k-1$. Suppose that at time k, we have the distribution $N_1 = n_1, \dots, N_k = n_k$. Let $n = (n_1, \dots, n_k)$ and let $\bar{n}$ be the state of the allelic partition at the previous time. Let

$$P_\theta(n) = \frac{k!}{\theta_{(k)}} \prod_{j=1}^{k} \frac{\left(\frac{\theta}{j}\right)^{n_j}}{n_j!}.$$

We will show that

$$\sum_{\bar{n}} P_\theta(\bar{n}) p(\bar{n}, n) = P_\theta(n), \quad (6.4.4)$$

where the transition probabilities $p(\bar{n}, n)$ are defined below.

We have several possibilities.

- If $\bar{n}_1 = n_1 - 1$: a new color has just been added. In this case, the transition probability for Hoppe's urn is

$$p(\bar{n}, n) = \frac{\theta}{\theta+k-1}.$$

  We also have by Ewens' formula that

$$\frac{P_\theta(n)}{P_\theta(\bar{n})} = \frac{k}{\theta+k-1} \frac{\theta}{n_1}.$$

- Suppose that for $1 \le j \le k$, we have $\bar{n}_j = n_j + 1$ and that $\bar{n}_{j+1} = n_{j+1} - 1$. An existing color, which was the color of j balls, has been chosen, and the number of balls of this color becomes $j+1$. In this case, the transition probability is then

$$p(\bar{n}, n) = \frac{j\bar{n}_j}{\theta + k - 1}.$$

The ratio of probabilities given by Ewens' formula is:

$$\frac{P_\theta(n)}{P_\theta(\bar{n})} = \frac{k}{\theta + k - 1} \cdot \frac{j\bar{n}_j}{(j+1)n_{j+1}}.$$

We now observe that

$$\sum_{\bar{n}} \frac{P_\theta(\bar{n})}{P_\theta(n)} p(\bar{n}, n) = \frac{\theta}{\theta + k - 1} \cdot \frac{\theta + k - 1}{k} \cdot \frac{n_1}{\theta} + \sum_{j=1}^{k-1} \frac{j\bar{n}_j}{\theta + k - 1} \cdot \frac{\theta + k - 1}{k} \cdot \frac{(j+1)n_{j+1}}{j\bar{n}_j}.$$

By simplifying the terms on the right-hand side, we finally obtain that

$$\sum_{\bar{n}} \frac{P_\theta(\bar{n})}{P_\theta(n)} p(\bar{n}, n) = \frac{n_1}{k} + \sum_{j=1}^{k-1} \frac{(j+1)n_{j+1}}{k} = 1,$$

since $\sum_{j=1}^{k} jn_j = k$. Thus, the probabilities defined by Ewens' formula satisfy (6.4.4). The distribution of the colors of Hoppe's urn also satisfies this property (6.4.4). Then by a recurrence assumption on n we deduce from (6.4.4) that $\mathbb{P}(\mathbf{N} = n)$ is given by Ewens' formula. □

As consequence, we obtain the following proposition.

Proposition 6.4.10 *Consider the distribution of* $(N_1, \ldots, N_h)$*, where* h *is a fixed number, while the sample size tends to infinity. We have*

$$\lim_{k \to \infty} \mathbb{P}(N_1 = n_1, \ldots, N_h = n_h) = \mathbb{P}(Y_1 = n_1, \ldots, Y_h = n_h),$$

where the random variables $Y_1, \ldots, Y_h$ *are independent and have a Poisson distribution with parameters respectively* $\theta, \frac{\theta}{2}, \ldots, \frac{\theta}{h}$.

6.4.4 A Branching Process With Immigration Point of View

We can link Hoppe's urn and the allelic distribution in the coalescent to a branching process with immigration. Individuals follow the reproduction rules of a Yule process: they never die and reproduce by giving birth to another individual after an exponential time of parameter 1. Migrants arrive in the population at the successive times of a Poisson process of parameter θ.

Proposition 6.4.11 *If each migrant is of a new type, and if offspring are of the same type as their parents, then the embedded Markov chain has the same distribution as that generated by Hoppe's urn.*

Proof If the population is composed of k individuals, a new arrival will have a new type with probability $\frac{\theta}{\theta+k}$ and it will take with probability $\frac{k}{k+\theta}$ the type of an individual chosen uniformly in the existing population. From this description, it is clear that the embedded Markov chain for this binary branching process with immigration has the same distribution as the color distribution in Hoppe's model. □

Let us combine the results of Propositions 6.4.11 and 5.4.7. We then obtain a new proof of Theorem 6.4.5.

Theorem 6.4.12 *Consider the coalescent starting from k lineages and stopping when there are ℓ lineages ($\ell \leq k$). Let $J_1, \ldots, J_\ell$ be the numbers of lineages in the sample from the k individuals, when they are randomly labeled. Then $(J_1, \ldots, J_\ell)$ is uniformly distributed over the vectors of $(\mathbb{N}^*)^\ell$ whose sum of coordinates equals k and for all $m \leq k - \ell + 1$,*

$$\mathbb{P}(J_i = m) = \binom{k-m-1}{\ell-2} \Big/ \binom{k-1}{\ell-1}.$$

Proof Let Z_t^i, $1 \leq i \leq \ell$, be independent copies of the Yule process. If $j_1, \ldots, j_\ell$ are positive integers summing to k, then by Proposition 5.4.7 applied with $a = 1$,

$$\mathbb{P}(Z_t^1 = j_1, \ldots, Z_t^\ell = j_\ell) = (1-p)^{k-\ell} p^\ell \quad \text{where} \quad p = \mathrm{e}^{-t}.$$

Since the term on the right depends only on the sum k and the number of terms ℓ, all possible vectors have the same probability. As already seen in the proof of Theorem 6.4.5, there are $\binom{k-1}{\ell-1}$ possible vectors $(j_1, \ldots, j_\ell)$ of positive integers summing to k. Thus it follows that

$$\mathbb{P}\left(Z_t^1 = j_1, \ldots, Z_t^\ell = j_\ell \,\middle|\, \sum_{j=1}^{\ell} Z_t^j = k\right) = 1/\binom{\ell-1}{k-1}.$$

The conditional distribution is therefore uniform over all possible vectors. Since the number of vectors $(j_2, \ldots, j_\ell)$ of positive integers summing to $k - j_1$ is $\binom{k-j_1-1}{\ell-2}$, we obtain

$$\mathbb{P}\left(Z_t^1 = j_1 \,\middle|\, \sum_{j=1}^{k} Z_t^j = \ell\right) = \binom{k-j_1-1}{\ell-2} / \binom{\ell-1}{\ell-1}.$$

This concludes the proof. □

6.5 Exercises

Exercise 6.5.1 (Genealogy of BGW conditioned to be of constant size)
We consider a genealogy of BGW conditioned to be of constant size N and with Poisson law of parameter m as reproduction law. This model is closely related to the Wright–Fisher model. We will see that in this case, the distribution of the number of descendants follows a multinomial law.

We denote by $Z_1,\dots,Z_N$ the number of descendants of individuals $1,\dots,N$. The random variables $Z_1,\dots,Z_N$ are therefore independent and identically distributed, with a Poisson law of parameter m.

1. What is the law of $Z_1+\cdots+Z_N$?
2. For all integers $k_1,\dots,k_N$ satisfying $k_1+\cdots+k_N=N$, compute the probability

$$\mathbb{P}(Z_1=k_1,\dots,Z_N=k_N \mid Z_1+\cdots+Z_N=N).$$

3. Conclude. (The solution of this exercise can be found in Section 3.3.)

Exercise 6.5.2 (The Moran Model)
We consider a population made up of N individuals of allele A or a. At time 0, there are x_0 individuals of allele A.
The individuals reproduce independently of each other at rate 1 and give birth to a unique individual. This individual carries the same allele as its parent. This descendant then replaces an individual drawn uniformly at random from the population, so that the population size remains constant.
We are interested in the dynamics of the number of individuals $(X_t, t\geq 0)$ carrying the allele A over time.

1. Show that the model is described by a Markovian jump process X with values in $\{0,\dots,N\}$ and give its generator.
2. Give the differential equation governing the mean and variance of the proportion of individuals with allele A over time.
3. What scaling limit would make it possible to find a Wright–Fisher diffusion process?

Exercise 6.5.3 (Diffusion approximation of a Moran model with selection)
The Moran model is a model of evolution of a fixed-size population N containing "resident" type individuals and "mutant" type individuals. When N is large, the proportion of mutants in the population can be approximated by the following diffusion process:

$$\mathrm{d}Y_t = s_N\, Y_t(1-Y_t)\,\mathrm{d}t + \sqrt{\frac{2+s_N}{N}\, Y_t(1-Y_t)}\,\mathrm{d}B_t,$$

where $s_N > -1$ is the selective advantage of the mutants.

1. What is the generator of the diffusion Y?
2. Compute the fixation probability of a subpopulation of Nx mutants.

3. Describe qualitatively the behavior of this diffusion approximation as N tends to infinity, depending on whether $s_N \longrightarrow s$, $s_N = \frac{s}{N}$ or $s_N = \frac{s}{N^2}$. Show that to obtain a non-trivial limit, it is necessary to make the time change $t \longrightarrow Nt$.

Exercise 6.5.4 (Kingman's coalescent and the number of polymorphic sites) Kingman's coalescent is a process in which we follow the lineages of an asexual genealogy in the backward sense of time and where each pair of lineages, independently, coalesces at a constant rate of 1. We call T_n the time it takes for n lineages to coalesce, that is, the time it takes for the Kingman coalescent starting from n lineages to reach the ancestral lineage, or the time that has elapsed since their most recent common ancestor.

Part I: Average time to the first common ancestor

1. Compute $\mathbb{E}(T_n)$.
2. What is the limit $\lim_{n\to\infty} \mathbb{E}(T_n)$? How to interpret it?

Part II: Distribution of the number of polymorphic sites

We assume that the Kingman coalescent represents the genealogical history of a chromosome. Conditionally on the coalescent, we throw a Poisson point cloud with parameter $\theta/2$ on its branches, each point corresponding to a mutation that at each time affects a new site of the chromosome DNA sequence. This means that

- on a length L of the genealogical tree, the number of mutations follows a Poisson distribution with parameter $\theta L/2$;
- the numbers $Z_1, \ldots, Z_k$ of mutations present on k *disjoint* portions of the tree are independent.

We denote by S_n the number of polymorphic sites of the sample, that is, existing in the ancestral state in at least one of the n sequences and in the mutant state in at least one other.

1. Compute the expectation and variance of the number S_n of polymorphic sites.
2. Denote by N_j the number of mutations that appeared on the genealogical tree during the period when the tree is composed of j ancestors. Compute the generating function of N_j. What is the law of N_j?
3. What is the generating function of S_n?

Part III: Estimating the mutation rate from the number of polymorphic sites

We seek to estimate the mutation rate θ from the number S_n of polymorphic sites in the sample.

1. Denote $\sum_{k=1}^{n-1} 1/k$ by H_n. Propose an estimator $\hat{\theta}_n$ of θ such that $\mathbb{E}(\hat{\theta}_n) = \theta$.
2. Compute the variance of $\hat{\theta}_n$. At what speed does $\mathbb{E}\big((\hat{\theta}_n/\theta - 1)^2\big)^{1/2}$ tend towards 0? Comment.
3. Compute the limit of $\log \mathbb{E}\Big(e^{it\sqrt{H_n(\hat{\theta}_n - \theta)}}\Big)$ as $n \longrightarrow \infty$.
4. Deduce that $\sqrt{\frac{\log n}{\hat{\theta}_n}}(\hat{\theta}_n - \theta) \stackrel{d}{\longrightarrow} Z$ as $n \longrightarrow \infty$, as Z follows a standard Gaussian distribution.

Chapter 7
Some Modern Developments in Ecology-Evolution

Models are only simplifications of a very complex reality.

Albert Einstein (1879–1955)

7.1 Survival and Growth of Meta-Populations Distributed on a Graph

This section is based on the work of Bansaye and Lambert [7].

Populations live in habitats with different characteristics (temperature, soil richness, resources, predation, etc.) and can be connected (by migration corridors of various kinds). The conditions of habitats and connections can evolve over time, following various events (climate changes, floods, pollution, deforestation, land development, etc.). We want to know how the characteristics of habitats and connections influence the probability of survival and the population growth.

7.1.1 First Approach: Multitype Galton–Watson Process

We consider K habitats, denoted by $i = 1, \dots, K$. Each individual in habitat i reproduces independently of others and gives an average number m_i of offspring. Each offspring born in habitat i has a probability p_{ij} of joining habitat j. The matrix

$$A = (m_i p_{ij})_{1 \le i, j \le K}$$

gives the average number of children of an individual from habitat i in habitat j.

S. Méléard, *Random Models in Biology, Ecology and Evolution*, Texts in Applied Mathematics 87, https://doi.org/10.1007/978-3-662-73483-4_7

Let $(Z_n)_n = (Z_n(j), j = 1, \dots, K, n \in \mathbb{N})$ be the process that describes the size of the subpopulations at each generation n. It takes values in $\mathbb{N}^K$ and $Z_n(j)$ is the number of individuals in habitat j at generation n. The process $(Z_n)_n$ is a multitype branching process (see Section 3.7). Assuming that Z_0 is given, the average of Z_n is

$$\mathbb{E}(Z_n) = Z_0\, A^n.$$

(The matrix A is the transpose matrix of M, using the notation of Section 3.7.) Suppose that A satisfies (3.7.1). By the Perron–Frobenius Theorem (see Theorem 3.7.7), there exists a maximum (simple) eigenvalue $\lambda > 0$ associated with positive eigenvectors u and v such that

$$A\,u = \lambda\, u; \qquad v\,A = \lambda\, v.$$

Moreover, we have the following asymptotic behaviors.

Theorem 7.1.1 *If $\lambda \leq 1$, the population goes extinct. If $\lambda > 1$, the population survives with positive probability and grows geometrically at rate λ. The asymptotic distribution in habitats is given by v. In the long run, the proportion of time spent by an individual in habitat i is $u_i\, v_i$.*

This approach is simple but the results are not very explicit. Indeed, it is difficult to interpret the results because we do not see the specific roles of dispersion and reproduction. Moreover, the graph structure is not explicit.

7.1.2 Second Approach – Markov Chain on a Graph

In this approach, we want to separate the movements from the reproductions and integrate more precisely the dynamics of movement from one habitat to another and the graph structure. We consider the Markov chain $(X_n)_n$ whose transition probabilities are defined by

$$\mathbb{P}(X_{n+1} = j \,|\, X_n = i) = p_{ij}.$$

We assume that the Markov chain is irreducible (see Chapter 2).

Theorem 7.1.2 *The population survives with positive probability if and only if*

$$m_1\, \mathbb{E}_1(m_{X_1} \cdots m_{X_{T-1}}) > 1, \tag{7.1.1}$$

where T is the return time in habitat 1*:*

$$T = \inf\{n \geq 1, X_n = 1\}.$$

T is well defined under the condition of irreducibility. We also note that even if 1 is assumed to be the initial habitat, the criterion does not depend on it, thanks to the irreducibility assumption. Thus, persistence will occur with strictly positive probability if there is a supercritical dynamic at a location on the graph.

The case of a completely mixing population

The case of a completely mixing population is represented by a matrix (p_{ij}), all of whose columns are equal to the same vector D ($\forall j = 1,\ldots,K,\ p_{ij} = D_j$): the migration probability to habitat j is independent of the parent's habitat. In this case, it is easy to compare the results of Theorems 7.1.1 and 7.1.2. We can easily compute the maximum eigenvalue λ of A. Indeed A is the rank K matrix equal to $m\,D^T$, where m is the column vector $(m_1,\ldots,m_K)^T$ and D^T the row vector $(D_1,\ldots,D_K)$. Thus, we can easily show that $A^n = (\sum_{i=1}^K m_i D_i)^{n-1} A$ and the number $\rho = \sum_{i=1}^K m_i D_i$ is therefore the largest eigenvalue of A.

On the other hand, the random variables X_n are independent and have the same law D. Thus the time T is geometrically distributed, with parameter D_1. By conditioning on T, we obtain

$$\begin{aligned}
m_1\,\mathbb{E}\Big(\prod_{n=1}^{T-1} m_{X_n}\Big) &= m_1 \sum_{k\geq 1} \mathbb{E}\Big(\prod_{n=1}^{T-1} m_{X_n}\,|T=k\Big)\mathbb{P}(T=k)\\
&= m_1 \sum_{k\geq 1} D_1(1-D_1)^{k-1}\Big(\sum_{j=2}^{K} m_j\frac{D_j}{1-D_1}\Big)^{k-1}\\
&= m_1 D_1 \sum_{k\geq 1}\Big(\sum_{j=2}^{K} m_j D_j\Big)^{k-1}.
\end{aligned}$$

If $\sum_{j=2}^K m_j D_j \geq 1$, then the right term is $+\infty$ and we have persistence with positive probability. If $\sum_{j=2}^K m_j D_j < 1$, we then obtain $m_1\,\mathbb{E}\Big(\prod_{n=1}^{T-1} m_{X_n}\Big) = \frac{m_1 D_1}{1-\sum_{j=2}^K m_j D_j}$. This quantity will be greater than 1 if and only if $\sum_{j=1}^K m_j D_j > 1$. We find the criterion given by the Perron–Frobenius Theorem.

Proof of Theorem 7.1.2 Consider an individual a located at time 0 in habitat 1. Let Y_1 be the number of a's children who are still in habitat 1, and similarly, let Y_n be the number of a's descendants at generation n living in habitat 1 all of whose parents (other than a) have lived outside this habitat. Let

$$Y = \sum_{n\geq 1} Y_n,$$

which is the number of a's descendants who are living in habitat 1 for the first time in their lineage (except a).

Let us compute $\mathbb{E}(Y)$. Let $Y_n^{(i)}$ be the number of descendants of 1 in patch i at generation n who have avoided patch 1 at generations $1,\ldots,n-1$. By recurrence, we can show that

$$\begin{aligned}\mathbb{E}(Y_n^{(i)}) &= \sum_{j=2}^{K} \mathbb{E}(Y_{n-1}^{(j)})\, m_j\, p_{ji} \\ &= \sum_{j_1,\cdots,j_{n-1}\in\{2,\cdots K\}} p_{1j_1}\, p_{j_1 j_2}\, p_{j_{n-2}j_{n-1}}\, p_{j_{n-1}i}\, m_1 m_{j_1}\cdots m_{j_{n-2}} m_{j_{n-1}}.\end{aligned}$$

Since $Y_n = Y_n^{(1)}$, we deduce that

$$\begin{aligned}\mathbb{E}(Y_n) &= \sum_{j_1,\ldots,j_{n-1}\in\{2,\ldots,K\}} p_{1j_1}\, p_{j_1 j_2}\, p_{j_{n-2}j_{n-1}}\, p_{j_{n-1}1}\, m_1 m_{j_1}\cdots m_{j_{n-2}} m_{j_{n-1}} \\ &= m_1\, \mathbb{E}(\mathbf{1}_{T=n}\, m_{X_1}\cdots m_{X_{n-1}}),\end{aligned}$$

and then

$$\mathbb{E}(Y) = \mathbb{E}\Big(\sum_{n\geq 1} Y_n\Big) = m_1 \sum_{n\geq 1} \mathbb{E}(\mathbf{1}_{T=n}\, m_{X_1}\cdots m_{X_{n-1}}) = m_1\, \mathbb{E}(m_{X_1}\cdots m_{X_{T-1}}).$$

This concludes the proof. □

The case of two types of habitats

The formula (7.1.1) separates the demographic effects from dispersion. For example, consider a graph composed of a source (the initial habitat) where the mean of the reproduction law is $m_1 > 1$ and of $K-1$ identical less favorable habitats (called sinks), where reproduction has the same mean $m < 1$. The condition (7.1.1) is written

$$m_1\, \mathbb{E}_1(m^{T-1}) > 1.$$

The expectation is the generating function of the random variable $T-1$, computed at the point m.

Let $p = \sum_{j=2}^{K} p_{1j}$ be the migration probability for an individual living in habitat 1. The average number of descendants per individual leaving the source at each generation is $m_1 p$. Let σ be the first visit time of a sink by a (fixed) walker coming from the source:

$$\sigma = \inf\{n \geq 0, X_n \neq 1\}.$$

It is easy to see that σ follows a geometric law with parameter p. Let S be the waiting time, after σ, before visiting the source:

$$S = \inf\{n \geq 0, X_{\sigma+n} = 1\}.$$

This duration S can be seen as the time spent in the wells between two consecutive visits to the source. We then have

$$\mathbb{E}\Big(\prod_{i=1}^{T-1} m_{X_n}\Big) = 1 - p + p\,\mathbb{E}(m^S).$$

Theorem 7.1.2 allows us to affirm that the population will persist with positive probability if and only if

$$m_1(1-p) + \delta m_1 p > 1, \tag{7.1.2}$$

where δ is the rate of depression due to passage through the wells:

$$\delta = \mathbb{E}(m^S) = \sum_{k\geq 1} m^k\,\mathbb{P}(S=k).$$

In the case where there are only two habitats ($K = 2$, a source and a well), the random variable S follows a geometric law with parameter $q = 1 - p$, hence $\delta = \frac{mq}{1-m(1-q)}$. Thus, the condition (7.1.2) is written

$$\frac{m_1 p}{1-m_1(1-p)} > \frac{1-m(1-q)}{mq}.$$

7.2 Abundance in a Random Environment

This section is based on an article by Evans, Hening and Schreiber [35].

In the Feller equation (5.6.10), the stochastic term is demographic in the sense that, as seen in Section 5.6.2, it comes from a multitude of births and deaths. It is also possible to model other sources of randomness, such as environmental stochasticity.

In this section, the process $(Y_t, t \geq 0)$ represents the abundance of a population in a random environment. It is governed by the stochastic differential equation

$$\mathrm{d}Y_t = Y_t\,(r - cY_t)\,\mathrm{d}t + \sigma Y_t\,\mathrm{d}W_t\ ,\ Y_0 > 0,$$

where $(W_t)_{t\geq 0}$ is a standard Brownian motion.

The growth rate has a stochastic component with σ^2 being the infinitesimal variance. The process is well defined and has an explicit form, which is easily shown using Itô's formula.

$$Y_t = \frac{Y_0 \exp\big((r-\frac{\sigma^2}{2})t + \sigma W_t\big)}{1 + cY_0\int_0^t \exp\big((r-\frac{\sigma^2}{2})s + \sigma W_s\big)\,\mathrm{d}s}. \tag{7.2.1}$$

We note that $Y_t \geq 0$ for all $t \geq 0$ almost surely.

It is then possible to deduce the long-term behavior of the process, which depends on the sign of $r - \frac{\sigma^2}{2}$.

Proposition 7.2.1

1. *If* $r - \frac{\sigma^2}{2} < 0$, *then* $\lim_{t\to\infty} Y_t = 0$ *almost surely.*
2. *If* $r - \frac{\sigma^2}{2} = 0$, *then almost surely,* $\liminf_{t\to\infty} Y_t = 0$ *and* $\limsup_{t\to\infty} Y_t = +\infty$.
3. *If* $r - \frac{\sigma^2}{2} > 0$, *then* $(Y_t)_t$ *has a unique stationary distribution, which is the law* $\Gamma(\frac{2r}{\sigma^2} - 1, \frac{\sigma^2}{2c}) = \Gamma(\kappa, \theta)$, *with density* $x \mapsto \dfrac{1}{\Gamma(\kappa)\,\theta^\kappa}\, x^{\kappa-1}\, \mathrm{e}^{-\frac{x}{\theta}}$.

Remark 7.2.2 We note that the behavior here is very different from the behavior of the Feller equation, which almost surely converges to 0 in the long term, regardless of the intrinsic growth rate. It is also very different from the behavior of the dynamical system that tends to 0 if the growth rate is negative and to the strictly positive limit c/r if the growth rate is strictly positive. The stochastic term modeling the environment has an unfavorable effect since r must be greater than $\sigma^2/2$ for there to be some stability of the population. In this case, the limit is random and its law depends on the competition parameter.

Proof We will provide some elements of the proof here. We refer to [35] for a complete and rigorous proof. It is first of all immediate, using the explicit formula (7.2.1) and the law of iterated logarithm (cf. Proposition 4.3.2), that if $r - \frac{\sigma^2}{2} < 0$, the process $(Y_t, t \geq 0)$ almost surely tends to 0 as t tends to infinity. Moreover, if $r - \frac{\sigma^2}{2} = 0$, $\liminf_{t\to\infty} Y_t = 0$ and $\limsup_{t\to\infty} Y_t = +\infty$ almost surely. Let's now focus on the case where $r - \frac{\sigma^2}{2} > 0$ and introduce, as in (4.6.20), the scale function

$$s(x) = \int_c^x \exp\left(-\int_a^y \frac{2b(z)}{\sigma^2(z)}\,\mathrm{d}z\right)\mathrm{d}y = \int_c^x \left(\frac{y}{a}\right)^{-\frac{2r}{\sigma^2}} \mathrm{e}^{\frac{2c}{\sigma^2}(y-a)}\,\mathrm{d}y,$$

where a and c are arbitrary in $\mathbb{R}_+^*$. We know by Itô's formula that $s(Y_t)$ defines a martingale and that

$$\mathrm{d}s(Y_t) = s'(Y_t)\sigma Y_t\,\mathrm{d}W_t = h(s(Y_t))\,\mathrm{d}W_t,$$

where $h(z) = \sigma\, s' \circ s^{-1}(z)\, s^{-1}(z)$.

Consider the measure $m(\mathrm{d}z) = \frac{1}{h^2(z)}\,\mathrm{d}z$. An explicit computation gives that

$$m(\mathbb{R}_+^*) = \frac{1}{\sigma^2 a^{2r/\sigma^2}} \int_0^{+\infty} u^{\frac{2r}{\sigma^2}-2}\, \mathrm{e}^{-\frac{2c}{\sigma^2}(u-a)}\,\mathrm{d}u.$$

Arguments of stochastic calculus in dimension 1, which are not developed in this course, show that the process Y has a stationary distribution if $s(\mathbb{R}_+^*) = (-\infty, +\infty)$ and if $m(\mathbb{R}_+^*) < \infty$, which is true if and only if $r - \frac{\sigma^2}{2} > 0$. In this case, the diffusion $s(Y)$ has a stationary distribution of density $f = \frac{1}{m(\mathbb{R}_+^*)h^2}$ (see [42] for these properties).

We deduce here that the stationary distribution of Y has density

$$\begin{aligned} g(x) = f(s(x))s'(x) &= \frac{1}{m(\mathbb{R}_+^*)h^2(s(x))} s'(x) \\ &= \frac{1}{m(\mathbb{R}_+^*)\sigma^2 x^2 s'(x)} \\ &= \frac{1}{m(\mathbb{R}_+^*)\sigma^2 x^2 \left(\frac{x}{a}\right)^{-\frac{2r}{\sigma^2}} \mathrm{e}^{\frac{2c}{\sigma^2}(x-a)}}. \end{aligned}$$

We recognize the density of a Gamma(κ,θ) law of parameters $\theta = \frac{\sigma^2}{2c}$ and $\kappa = \frac{2r}{\sigma^2} - 1$. □

7.3 Study of the Invasion of a Mutant in a Large Resident Population at Equilibrium

In this section, we model one of the steps of the invasion mechanism of a mutant individual in a population at equilibrium. This study is part of research on the random modeling of adaptive dynamics, which can be found for example in Champagnat–Ferrière–Méléard [19], [18] or in Méléard [60].

Let us consider a large population at equilibrium, assuming that all individuals are of the same genetic type 1. The demographic parameters depend on the genetic type. Individuals of type 1 reproduce at rate b_1 and give birth to a single offspring. By heredity, an individual transmits his type to his descendants, unless a mutation occurs. Individuals of type 1 die at rate d_1 or due to competition with other present individuals. We denote by c_{11} the competition pressure between two individuals of type 1. We assume that without competition, the birth and death process would be supercritical: $b_1 - d_1 > 0$.

In Section 5.6.1 (Theorem 5.6.1), we proved that the size of this population is approximated (under a large population assumption) by the solution $(x(t), t \geq 0)$ of the differential equation

$$x'(t) = x(t)\,(b_1 - d_1 - c_{11}x(t)).$$

The stable equilibrium of this equation, which is given by

$$x_1^* = \frac{b_1 - d_1}{c_{11}},$$

therefore represents the size of the type 1 population at equilibrium.

Let us assume that a mutant individual, of type 2, appears during a reproduction event. Even if this individual is unique at the time of its appearance, it develops by heredity a sub-population of type 2 individuals. We assume that the demographic parameters of type 2 individuals are b_2 and d_2, with $b_2 > d_2$. The population of type 2 individuals is initially small; it is therefore modeled by a birth and death process. For a while, the size of this population is negligible compared to that of type 1 individuals. A type 2 individual is therefore primarily impacted by the presence of type 1 individuals in its struggle for survival (resource sharing), and type 1 individuals do not feel the influence of type 2 individuals. We introduce the parameter c_{21}, which describes the competition pressure of a type 1 individual on a type 2 individual. The death rate of a type 2 individual is therefore approximated by

$$d_2 + c_{21} x_1^* = d_2 + c_{21} \frac{b_1 - d_1}{c_{11}}.$$

For type 2 individuals to be able to establish themselves in the population, it is necessary that the birth and death process associated with type 2 survives in the long term. In Chapter 5 (see Theorem 5.5.5 and the application to the binary branching process with a single ancestor), we obtained the conditions for such survival and in the case of possible survival, we computed the probability of this event. By adapting these conditions, we deduce the following results.

- If $b_2 \leq d_2 + c_{21} x_1^*$, the population of type 2 individuals resulting from the mutant individual will go extinct and the population will return, if we wait long enough, to a type 1 population. (We assume in this interpretation that mutations are rare and that a new mutation will only occur well after this phase of development and the disappearance of the type 2 population.)
- On the other hand, if $b_2 > d_2 + c_{21} x_1^*$, the population of type 2 individuals develops according to a supercritical process and the survival probability is non-zero, given by
$$\frac{b_2 - d_2 - c_{21} x_1^*}{b_2}.$$
Under this condition, we can hope that the mutant establishes itself in the population.

The Survival Condition

$$F_{21} = b_2 - d_2 - c_{21} \frac{b_1 - d_1}{c_{11}} > 0 \qquad (7.3.1)$$

corresponds to the growth rate of a small type 2 population within a resident population at equilibrium of type 1. This number F_{21} is called the *invasion fitness of a type 2 population in a resident population of type 1*. Note that this fitness is zero if the two types are identical. If the competition parameter is constant (resources are consumed uniformly by individuals of both types), then the survival condition becomes

$$b_2 - d_2 > b_1 - d_1. \qquad (7.3.2)$$

This means that the intrinsic growth rate (without competition) of the type 2 population is higher than that of the type 1 population.

On the survival event, the size of the type 2 population will reach at a certain time a sufficiently large size to be approximated by its deterministic approximation. At that point, we must take into account the two sub-populations and their interactions. We introduce the competition parameters c_{22} between two type 2 individuals and c_{12} of a type 2 individual on a type 1 individual. We saw in Section 5.6.3 that the deterministic system summarizing the dynamics of the coupled system of the two sub-populations is a competitive Lotka–Volterra system

$$\begin{aligned} x_1'(t) &= x_1(t)\,(b_1 - d_1 - c_{11}x_1(t) - c_{12}x_2(t)), \\ x_2'(t) &= x_2(t)\,(b_2 - d_2 - c_{21}x_1(t) - c_{22}x_2(t)). \end{aligned} \tag{7.3.3}$$

A study of this system (see Istas [43]) gives the following two possible conclusions. If in addition to condition (7.3.1), the demographic parameters also satisfy $F_{12} < 0$, then the equilibrium reached by the system (7.3.3) will be $(0, x_2^*)$. The population of type 1 individuals will go extinct and after a certain time, only type 2 individuals will remain. We say that type 2 has fixed in the population.

If on the other hand $F_{12} > 0$, the equilibrium reached by the system will be a non-trivial equilibrium such that both coordinates are non-zero. In this case, we will have coexistence of types 1 and 2 in the long term. Let $r_i = b_i - d_i$ for $i = 1,2$. Note that the conditions $F_{12} > 0$ and $F_{21} > 0$ necessarily imply that

$$(c_{11}r_2 - c_{21}r_1)(c_{22}r_1 - c_{12}r_2) > 0$$

and we see that then, $D = c_{11}c_{22} - c_{21}c_{12} \neq 0$. Indeed, otherwise, we would have

$$(c_{11}r_2 - c_{21}r_1)(c_{22}r_1 - c_{12}r_2) = -\frac{c_{22}c_{21}}{c_{12}}(r_1 - r_2\frac{c_{12}}{c_{22}})^2 < 0.$$

It is then easy to compute the sizes x_{12}^* and x_{21}^* of the two type 1 and 2 sub-populations at equilibrium:

$$x_{12}^* = \frac{c_{22}r_1 - r_2c_{12}}{D}; \quad x_{21}^* = \frac{c_{11}r_2 - r_1c_{21}}{D}.$$

Since $D \neq 0$, the case of coexistence is impossible if the competition coefficients are equal (see also (7.3.2)). Thus, the impact of the environment, for example on resources, can change the equilibria.

7.4 A Stochastic Model for the Self-Incompatibility of Flowering Plants

This section is based on Billiard and Tran [9] and Lafitte-Godillon, Raschel, and Tran [54]. It concerns the study of a population of flowering plants whose reproduction is self-incompatible. Self-incompatible reproduction prevents the fertilization of a plant's stigma by its own pollen. The individuals are diploid and can carry the alleles A or a at the locus S, which determines the partners with which the plant can reproduce. The possible genotypes at this locus are therefore AA, Aa, and aa. In this model, the A allele is dominant and the a allele is recessive. The phenotype corresponding to AA, Aa, or aa (i.e., the type of protein carried by the pollen and the stigma) will therefore be A, A, or a, respectively. Only pollen and stigmas carrying different phenotypes can produce viable seeds. Thus, AA cannot be created, since the genotype of an individual with phenotype a is necessarily aa. Due to self-incompatibility, the only possibilities for reproduction will therefore be between individuals of genotypes AA and aa or Aa and aa.

We will therefore study the birth and death bi-type process $(N_t, t \geq 0) = ((N_t^{Aa}, N_t^{aa}), t \geq 0)$ which describes the dynamics over time of the number of individuals carrying the genotypes Aa and aa. We assume that the ovules reproduce in continuous time at rate $r > 0$ and that each ovule is fertilized provided there is compatible pollen in the population. The death rate of each individual is $d > 0$. Note that

- AA and aa will produce an individual of genotype Aa with probability 1,
- Aa and aa will produce an individual of genotype Aa with probability $1/2$ and an individual of genotype aa with probability $1/2$.

Thus the generator Q will be given for $(i, j) \in (\mathbb{N}^*)^2$ by

$$\begin{aligned}
(i,j) &\longrightarrow (i+1,j) \ \text{ at rate } \frac{r(i+j)}{2} \\
(i,j) &\longrightarrow (i,j+1) \ \text{ at rate } \frac{r(i+j)}{2} \\
(i,j) &\longrightarrow (i-1,j) \ \text{ at rate } di \\
(i,j) &\longrightarrow (i,j-1) \ \text{ at rate } dj.
\end{aligned} \tag{7.4.1}$$

The other terms of Q are null.

When one of the two phenotypes A or a disappears, reproduction becomes impossible and the system goes extinct. We are looking for the probability of population extinction.

Let us define

$$\tau_0 = \inf\{t > 0, N_t^{Aa} = 0 \ or \ N_t^{aa} = 0\}$$

and for all $(i, j) \in \mathbb{N}^2$,

$$p_{i,j} = \mathbb{P}_{i,j}(\tau_0 < +\infty).$$

Proposition 7.4.1 *The sequence $(p_{i,j})_{(i,j)\in\mathbb{N}^2}$ is the smallest positive bounded solution of the following system: for all $(i, j) \in \mathbb{N}^2$,*

$$\begin{aligned}
p_{i,j} &= \frac{di}{(r+d)(i+j)} p_{i-1,j} + \frac{dj}{(r+d)(i+j)} p_{i,j-1} + \frac{r}{2(r+d)} p_{i,j+1} + \frac{r}{2(r+d)} p_{i+1,j}\,; \\
p_{i,0} &= p_{0,j} = 1.
\end{aligned} \tag{7.4.2}$$

To prove this proposition, we will use a very important result in stochastic analysis.

Theorem 7.4.2 (Dynkin's Criterion) *Let $(X_t)_t, t \geq 0)$ be a Markov jump process with values in a countable space E, with infinitesimal generator Q. Let $\mathcal{F}_t = \sigma(X_s, s \leq t)$. For any continuous bounded function on E such that Qf is bounded,*

$$f(X_t) - f(X_0) - \int_0^t Qf(X_u)\,\mathrm{d}u$$

is a continuous $\mathcal{F}_t$-martingale to the right and limited to the left.

Proof of Theorem 7.4.2 We obtained in Chapter 5 the Kolmogorov equations, in particular the backward Kolmogorov formula (5.5.3): for all $x \in E$,

$$P_t f(x) - f(x) = \int_0^t P_u Q f(x)\,\mathrm{d}u.$$

As the functions f and Qf are bounded, we deduce from Fubini's Theorem that for all $x \in E$,

$$\mathbb{E}_x\Big(f(X_t) - f(X_0) - \int_0^t Qf(X_u)\,\mathrm{d}u\Big) = 0. \tag{7.4.3}$$

Let's show that $M_t = f(X_t) - f(X_0) - \int_0^t Qf(X_u)\,\mathrm{d}u$ defines a martingale. For all $t > 0$, M_t is integrable because f and Qf are bounded. Let $t, s > 0$. Using the Markov property, we can write

$$\mathbb{E}_x(f(X_{t+s}|\mathcal{F}_t) = E_{X_t}(f(X_s));$$

$$\mathbb{E}_x\Big(\int_t^{t+s} Qf(X_u)\,\mathrm{d}u|\mathcal{F}_t\Big) = \mathbb{E}_x\Big(\int_0^s Qf(X_{t+v})\,\mathrm{d}v|\mathcal{F}_t\Big) = \mathbb{E}_{X_t}\Big(\int_0^s Qf(X_v)\,\mathrm{d}v\Big).$$

Finally, we have

$$\mathbb{E}_x(M_{t+s} - M_t|\mathcal{F}_t) = E_{X_t}\Big(f(X_s) - f(X_0) - \int_0^s Qf(X_v)\,\mathrm{d}v\Big) = 0,$$

by (7.4.3). We deduce that $(M_t, t \geq 0)$ is an $\mathcal{F}_t$-martingale. □

Remark 7.4.3 This theorem has already been proven as a consequence of the Itô formula for the solution $(X_t, t \geq 0)$ of a stochastic differential equation and f of class C_b^2.

Proof of Proposition 7.4.1 Let's first note that the constant sequence equal to 1 is a solution of (7.4.2). We can easily adapt (5.5.3) and (5.5.4) and show that the sequence $(p_{i,j})_{(i,j)\in\mathbb{N}^2}$ is also a solution of (7.4.2). Let's show that it is in fact the smallest positive bounded solution of the latter.

Let f be another positive and bounded solution. It is easy to show that this is equivalent to $Qf(i,j) = 0$ for all (i,j), where Q was defined in (7.4.1) and Qf in (5.3.2). Theorem 7.4.2 then implies that $(M_t)_t = (f(N_t^{Aa}, N_t^{aa}))_t$ is a martingale, as is $(M_{t\wedge\tau_0})_t$ by Proposition 4.5.16. Moreover, $M_{t\wedge\tau_0}$ converges on $\tau_0 < +\infty$ to

$f(N^{Aa}_{\tau_0}, N^{aa}_{\tau_0}) = 1$ under the boundary conditions of (7.4.2). Using the positivity of f and Fatou's Lemma, we deduce that for all $(i, j) \in \mathbb{N}^2$,

$$\begin{aligned} f(i,j) &= \mathbb{E}_{(i,j)}(M_0) = \lim_{t\to+\infty} \mathbb{E}_{(i,j)}(M_{t\wedge\tau_0}) \\ &\geq \mathbb{E}_{(i,j)}(\liminf_{t\to+\infty} M_{t\wedge\tau_0} \mathbf{1}_{\tau_0<+\infty}) = \mathbb{E}_{(i,j)}(\mathbf{1}_{\tau_0<+\infty}) = p_{i,j}. \end{aligned}$$ □

This solution is unfortunately impossible to compute with classical tools. In [9, Prop. 9], upper and lower bounds for these probabilities are given, using sophisticated comparisons of stochastic processes.

Proposition 7.4.4

- *If $r \leq d$, then $p_{ij} = 1$ for all i, j.*
- *If $r > d$, then*

$$\left(\frac{d}{r}\right)^{i+j} \leq p_{ij} \leq \left(\frac{d}{r}\right)^{i} + \left(\frac{d}{r}\right)^{j} - \left(\frac{d}{r}\right)^{i+j}.$$

Proof We will provide elements of the proof. The basic idea is to compare the process $((N^{Aa}_t, N^{aa}_t), t \geq 0)$ to processes that are easier to study. First, we will compare the process $((N^{Aa}_t, N^{aa}_t), t \geq 0)$ to the process $((\widetilde{N}^{Aa}_t, \widetilde{N}^{aa}_t), t \geq 0)$ defined on the same probability space, resulting from the same initial condition (i, j) and whose generator is given by the same system as (7.4.1), but where the value of the reproduction rates extends to the case where i or j can be zero. As the death rates are identical, it is clear that $((\widetilde{N}^{Aa}_t, \widetilde{N}^{aa}_t), t \geq 0)$ stochastically dominates $((N^{Aa}_t, N^{aa}_t), t \geq 0)$, in the sense that each of the coordinates dominates the corresponding coordinate. The process $((\widetilde{N}^{Aa}_t, \widetilde{N}^{aa}_t), t \geq 0)$ is a bi-type branching process whose dynamics can be described as follows. Each individual lives for an exponential time of parameter $r+d$. At its death, there are no descendants with probability $d/(r+d)$. There are two descendants of the same type as the ancestor with probability $r/2(r+d)$ and two descendants carrying the two different types with probability $r/2(r+d)$. In particular, the process $(\widetilde{N}_t)_t = (\widetilde{N}^{Aa}_t + \widetilde{N}^{aa}_t)_t$ is a binary branching process with birth rate r and death rate d. The extinction probability of this process was studied in Section 5.4.5. In the case where $r \leq d$, the process $(\widetilde{N}_t)_t$ almost surely goes extinct and the same is therefore true for the process $((N^{Aa}_t, N^{aa}_t), t \geq 0)$. In the case where $r > d$, the process $(\widetilde{N}_t)_t$ goes extinct with probability $\left(\frac{d}{r}\right)^{i+j}$, because (i, j) is the initial condition of $(N^{Aa}_t, N^{aa}_t)_t$ and therefore of $(\widetilde{N}^{Aa}_t, \widetilde{N}^{aa}_t)_t$. The domination property then leads to

$$\left(\frac{d}{r}\right)^{i+j} \leq p_{i,j}.$$

The upper bound of $p_{i,j}$ is more subtle and we refer the reader to [9] for the proof.□

7.5 Modeling a Diploid Population

This study is part of research on the random modeling of the dynamics of diploid populations and is based on the articles by Coron [27], Collet–Méléard–Metz [25] and Coron [28].

In the previous section, as in Chapters 5 and 6, we assumed that populations have haploid reproduction, in the sense that we did not take into account the segregation rules (called *Mendel's laws*) of diploid reproduction. In this section, we will propose and study a model of a multi-type birth and death process describing the dynamics of a diploid population with competition and Mendelian reproduction.

7.5.1 A Multi-Type Birth and Death Process

We consider a population of hermaphroditic diploid organisms, characterized by their genotype composed of two DNA strands. We assume here that individuals differ only by a gene that carries two alleles A and a. Thus, the three possible genotypes are AA, Aa and aa. The population dynamics is then represented by the Markov process

$$Z : t \longrightarrow Z_t = (k_1(t), k_2(t), k_3(t)),$$

which gives respectively the number of individuals of genotype AA, Aa and aa at time t. For simplicity, we will denote these types by $1, 2$ and 3.

The proportion of allele A is then given by the formula

$$p_A(t) = \frac{2k_1(t) + k_2(t)}{2(k_1(t) + k_2(t) + k_3(t))}.$$

Each individual reproduces at rate $b > 0$, which does not depend on the genotype of the individual, and chooses a reproduction partner uniformly at random among the other individuals in the population.

The process $(Z_t, t \geq 0)$ is a birth and death process whose individual birth rates of type AA, resp. Aa, aa, are given, by virtue of Mendelian laws and for $k = (k_1, k_2, k_3)$ and $N = k_1 + k_2 + k_3$, by

$$\begin{aligned} b_1(k) &= b\left(k_1 + \frac{1}{2}k_2\right) p_A = \frac{b}{N}\left(k_1 + \frac{1}{2}k_2\right)^2 ; \\ b_2(k) &= b\left(k_1 + \frac{1}{2}k_2\right)(1 - p_A) + b\left(k_3 + \frac{1}{2}k_2\right) p_A = 2\frac{b}{N}\left(k_1 + \frac{1}{2}k_2\right)\left(k_3 + \frac{1}{2}k_2\right); \\ b_3(k) &= b\left(k_3 + \frac{1}{2}k_2\right)(1 - p_A) = \frac{b}{N}\left(k_3 + \frac{1}{2}k_2\right)^2 . \end{aligned}$$

Indeed, let us detail for example the case of $b_1(k)$. The total reproduction rate is bN. As each individual chooses a partner uniformly at random, the probability of obtaining the type AA in reproduction will be given by

$$\frac{k_1^2}{N^2} + \frac{1}{2}\left(\frac{2k_1k_2}{N^2}\right) + \frac{1}{4}\frac{k_2^2}{N^2},$$

depending on whether the parents have the types AA, AA or AA, Aa or Aa, Aa. Note that

$$b_1(k) + b_2(k) + b_3(k) = b\,(k_1 + k_2 + k_3). \tag{7.5.1}$$

The death rates of individuals of genotype AA, Aa and aa are given by

$$\begin{aligned} d_1(k) &= k_1\big(d_1 + (c_{11}k_1 + c_{12}k_2 + c_{13}k_3)\big); \\ d_2(k) &= k_2\big(d_2 + (c_{21}k_1 + c_{22}k_2 + c_{23}k_3)\big); \\ d_1(k) &= k_3\big(d_3 + (c_{31}k_1 + c_{32}k_2 + c_{33}k_3)\big), \end{aligned}$$

where d_1, d_2, d_3 are the intrinsic death rates of the different types and the parameters c_{ij}, for $i, j \in \{1,2,3\}$, are the competition parameters between individuals of types i and j. We assume that

$$\underline{c} = \inf_{i,j\in\{1,2,3\}} c_{ij} > 0.$$

We then have that

$$d_1(k) + d_2(k) + d_3(k) \geq \underline{c}\,(k_1 + k_2 + k_3)^2. \tag{7.5.2}$$

Extinction

Thanks to (7.5.1) and (7.5.2), we see that the process

$$(N_t = k_1(t) + k_2(t) + k_3(t), t \geq 0),$$

giving the size of the population at time t, is stochastically dominated by a birth and death process $(Y_t, t \geq 0)$ with birth rate by and death rate $\underline{c}\,y^2$, if $Y_t = y$. This domination is expressed by the fact that we can construct the two processes on the same probability space and then, for all $t \geq 0$, N_t is almost surely less than Y_t. The process N will therefore almost surely extinguish like Y, by virtue of Theorem 5.5.5 applied to the logistic birth and death process.

Fixation of one of the alleles in the neutral case

Suppose that the parameters d_i are equal to d and that the parameters c_{ij} are all equal to c. We want to avoid the extinction of the process in order to observe the eventual fixation of one of the alleles. To do this we will assume that if there is only one individual left in the population, it cannot die, by imposing that

$$d_i(k) = 0 \text{ if } k_1 + k_2 + k_3 = 1. \tag{7.5.3}$$

The process N then takes its values in $\mathbb{N}^*$.

Proposition 7.5.1 *Let us place ourselves in the neutral case and under the additional assumption* (7.5.3). *Then, if the process originates from* (k_1, k_2, k_3)*, the probability of fixation of* a *is*

$$u(k_1, k_2, k_3) = \frac{k_2 + 2k_3}{2(k_1 + k_2 + k_3)},$$

which is the initial proportion of alleles a.

Proof Let's denote by $(\mathcal{F}_t, t \geq 0)$ the filtration of the process Z. We introduce the stopping time

$$T_a = \inf\{t \geq 0, Z_t \in \{0\} \times \{0\} \times \mathbb{N}^*\},$$

which is the time of fixation of the allele a. Let $p_a(k)$ be the proportion of allele a in the population $k = (k_1, k_2, k_3)$:

$$p_a(k_1, k_2, k_3) = \frac{k_2 + 2k_3}{2(k_1 + k_2 + k_3)}.$$

We will consider the embedded Markov chain associated to the process Z, namely the Markov chain $(\mathcal{Z}_\ell, \ell \in \mathbb{N})$, with jump $\mathcal{Z}_\ell - \mathcal{Z}_{\ell-1}$ equal to the jump amplitude of the ℓ-th jump time of Z. The sequence $p_a(\mathcal{Z}_\ell)$ therefore represents the sequence of successive proportions of a in the population. It is clear that 0 and 1 are absorbing points for this process. We will prove that if T_ℓ is the ℓ-th jump time of Z and if $\widetilde{\mathcal{F}}_\ell = \mathcal{F}_{T_\ell}$, then $p_a(\mathcal{Z}_\ell)$ is a bounded $\widetilde{\mathcal{F}}_\ell$-martingale. Indeed, if $\mathcal{N}_\ell$ denotes the size of $\mathcal{Z}_\ell$, and for $\mathcal{N}_\ell \geq 2$,

$$\begin{aligned}
&\mathbb{E}(p_a(\mathcal{Z}_{\ell+1}) \mid \widetilde{\mathcal{F}}_\ell) \\
&\quad = \frac{2\mathcal{N}_\ell p_a(\mathcal{Z}_\ell) - 2}{2\mathcal{N}_\ell - 2} \mathbb{P}(\text{death of } aa) + \frac{2\mathcal{N}_\ell p_a(\mathcal{Z}_\ell) - 1}{2\mathcal{N}_\ell - 2} \mathbb{P}(\text{death of } Aa) \\
&\quad + \frac{2\mathcal{N}_\ell p_a(\mathcal{Z}_\ell)}{2\mathcal{N}_\ell - 2} \mathbb{P}(\text{death of } AA) + \frac{2\mathcal{N}_\ell p_a(\mathcal{Z}_\ell) + 2}{2\mathcal{N}_\ell + 2} \mathbb{P}(\text{birth of } aa) \\
&\quad + \frac{2\mathcal{N}_\ell p_a(\mathcal{Z}_\ell) + 1}{2\mathcal{N}_\ell + 2} \mathbb{P}(\text{birth of } Aa) + \frac{2\mathcal{N}_\ell p_a(\mathcal{Z}_\ell)}{2\mathcal{N}_\ell + 2} \mathbb{P}(\text{birth of } AA) \\
&\quad = p_a(\mathcal{Z}_\ell).
\end{aligned}$$

(More precise computations are left to the reader.)

A similar result is true if $\mathcal{N}_\ell = 1$. We can then apply the stopping theorem to the bounded martingale $p_a(\mathcal{Z}_.)$ at the stopping time $T = T_a \wedge T_A$. We deduce that

$$\mathbb{E}_{(k_1,k_2,k_3)}(p_a(\mathcal{Z}_T)) = \mathbb{E}_{(k_1,k_2,k_3)}(p_a(\mathcal{Z}_0)) = \frac{k_2+2k_3}{2(k_1+k_2+k_3)}.$$

Furthermore,

$$\mathbb{E}_{(k_1,k_2,k_3)}(p_a(\mathcal{Z}_T)) = \mathbb{E}_{(k_1,k_2,k_3)}(T_a < T_A) + \mathbb{E}_{(k_1,k_2,k_3)}(p_a(\mathcal{Z}_T)\mathbb{1}_{T=\infty}).$$

Since the martingale $(p_a(\mathcal{Z}_\ell))_\ell$ is non-negative and bounded, it converges almost surely as ℓ tends to infinity. Reasoning as in the proof of Theorem 6.2.6, we can prove that 0 and 1 are the unique possible limits and then $T < +\infty$ almost surely. We deduce the value of the fixation probability. □

7.5.2 Large Population Approximations

Deterministic approximation

As in Chapter 5, we are interested in large population approximation of the three-dimensional process Z. Introducing a resource parameter K, we now assume (as in Chapter 5) that the competition parameters are of the form c_{ij}/K. We denote by Z^K the process defined as above with these new competition parameters and we assume that Z_0^K/K converges to a deterministic vector $(z_1(0), z_2(0), z_3(0))$ as K tends to infinity. Then by adapting the proof of Theorem 5.6.1, we can prove that

Proposition 7.5.2 *The process* $\left(\frac{Z_t^K}{K}, t \geq 0\right)$ *converges in law, as K tends to infinity, to the solution* $\left((z_1(t), z_2(t), z_3(t)), t \geq 0\right)$ *of*

$$\frac{\mathrm{d}}{\mathrm{d}t}\begin{pmatrix} z_1(t) \\ z_2(t) \\ z_3(t) \end{pmatrix} = \psi\big(z_1(t), z_2(t), z_3(t)\big), \tag{7.5.4}$$

where

$$\psi(z_1,z_2,z_3) = \begin{pmatrix} \tilde{b}_1(z_1,z_2,z_3) - \tilde{d}_1(z_1,z_2,z_3) \\ \tilde{b}_2(z_1,z_2,z_3) - \tilde{d}_2(z_1,z_2,z_3) \\ \tilde{b}_3(z_1,z_2,z_3) - \tilde{d}_3(z_1,z_2,z_3) \end{pmatrix},$$

with for $z = (z_1, z_2, z_3)$,

$$\tilde{b}_1(z) = b\,\frac{(z_1 + \frac{1}{2}z_2)^2}{z_1+z_2+z_3}\;;\; b_2(z) = 2b\,\frac{(z_1 + \frac{1}{2}z_2)(z_3 + \frac{1}{2}z_2)}{z_1+z_2+z_3}\;;\; b_3(z) = b\,\frac{(z_3 + \frac{1}{2}z_2)^2}{z_1+z_2+z_3},$$

$$\tilde{d}_1(z) = (d_1 + c_{11}\,z_1 + c_{12}\,z_2 + c_{13}\,z_3)\,z_1,$$

and a similar expression for the other terms.

It can be shown that the vector field ψ has two fixed points $(z_1^*, 0, 0)$ and $(0, 0, z_3^*)$, where

$$z_1^* = \frac{b - d_1}{c_{11}} \quad \text{and} \quad z_3^* = \frac{b - d_3}{c_{33}}.$$

These points correspond to the homozygous populations.

Suppose that for all $i \in \{1,2,3\}$, $d_i = d$ and for all i, j, $c_{ij} = c$. A solution of the system is not easily obtained, even in this simple case. To solve the system, we introduce a change of variables ϕ related to the quantities of interest: the population size, the proportion of allele A and a distance to Hardy–Weinberg equilibrium (distance to independence). Let for all $z \in (\mathbb{R}_+)^3 \backslash \{(0,0,0\}$, $\phi(z) = (\phi_1(z), \phi_2(z), \phi_3(z))$ with

$$\phi_1(z) = z_1 + z_2 + z_3; \qquad \phi_2(z) = \frac{2z_1 + z_2}{2(z_1 + z_2 + z_3)}; \qquad \phi_3(z) = \frac{4z_1 z_3 - (z_2)^2}{4(z_1 + z_2 + z_3)}.$$

We see that

$$\phi_3(z) = \phi_1(z)(p^{AA}(z) - (p^A)^2(z)) = \phi_1(z)(p^{aa}(z) - (p^a)^2(z)).$$

Let $\phi(z_1(t), z_2(t), z_3(t)) = (\mathcal{N}_t, \mathcal{X}_t, \mathcal{Y}_t)$ and make the change of variables in (7.5.4). We obtain

$$\frac{\mathrm{d}}{\mathrm{d}t} \begin{pmatrix} \mathcal{N}_t \\ \mathcal{X}_t \\ \mathcal{Y}_t \end{pmatrix} = \begin{pmatrix} (b - d - c\mathcal{N}_t)\mathcal{N}_t \\ 0 \\ -(d + c\mathcal{N}_t)\mathcal{Y}_t \end{pmatrix}.$$

We can explicitly solve this system by noticing that the first equation is the logistic equation that we have already encountered in Chapter 5. The proof of the following proposition is elementary and left to the reader.

Proposition 7.5.3 *Let $t \geq 0$.*

(i) *If $b = d$, then $\mathcal{N}_t = \dfrac{\mathcal{N}_0}{c\,\mathcal{N}_0 t + 1}$. Otherwise,*

$$\mathcal{N}_t = \frac{(b-d)\mathcal{N}_0 \mathrm{e}^{(b-d)t}}{(b-d) + c\mathcal{N}_0(\mathrm{e}^{(b-d)t} - 1)}.$$

(ii) $\mathcal{X}_t = \mathcal{X}_0$.

(iii) *If $c = 0$, then $\mathcal{Y}_t = \mathcal{Y}_0 \mathrm{e}^{-dt}$.*

If $c \neq 0$ and $b = d$, then $\mathcal{Y}_t = \dfrac{\mathcal{Y}_0}{c\mathcal{N}_0 t + 1}\mathrm{e}^{-dt}$.

If $c \neq 0$, $b \neq d$ and $\mathcal{N}_0 = \dfrac{b-d}{c}$, then $\mathcal{N}_t = \dfrac{b-d}{c}$ and $\mathcal{Y}_t = \mathcal{Y}_0 \mathrm{e}^{-bt}$.

If $c \neq 0$, $b \neq d$ and $\mathcal{N}_0 \neq \dfrac{b-d}{c}$, then by setting $C = \dfrac{\mathcal{Y}_0}{1 - \frac{c\mathcal{N}_0}{b-d}}$,

$$\mathcal{Y}_t = C\mathrm{e}^{-dt}\left(1 - \frac{c\mathcal{N}_0 \mathrm{e}^{(b-d)t}}{(b-d) + c\mathcal{N}_0(\mathrm{e}^{(b-d)t} - 1)}\right).$$

We note that the population does not go extinct in finite time and that the proportion of alleles A remains constant. Moreover, $\mathcal{Y}_t$ tends to 0 as t tends to infinity, which results in a convergence of the population over a long period of time towards the Hardy–Weinberg equilibrium. Therefore, there is conservation of biodiversity but no possibility of evolution in the Darwinian sense since none of the alleles will disappear. This is due to the fact that the population is neutral (the demographic coefficients are identical regardless of the genotype of the individuals) and that we are under a large population assumption: selection does not operate.

Slow-fast diffusive dynamics

This case generalizes Section 5.6.2 to the diploid case. We now assume an affine dependence of the demographic parameters on the scaling parameter K and highlight a limit dynamics of slow-fast stochastic diffusion, different from the deterministic dynamics obtained in Proposition 7.5.2. The difference between the limiting behaviors obtained in Proposition 7.5.2 and Proposition 7.5.4 shows the importance of the choice of scales between the coefficients depending on the biological problem studied.

We study the neutral model where

$$b = b_K = \gamma K + \beta; \qquad d = d_K = \gamma K + \delta$$

and the competition parameters are all equal to $\frac{c}{K}$.

We also assume that Z_0^K/K converges in law towards a random vector Z_0. Let us consider again the variable change ϕ introduced above and let us set

$$\phi\left(\frac{Z_t^K}{K}\right) = (N_t^K, X_t^K, Y_t^K).$$

We present here the results obtained in Coron [28].

Proposition 7.5.4

(1) *The process $(Y_t^K, t \geq 0)$ is a fast variable that converges to 0 as K tends to infinity, in the sense that for all $t, s > 0$, $\sup_{t \leq u \leq t+s} \mathbb{E}((Y_u^K)^2) \to_{K\to\infty} 0$.*

(2) *The pair $(N_t^K, X_t^K)_{t\geq 0}$ converges in law towards the solution (N, X) of the stochastic differential system*

$$\begin{aligned} \mathrm{d}N_t &= \sqrt{2\gamma N_t}\,\mathrm{d}B_t^1 + N_t\,(b - d - cN_t)\,\mathrm{d}t; \\ \mathrm{d}X_t &= \sqrt{\frac{\gamma X_t(1-X_t)}{N_t}}\,\mathrm{d}B_t^2, \end{aligned}$$

where B^1 and B^2 are independent Brownian motions.

We observe that the process $(N_t, t \geq 0)$ tends to 0 as t tends to infinity, as proved in Section 4.7.2. The system of equations is well defined until the stopping time $\inf\{t \geq 0, N_t = \varepsilon\}$, for all $\varepsilon > 0$.

Remark 7.5.5 The equation satisfied by the process X is a generalization of the Wright–Fisher diffusion equation obtained in Chapter 6, in the case where the population size is variable and the population is diploid. A haploid bi-type model, with two alleles A and a but without segregation at birth, would lead to the same type of equations for the size and proportion of allele A, but with γ changed to 2γ in the equation for X. This simple term can change certain behaviors, such as the quasi-stationary behavior of the process (see [28]).

Proof The proof of item (2) is based on a tightness-identification argument and uniqueness of the limit, as explained in Chapter 5. Here, we will only compute the generator of the process $(N_t^K, X_t^K, Y_t^K)_{t\geq 0}$, which will allow us to understand the form of the limit. Let f be a function from $\mathbb{R}^3$ to $\mathbb{R}$ and L^K the generator of the process $(\frac{Z_t^K}{K})$. As before, let $\phi(z) = (n, x, y)$ and let e_1, e_2, e_3 be the unit vectors. The generator decomposes as follows (noting for example that $y = z_1 - nx^2$):

$$\begin{aligned}
L^K((z) = {} & \gamma K^2 y \left(f(z-\frac{e_1}{K}) - 2f(z-\frac{e_2}{K}) + f(z-\frac{e_3}{K})\right) \\
& + \gamma K^2 n x^2 \left(f(z+\frac{e_1}{K}) + f(z-\frac{e_1}{K}) - 2f(z)\right) \\
& + \gamma K^2 2nx(1-x) \left(f(z+\frac{e_2}{K}) + f(z-\frac{e_2}{K}) - 2f(z)\right) \\
& + \gamma K^2 n(1-x)^2 \left(f(z+\frac{e_3}{K}) + f(z-\frac{e_3}{K}) - 2f(z)\right) \\
& + \beta K n x^2 \left(f(z+\frac{e_1}{K}) - f(z)\right) \\
& + \beta K 2nx(1-x) \left(f(z+\frac{e_2}{K}) - f(z)\right) \\
& + \beta K n(1-x)^2 \left(f(z+\frac{e_3}{K}) - f(z)\right) \\
& + K \sum_{i\in\{1,2,3\}} z_i(\delta + cn) \left(f(z-\frac{e_i}{K}) - f(z)\right).
\end{aligned}$$

By applying the generator to $f = (\phi_3)^2$, we can show that there exists a constant $C > 0$ such that

$$\frac{\mathrm{d}\mathbb{E}((Y_t^K)^2)}{\mathrm{d}t} \leq -2\gamma K \mathbb{E}((Y_t^K)^2) + C,$$

which leads to item (1).

We can then apply the generator to $f = \phi_1$ and let K tend to infinity to obtain the diffusive limit behavior given by the generator

$$Lg(n) = \gamma n g''(n) + n(\beta - \delta - cn) g'(n),$$

for any real function g of a real variable n. We can similarly study the joint generator of $(n,x) = (\phi_1,\phi_2)(z)$ and deduce the stochastic differential equation that drives X.□

7.6 Genealogical Trees of Sexual Populations

This section is based on Sainudiin, Thatte, and Véber [73].

We are interested in the genetic diversity of a population of size N. As we saw in Chapter 6, it is an important issue for biologists to study the future of this genetic diversity, the fixation of an allele, or the maintenance of a polymorphism with several coexisting alleles. The genetic heritage of an individual is inherited from his ancestors. Thus, the genetic diversity of the population is linked to the genealogical relationships between individuals: the closer they are (in the genealogical sense), the less time there will be for mutations to appear and differentiate the individuals. Comparing genealogical trees will thus provide indications of genetic inheritance.

Let us first recall that in Chapter 6, we introduced the simple Wright–Fisher model, and studied its genealogy. Let τ_2 be the minimum number of generations to go back to find the first common ancestor of two individuals chosen at random in a population of size N.

Proposition 7.6.1 *The random variable τ_2 has a geometric distribution with parameter $\frac{1}{N}$.*

Proof Using the definition of the Wright–Fisher model, we immediately have that

$$\mathbb{P}(\tau_2 = 1) = \sum_{i=1}^{N} \frac{1}{N^2} = \frac{1}{N}.$$

Furthermore, by reasoning identical to that of the proof of Proposition 6.2.8 developed in Remark 6.2.9, for all $k \in \mathbb{N}^*$,

$$\mathbb{P}(\tau_2 = k) = \left(1 - \frac{1}{N}\right)^{k-1} \frac{1}{N}.$$ □

In particular, $\mathbb{E}(\tau_2) = N$. As the population is *a priori* large, two individuals need to go very far back in time to find a common ancestor. We have also seen in Proposition 6.2.10 that the average time for an allele to become fixed is also of the order of N generations.

7.6.1 Diploid Wright–Fisher Model With Recombination

The Wright–Fisher model is widely used by biologists, but it assumes that the portion of DNA considered comes from the genes of a single parent. However, in sexually reproducing populations, the process of recombination that occurs during reproduction allows an individual to inherit a mosaic of parental genes. To understand the origin of an individual's genetic information, we must reconstruct the genealogy of their two parents if there was recombination, or only the single parent from whom they inherited their genes in the absence of recombination. We will therefore define a Wright–Fisher process with recombination.

We assume here that each individual has two parents from whom they inherit their genetic heritage. The population is of fixed size N and the probability of recombination is fixed, equal to $r \in [0, 1]$. In each generation, an individual's genetic heritage, independent of others,

- with probability r, is inherited from two parents chosen uniformly at random from the previous generation.
- with probability $1-r$, is inherited from a single parent chosen uniformly at random from the previous generation.

Thus, although each individual physically has two parents, for the portion of the genome we are interested in, only one of their parents can transmit all of their genes. To simplify the terminology, we will then say that this individual has only one parent (genetically). Note that if $r = 1$, the physical and genetic parentage coincide and in this case, the model gives us the complete pedigree (the genealogical tree) of the chosen individuals. We can then construct parental/genealogical relationships for $r < 1$ by removing links when there is no recombination. The genealogical tree for $r < 1$ is therefore naturally included in a tree with $r' = 1$.

Also note that when N is large, choosing a pair of individuals at random (the two parents) is equivalent to drawing two individuals with replacement (as the hypergeometric law is in this case very close to a binomial law).

Consider a fixed individual $i \in \{1, \dots, N\}$, belonging to generation 0. We wish to know the law of the process that describes the dynamics of the size of the population descended from this individual over time.

We define G^i_n, the number of descendants of individual i at generation n.

Proposition 7.6.2 *Let $i \in \{1, \dots, N\}$. The process $(G^i_n)_n$ is a Markov chain such that, conditionally on G^i_n, the random variable G^i_{n+1} has a binomial distribution*

$$\mathcal{B}(N, (1+r)\frac{G^i_n}{N} - r(\frac{G^i_n}{N})^2).$$

Furthermore, if B^i_n denotes the number of individuals which are not descendants of i, then the process $(B^i_n)_n$ is a Markov chain such that, conditionally on B^i_n, the random variable B^i_{n+1} has a binomial distribution

$$\mathcal{B}(N, (1-r)\frac{B^i_n}{N} + r(\frac{B^i_n}{N})^2).$$

Proof Let's fix an individual from the generation $n+1$ and compute the probability that it comes from individual i. Let's denote by $\mathcal{R}$ the event "There was a recombination to produce this individual, from generation n to generation $n+1$". The probability is

$$\begin{aligned}&(1-r)\,\mathbb{P}(\text{parent comes from } i\,|\,\mathcal{R}^C)\\&\quad+r\,\mathbb{P}(\text{at least one of the two parents comes from } i\,|\,\mathcal{R})\\&\quad=(1-r)\frac{G_n^i}{N}+r\Big(1-\Big(1-\frac{G_n^i}{N}\Big)^2\Big)\\&\quad=(1+r)\frac{G_n^i}{N}-r\Big(\frac{G_n^i}{N}\Big)^2.\end{aligned}$$

The second assertion immediately follows from the first. □

From now on, we fix an individual $i\in\{1,\ldots,N\}$ and consider only the offspring of this individual. Let us denote by G_n the size of the population starting from i at time n. We have $G_0=1$.

If G_n is small compared to N, the binomial law obtained in Proposition 7.6.2 is close (as N tends to infinity) to a Poisson distribution of parameter $(1+r)G_n$. In other words, despite the dependence between the sizes of different families, the initial development of (G_n) is very close to a Galton–Watson process with reproduction law $\mathcal{P}(1+r)$. In the same way, as N tends to infinity and B_n is small compared to N, the process $(B_n, n\in\mathbb{N})$ behaves like a Galton–Watson process with reproduction law $\mathcal{P}(1-r)$.

The following proposition shows more precisely that the probabilities of reaching points by G_n are approximately close to those of a Galton–Watson process with reproduction law $\mathcal{P}(1+r)$. (A similar result occurs for B_n.)

Lemma 7.6.3 *(Adaptation of Chang [21, Lemma 3].) Let $(Y_n)_n$ be a Galton–Watson process with reproduction law a Poisson distribution of parameter $1+r$. For all $b\geq 0$, let us define*

$$\tau_b^Y=\inf\{n, Y_n\geq b\};\qquad \tau_{0,b}^Y=\inf\{n, Y_n\geq b \text{ or } Y_t=0\},$$

and similarly for the process $(G_n)_n$.

Then, if $m\geq 0$ and b grow with N such that $mb^2=o(N)$, as $N\to\infty$,

$$\mathbb{P}(\tau_b^G>m)=\mathbb{P}(\tau_b^Y>m)(1+o(1));\qquad \mathbb{P}(\tau_{0,b}^G>m)=\mathbb{P}(\tau_{0,b}^Y>m)(1+o(1)).$$

Proof Let $x>0$ and $y\geq 0$ or $x=y=0$. Let's introduce the likelihood ratio

$$\begin{aligned}L(y\,|\,x)&=\frac{\mathbb{P}(G_{n+1}=y\,|\,G_n=x)}{\mathbb{P}(Y_{n+1}=y\,|\,Y_n=x)}\\&=\frac{\mathbb{P}(\mathcal{B}(N,(1+r)\frac{x}{N}-r(\frac{x}{N})^2)=y)}{\mathbb{P}(\mathcal{P}((1+r)x)=y)}\\&\leq \mathrm{e}^{(1+r)x}\Big(1-(1+r)\frac{x}{N}+r(\frac{x}{N})^2\Big)^{N-y}.\end{aligned}$$

For $x < b$, $y < b$ and N sufficiently large,

$$\log L(y \mid x) \le \frac{2(1+r)b^2}{N},$$

and by a similar computation, we also obtain

$$\log L(y \mid x) \ge -\frac{2(1+r)b^2}{N}.$$

Let then $x_1, \ldots, x_m$ be positive and less than b. We deduce that

$$\begin{aligned}\mathbb{P}(G_1 = x_1, \ldots, G_m = x_m) &= \mathbb{P}(G_1 = x_1 \mid G_0 = 1) \cdots \mathbb{P}(G_m = x_m \mid G_{m-1} = x_{m-1}) \\ &= \mathbb{P}(Y_1 = x_1, \ldots, Y_m = x_m) L(x_1 \mid 1) \cdots L(x_m \mid x_{m-1}) \\ &\le \mathbb{P}(Y_1 = x_1, \ldots, Y_m = x_m)\, \mathrm{e}^{2(1+r)mb^2/N},\end{aligned}$$

and

$$\mathbb{P}(G_1 = x_1, \ldots, G_m = x_m) \ge \mathbb{P}(Y_1 = x_1, \ldots, Y_m = x_m)\, \mathrm{e}^{-2(1+r)mb^2/N}.$$

Thus,

$$\begin{aligned}\mathbb{P}(\tau_b^G > m) &= \sum_{0 \le x_1 < b} \cdots \sum_{0 \le x_m < b} \mathbb{P}(G_1 = x_1, \ldots, G_m = x_m) \\ &\le \sum_{0 \le x_1 < b} \cdots \sum_{0 \le x_m < b} \mathbb{P}(Y_1 = x_1, \ldots, Y_m = x_m)\, \mathrm{e}^{2(1+r)mb^2/N} \\ &\le \mathbb{P}(\tau_b^Y > m)\, \mathrm{e}^{2(1+r)mb^2/N}, \end{aligned} \tag{7.6.1}$$

and similarly,

$$\mathbb{P}(\tau_b^G > m) \ge \mathbb{P}(\tau_b^Y > m)\, \mathrm{e}^{-2(1+r)mb^2/N}.$$

The hypothesis $m\, b^2 = o(N)$ then implies that

$$\mathbb{P}(\tau_b^G > m) = \mathbb{P}(\tau_b^Y > m)(1 + o(1)).$$

The proof of the second assertion follows exactly the same reasoning with the summations of (7.6.1) taken over $0 < x_n < b$ instead of $0 \le x_n < b$. □

We have seen that if G_n is small compared to N, the process $(G_n, n \in \mathbb{N})$ behaves, as N tends to infinity, like a Galton–Watson process with reproduction law the Poisson distribution $\mathcal{P}(1+r)$. On the other hand, if G_n is of the order of N, we know by the Law of Large Numbers that G_{n+1} is close to its expectation. In the same way, as N tends to infinity and if B_n is small compared to N, the process $(B_n, n \in \mathbb{N})$ behaves like a Galton–Watson process with the Poisson reproduction law $\mathcal{P}(1-r)$, whereas if B_n is of the order of N, the process is close to its expectation. We even have precise inequalities that allow us to control this approximation, known as Bernstein's inequalities (recalled below).

Lemma 7.6.4 *(See [73, Lemma 7].) Let X be a random variable of binomial law* $\mathcal{B}(N,p)$*, where* $p \in (0,1)$*, and* $x > 0$*. Then*

$$\mathbb{P}(X \geq Np+x) \leq \exp\left(\frac{-x^2}{2Np(1-p)+(2/3)x}\right);$$
$$\mathbb{P}(X \leq Np-x) \leq \exp\left(\frac{-x^2}{2Np(1-p)+(2/3)x}\right).$$

7.6.2 Number of Generations to Reach the Common Ancestor

With these preliminaries established, we will be able, as for the haploid Wright–Fisher process, to determine the order of magnitude of the number of generations it takes to find at least one common ancestor for the entire current population, of size N.

Let us denote by τ_N this number of generations.

Theorem 7.6.5 *Suppose that* $r \in]0,1[$*. For all* $\varepsilon > 0$*,*

$$\lim_{N\to+\infty} \mathbb{P}((1-\varepsilon)C(r)\log N \leq \tau_N \leq (1+\varepsilon)C(r)\log N) = 1,$$

where

$$C(r) = \frac{1}{\log(1+r)} - \frac{1}{\log(1-r)}.$$

It follows that the sequence $\left(\frac{\tau_N}{C(r)\log N}\right)_N$ *converges in probability to 1, as N tends to infinity.*

Note that the genealogical tree of the population is much shorter here than in the haploid Wright–Fisher model ($r = 0$). Recombination, however small r may be, allows for much faster genetic mixing. The case where $r = 1$ was treated in [21], where the author shows that the sequence $\left(\frac{\log 2}{\log N}\tau_N\right)_N$ converges in probability to 1 as N tends to infinity.

Proof We will give a broad outline of the proof.

As the reproduction process starts at generation 0, we consider its dynamics in the forward direction of time. By Proposition 7.6.2, we know the law of the size of each subpopulation resulting from one of the individuals of the initial generation.

The proof of Theorem 7.6.5 then unfolds in several steps, described below.

We take $\varepsilon \in (0,1/2)$.

1) With a probability that tends to 1 as N tends to infinity, there is an individual i in generation 0 for which the offspring $(G^i_n, n \in \mathbb{N})$ reaches $(\log N)^2$ after $\tau_N^{(G1)}$ generations, where $\tau_N^{(G1)} = O\big((\log\log N)^2\big)$ with high probability.

2) The number of descendants of i grows from $(\log N)^2$ to $g_2 N$ in $\tau_N^{(G2)}$ generations, where $g_2 \in (0,1/2)$ is a well-chosen constant, and $\tau_N^{(G2)}$ is of the order of $\frac{\log N}{\log(1+r)}$. More precisely, the probabilities

$$\mathbb{P}\Big(\tau_N^{(G2)} > \Big(1+\frac{\varepsilon}{2}\Big)\frac{\log N}{\log(1+r)}\Big) \quad \text{and} \quad \mathbb{P}\Big(\tau_N^{(G2)} < \Big(1-\frac{\varepsilon}{2}\Big)\frac{\log N}{\log(1+r)}\Big)$$

are of the order of $o(\frac{1}{N})$.

3) The number of descendants of i grows from $g_2 N$ to $\frac{N}{2}$ in $\tau_N^{(G3)}$ generations, where $\tau_N^{(G3)} \leq \log\log N$ with probability $1-o(\frac{1}{N})$.

4) The number of non-descendants of i decreases from at most $\frac{N}{2}$ to at most $b_1 N$ in $\tau_N^{(B1)}$ generations, where $b_1 \in (0,1/2)$ is a well-chosen constant, and $\tau_N^{(B1)} \leq \log\log N$ with probability $1-o(\frac{1}{N})$.

5) The number of non-descendants of i decreases from at most $b_1 N$ to $(\log N)^2$ in $\tau_N^{(B2)}$ generations, where $\tau_N^{(B2)}$ is approximately $-\log N/\log(1-r)$ generations. More precisely, the probabilities

$$\mathbb{P}\Big(\tau_N^{(B2)} > \Big(1+\frac{\varepsilon}{2}\Big)\frac{-\log N}{\log(1-r)}\Big) \quad \text{and} \quad \mathbb{P}\Big(\tau_N^{(B2)} < \Big(1-\frac{\varepsilon}{2}\Big)\frac{-\log N}{\log(1-r)}\Big)$$

are of the order of $o(\frac{1}{N})$.

6) The number of non-descendants of i decreases from at most $(\log N)^2$ to 0 in $\tau_N^{(B3)}$ generations, where $\tau_N^{(B3)} = O\big((\log\log N)^2\big)$ with high probability.

All these results combined show that with probability tending to 1, the first time at which an individual becomes the common ancestor of the entire population is bounded above by

$$O\big((\log\log N)^2\big) + \Big(1+\frac{\varepsilon}{2}\Big)\log N\Big(\frac{1}{\log(1+r)} - \frac{1}{\log(1-r)}\Big),$$

which leads to the upper bound given in Theorem 7.6.5.

The lower bound is the one that requires that the probabilities corresponding to the phases in $O(\log N)$ are of the order of $o(\frac{1}{N})$. Indeed, this ensures that with a probability tending to 1, none of the N families born from the initial individuals can reach $g_2 N$ in less than $\Big(1-\frac{\varepsilon}{2}\Big)\frac{\log N}{\log(1+r)}$ generations, and that no family reaching $N/2$ can reach $N-(\log N)^2$ in less than $-\Big(1-\frac{\varepsilon}{2}\Big)\frac{\log N}{\log(1-r)}$ generations. Thus, with a probability tending to 1, an individual needs at least

$$\Big(1-\frac{\varepsilon}{2}\Big)\log N\Big(\frac{1}{\log(1+r)} - \frac{1}{\log(1-r)}\Big)$$

generations to become a common ancestor of the entire population.

We will not detail all the steps of this proof, which mainly use the two previous lemmas and largely rely on ad hoc results for Galton–Watson processes whose reproduction law is a Poisson distribution.

Let us give the proof of step 1. We apply Lemma 7.6.3 with

$$b_N = (\log N)^2 \quad \text{and} \quad m_N = \frac{3}{\log(1+r)} \log\log N.$$

So the probability that the process $(G_n)_n$ starting from 1 reaches $(\log N)^2$ in less than m_N generations is asymptotically equivalent to that of a Galton–Watson process $(Y_n)_n$ with reproduction law $\mathcal{P}(1+r)$. Since the expectation of the reproduction law is $1+r$, the process $(Y_n)_n$ is supercritical, and by Theorem 3.2.10, we know that the process $(\frac{Y_n}{(1+r)^n})_n$ is a positive martingale M_∞ and almost surely converges to a strictly positive random variable on the persistence set of the process. Moreover, if we denote by ρ the extinction probability of this process, then

$$\mathbb{P}(M_\infty = 0) = \rho < 1.$$

We have

$$\mathbb{P}(\tau^Y_{b_N} > m_N) \le \mathbb{P}(Y_{m_N} \le b_N) = \mathbb{P}(M_{m_N} \le b_N\,(1+r)^{-m_N}).$$

Using Fatou's lemma and the fact that by hypothesis, $b_N\,(1+r)^{-m_N}$ tends to 0 as $N \to \infty$, we obtain

$$\begin{aligned}\limsup_N \mathbb{P}(\tau^Y_{b_N} > m_N) &\le \limsup_N \mathbb{P}(M_{m_N} \le b_N\,(1+r)^{-m_N})\\ &\le \mathbb{P}(\limsup_N M_{m_N} \le b_N\,(1+r)^{-m_N})\\ &= \mathbb{P}(M_{m_N} \le b_N\,(1+r)^{-m_N} \text{ infinitely often})\\ &\le \mathbb{P}(M_\infty = 0) = \rho < 1.\end{aligned}$$

By Lemma 7.6.3,

$$\mathbb{P}(\tau^G_{b_N} > m_N) = \mathbb{P}(\tau^Y_{b_N} > m_N)(1+o(1)) \qquad \text{as } N \to \infty.$$

We deduce that

$$\limsup_N \mathbb{P}(\tau^G_{b_N} > m_N) \le \rho < 1,$$

and therefore

$$\liminf_N \mathbb{P}(\tau^G_{b_N} \le m_N) \ge 1-\rho > 0.$$

Thus, if every m_N generations, we test whether the individual labeled 1 has at least $(\log N)^2$ descendants m_N generations later, we obtain a geometric-type sequence with a success probability in a finite number of steps equal to 1. In particular, this implies that the probability of having no individual from generation 0 with at least

$(\log N)^2$ descendants after $(\log\log N)\, m_N = o(\log N)$ generations is bounded by $(1-\rho+\delta)^{\log\log N}$, as $N \to \infty$, for any $\delta \in (0,\rho)$. This gives

$$\lim_{N\to\infty} \mathbb{P}\Big(\tau_N^{(G1)} > \frac{3}{\log(1+r)}(\log\log N)^2\Big) = 0.$$

This reasoning can also be adapted to prove step 6. □

Appendix A
Measures, Integration and Probability Measures

A.1 σ-Fields and Measures

For more details, the reader can, for example, consult the books of Billingsley [10] or Rudin [72].

The starting point of Lebesgue's integration theory is to observe that every bounded function is a uniform limit of a sequence of functions *taking only a finite number of values*. These latter functions are of the form

$$\sum_{i=1}^{k} \lambda_i \, \mathbf{1}_{\{x \in [a,b] : f(x) = \lambda_i\}}.$$

From this point of view, linearity means knowing how to calculate, first and foremost, the integral of a characteristic function of a part A of $[a, b]$; this integral can be seen as a measure of A. This point of view can easily be generalized to arbitrary sets, much more general than $\mathbb{R}$ or $\mathbb{R}^d$, and will be particularly well suited to probability theory.

A.1.1 σ-Fields and Measurable Space

To define a coherent concept of measuring sets, we use the notion of a σ-field.

Definition A.1.1 A *σ-field* (or *σ-algebra*) on a set Ω is a set $\mathcal{A}$ of subsets of Ω satisfying the following axioms:

- The empty set $\emptyset$ belongs to $\mathcal{A}$.
- The union of countably many sets in $\mathcal{A}$ belongs to $\mathcal{A}$.
- The complement of any set in $\mathcal{A}$ (with respect to Ω) belongs to $\mathcal{A}$.

The pair $(\Omega, \mathcal{A})$ is called a *measurable space*.

S. Méléard, *Random Models in Biology, Ecology and Evolution*, Texts in Applied Mathematics 87, https://doi.org/10.1007/978-3-662-73483-4

In short: a σ-field of subsets of Ω is a set of subsets of Ω containing the empty set and Ω itself, and stable under *countable* union and complementation.

In a context where the σ-field $\mathcal{A}$ on a set Ω is fixed, a set is said to be *measurable* if it belongs to $\mathcal{A}$.

Elementary properties

It follows from the axioms that a σ-field is stable under two important set operations, difference and symmetric difference. Recall that if A and B are subsets of Ω, the *difference* of A and B is

$$A \setminus B = \{x \in \Omega : x \in A \text{ but } x \notin B\},$$

and their *symmetric difference* is

$$A \,\Delta\, B = (A \cup B) \setminus (A \cap B) = (A \setminus B) \cup (B \setminus A).$$

A σ-field is also stable under more subtle set-theoretic operations: the limit superior and limit inferior of a sequence of subsets. If $\{A_n\}_{n\geq 0}$ is a sequence of subsets of Ω, the *limit superior*

$$\overline{\lim}_{n\to\infty} A_n = \cap_k \cup_{n\geq k} A_n$$

of $\{A_n\}_{n\geq 0}$ is the set of points that belong to A_n for an infinite number of indices n, and the *limit inferior*

$$\underline{\lim}_{n\to\infty} A_n = \cup_k \cap_{n\geq k} A_n$$

of $\{A_n\}_{n\geq 0}$ is the set of points that belong to all A_n from a certain rank onwards.

Example The pair $\{\varnothing; X\}$ and the set $\mathcal{P}(\Omega)$ of all subsets of Ω are trivial σ-fields on Ω.

σ-fields generated by a set

There is an order relation for σ-fields on a fixed set Ω: we say that a σ-field $\mathcal{A}$ contains a σ-field $\mathcal{B}$ if every subset of Ω that is in $\mathcal{B}$ is also in $\mathcal{A}$. For any set $\mathcal{E}$ of subsets of Ω, we can then define the smallest σ-field on Ω containing $\mathcal{E}$, which by definition is called the *σ-field generated by $\mathcal{E}$*.

Definition A.1.2 The σ-field generated by $\mathcal{E}$ is the intersection of all σ-fields containing $\mathcal{E}$.

Indeed, if we consider a family $(\mathcal{A}_i)_{i\in I}$ of σ-fields on Ω, we can show that $\bigcap_{i\in I} \mathcal{A}_i$ is a σ-field of Ω. Since the empty set is in all σ-fields $\mathcal{A}_i$, it is in $\bigcap_{i\in I} \mathcal{A}_i$; if $A \in \bigcap_{i\in I} \mathcal{A}_i$, then by stability through passing to the complement of the σ-fields, we have $\Omega \setminus A \in \mathcal{A}_i$ for all $i \in I$, i.e. $\Omega \setminus A \in \bigcap_{i \in I} \mathcal{A}_i$, and the stability of $\bigcap_{i \in I} \mathcal{A}_i$ under countable union can be seen in a similar way.

Although this definition is very simple, it is not constructive. In general, a generated σ-field contains many parts, and one should not try to represent it in its entirety.

Note that there is always at least one σ-field containing $\mathcal{E}$, namely $\mathcal{P}(\Omega)$.

Borel sets of a metric space

Let us now consider the special case where the set Ω is a metric space equipped with a distance d.

Definition A.1.3 Let (Ω, d) be a metric space. The *Borel σ-field* of Ω is the σ-field generated by the topology of Ω, that is, by the collection of open sets in Ω. The Borel σ-field is denoted by $\mathcal{B}(\Omega, d)$, or $\mathcal{B}(\Omega)$ if there is no risk of confusion.

A *Borel set* of Ω is an element of the Borel σ-field of Ω.

By definition, the open sets of Ω are Borel sets, and by the stability of a σ-field under complementation, the closed sets of Ω are also Borel sets.

Recall that in $\mathbb{R}^d$, every open set is a *countable union of open rectangles*. In particular, it is equivalent to define the Borel σ-field of $\mathbb{R}$, denoted by $\mathcal{B}(\mathbb{R})$ (respectively of $\mathbb{R}^d$, denoted by $\mathcal{B}(\mathbb{R}^d)$), as the σ-field generated by the open intervals of $\mathbb{R}$ (respectively, the open rectangles of $\mathbb{R}^d$).

Remark A.1.4 It can be proven that there are many more subsets of $\mathbb{R}$ than Borel sets, but paradoxically, the question of exhibiting a non-Borel subset of $\mathbb{R}$ is delicate.

Inverse image σ-fields

We want to know how the structure of a σ-field behaves with respect to mappings. We introduce the notion of an inverse image σ-field.

Proposition A.1.5 *Let $f : \Omega_1 \to \Omega_2$ be a mapping from a set Ω_1 to a set Ω_2.*

- *The inverse image, under f, of any σ-field of Ω_2 is a σ-field of Ω_1.*
- *If $\mathcal{A}_1$ is a σ-field of Ω_1, then the set $\mathcal{A}_2$ of subsets of Ω_2 whose inverse image is in $\mathcal{A}_1$ is a σ-field of Ω_2.*

Proof We prove the second point, the proof being similar for the first one. The inverse image, under f, of the empty subset of Ω_2 is the empty subset of Ω_1, which is in $\mathcal{A}_1$; thus $\varnothing \in \mathcal{A}_2$. Let $A \in \mathcal{A}_2$; then the complement $\Omega_2 \setminus A$ satisfies $f^{-1}(\Omega_2 \setminus A) = \Omega_1 \setminus f^{-1}(A)$. Now $f^{-1}(A) \in \mathcal{A}_1$ by definition of $\mathcal{A}_2$, and therefore $\Omega_1 \setminus f^{-1}(A) \in \mathcal{A}_1$ by stability of a σ-field by passage to the complement; thus $\Omega_2 \setminus A \in \mathcal{A}_2$. The stability of $\mathcal{A}_2$ under countable union comes from the fact that the inverse image of a union is the union of the inverse images. □

A.1.2 Positive Measures on σ-Fields

The notion of measure and basic properties

Definition A.1.6 Let $(\Omega, \mathcal{A})$ be a measurable space. A *positive measure* on $(\Omega, \mathcal{A})$ is a function $\mu : \mathcal{A} \to \overline{\mathbb{R}}_+$ satisfying the following axioms:

- Measure of the empty set: $\mu(\emptyset) = 0$.
- Countable additivity: if $\{A_n\}_{n \geq 0}$ is a sequence (finite or countable) of elements of $\mathcal{A}$ that are pairwise disjoint, then:

$$\mu\Big(\bigcup_{n \geq 0} A_n\Big) = \sum_{n \geq 0} \mu(A_n).$$

The quantity $\mu(\Omega)$ is called the *total mass*. The measure μ is said to be *finite* (or *bounded*) if $\mu(\Omega) < +\infty$. The measure μ is called a *probability measure* if $\mu(\Omega) = 1$ (see Section A.2).

If there exists a sequence $(A_n)_n$ of elements of $\mathcal{A}$ such that $\cup_n A_n = \Omega$, and if for all $n \in \mathbb{N}$, $\mu(A_n)$ is finite, then we say that μ is a *σ-finite measure*.

Definition A.1.7 A *measured space* is any triplet $(\Omega, \mathcal{A}, \mu)$ where $(\Omega, \mathcal{A})$ is a measurable space and μ is a positive measure on $(\Omega, \mathcal{A})$.

Definition A.1.8 For a measured space $(\Omega, \mathcal{A}, \mu)$, a subset of Ω is said to be *negligible* (or *μ-negligible*) if it is contained in a measurable subset of measure zero.

Definition A.1.9 A property is true *μ-almost everywhere* (a.e.) if it is true outside a μ-negligible set.

Examples of measures

We give a few examples of measures, which illustrate the wide variety of possibilities.

- The *coarse measure* on a σ-field $\mathcal{A}$ is the measure that is $+\infty$ on all sets in $\mathcal{A}$ other than $\emptyset$.
- The *zero measure* assigns the measure 0 to all sets.
- The *Dirac measure* at a point $a \in X$ is the measure that is equal to 1 on all $A \in \mathcal{A}$ such that $a \in A$, and 0 otherwise.
- The *counting measure* on Ω is the measure – defined on the entire $\mathcal{P}(\Omega)$ – which is equal to the cardinality of $A \subset X$ if A is finite, and $+\infty$ otherwise.

A fundamental example that leads to the correct generalization of the Riemann integral is the Lebesgue measure.

A.1.3 Definition of the Lebesgue Measure

Definition A.1.10 There exists a unique measure λ_d on the Borel σ-field of $\mathbb{R}^d$ such that the measure of any rectangle $A = [a_1, b_1] \times \cdots \times [a_d, b_d]$ is $\lambda_d(A) = \prod_{i=1}^{d} (b_i - a_i)$. This measure is called the *Lebesgue measure* of $\mathbb{R}^d$.

The existence and uniqueness of this measure are not at all obvious.

The Lebesgue measure is the only measure defined on the Borel σ-field of $\mathbb{R}^d$ that possesses the following properties:

- It is invariant under any translation of $\mathbb{R}^d$.
- It assigns the measure 1 to the unit cube.

Some practical properties:

- For any Borel set A of $\mathbb{R}^d$, we have $\lambda_d(A) = \inf_U \lambda_d(U)$, where U runs through the set of open sets containing A.
- For any Borel set A of $\mathbb{R}^d$, we have $\lambda_d(A) = \sup_K \lambda_d(K)$, where K runs through the set of compact sets contained in A.
- For any point $x \in \mathbb{R}^d$, $\lambda_d(\{x\}) = 0$.
- For real numbers $a < b$, we have $\lambda_1([a,b]) = \lambda_1(]a,b]) = \lambda_1(]a,b[) = \lambda_1([a,b[) = b - a$.

Properties of measures

Let us state some properties of measures that follow directly from the axioms.

In the following statements, $\mu : \mathcal{A} \to \overline{\mathbb{R}}_+$ denotes a positive measure on a measurable space $(\Omega, \mathcal{A})$.

Lemma A.1.11 (Growth of measures) *A measure $\mu : \mathcal{A} \to \overline{\mathbb{R}}_+$ is an increasing function for the order relation given by inclusion on the sets of the σ-field $\mathcal{A}$; in other words, if A and B in $\mathcal{A}$ are such that $A \subset B$, then $\mu(A) \leq \mu(B)$.*

Proof We write A as the disjoint union of A and $B \setminus A$ (noting that $B \setminus A$ is still in $\mathcal{A}$) and apply the additivity property to the finite sequence $\{A; B \setminus A\}$ of elements of $\mathcal{A}$. □

Note that the calculation conventions in $\overline{\mathbb{R}}_+$ allow us to write that

$$\mu(B \setminus A) = \mu(B) - \mu(A) \quad \text{if} \quad \mu(A) < +\infty,$$

but that we cannot conclude anything of the sort if A is of infinite μ-measure.

Lemma A.1.12 (Strong additivity of measures) *If A and B are any measurable sets, then we have*

$$\mu(A \cup B) + \mu(A \cap B) = \mu(A) + \mu(B).$$

Proof We write $A \cup B$ as the disjoint union of the three (measurable) sets $A \setminus B$, $B \setminus A$, and $A \cap B$. Applying the axiom of additivity of measures, we obtain:

$$\mu(A \cup B) = \mu(A \setminus B) + \mu(B \setminus A) + \mu(A \cap B).$$

We still need to add the term $\mu(A \cap B)$ to both sides of the equation and use, as in the previous proof, that $\mu(A \setminus B) + \mu(A \cap B) = \mu(A)$ and $\mu(B \setminus A) + \mu(A \cap B) = \mu(B)$. □

In measure theory, the letter σ is often used as a reference to countability, as in the following result. The difference with the second axiom of measures is that we do not assume that the A_n are pairwise disjoint, but the result is an inequality.

Lemma A.1.13 (σ-subadditivity of measures) *For any (at most) countable sequence of measurable sets $(A_n)_{n\geq 0}$ we have*

$$\mu\Big(\bigcup_{n\geq 0} A_n\Big) \leq \sum_{n\geq 0} \mu(A_n).$$

Proof The result is clear if one of the terms A_n has infinite μ-measure, which we now exclude. We then perform a natural manipulation to produce, from $(A_n)_{n\geq 0}$, a sequence $(B_n)_{n\geq 0}$ satisfying the disjointness condition of the second axiom. Specifically, we set $B_1 = A_1$ and $B_n = A_n \setminus \bigcup_{k=1}^{n-1} A_k$ for $n \geq 2$. This gives:

$$\mu\Big(\bigcup_{n\geq 0} A_n\Big) = \mu\Big(\bigcup_{n\geq 0} B_n\Big) = \sum_{n\geq 0} \mu(B_n) \leq \sum_{n\geq 0} \mu(A_n).$$

The lemma is proven. □

Lemma A.1.14 (Increasing sequences of measurable sets) *For any increasing sequence of measurable sets $(A_n)_{n\geq 0}$ and limit (i.e., union) A, we have*

$$\mu(A) = \lim_{n\to\infty} \mu(A_n) \leq +\infty.$$

Proof This is the same manipulation. We set $B_1 = A_1$ and $B_n = A_n \setminus A_{n-1}$ for $n \geq 2$. The sequence $(B_n)_{n\geq 0}$ consists of measurable sets that are pairwise disjoint, so we have

$$\lim_{n\to\infty} \mu(A_n) = \sum_{m=1}^{\infty} \mu(B_m) = \mu\Big(\bigcup_{m=1}^{\infty} B_m\Big) = \mu(A).$$ □

We can deduce the following result by passing to the complement.

Lemma A.1.15 (Decreasing sequences of measurable sets) *If $(A_n)_{n\geq 0}$ is a decreasing sequence of measurable sets with limit (i.e., intersection) A, and if one of its terms has finite μ-measure, then we have*

$$\mu(A) = \lim_{n\to\infty} \mu(A_n) < +\infty.$$

Proof By hypothesis, for an index p, we have $\mu(A_p) < +\infty$. The previous lemma for the increasing sequence $(A_p \setminus A_n)_{n \geq p}$ then provides

$$\mu(A_p) - \mu(A) = \mu(A_p \setminus A) = \lim_{n \to \infty} \mu(A_p \setminus A_n) = \mu(A_p) - \lim_{n \to \infty} \mu(A_n),$$

and the result follows since $\mu(A_p) < +\infty$. □

Remark A.1.16 The finiteness condition is essential. To see this, take μ to be the counting measure on $\mathbb{N}$ and the sequence given by $A_n = \{\text{integers} \geq n\}$; in this case, $A = \bigcap_n A_n = \varnothing$, and therefore $\mu(A) = 0$, while $\mu(A_n) = +\infty$ for all $n \geq 0$.

A.1.4 Measurable Functions

The definition of a measurable function is fundamental to the construction of the Lebesgue integral.

Definition A.1.17 If $(\Omega_1, \mathcal{A}_1)$ and $(\Omega_2, \mathcal{A}_2)$ are measurable spaces, we say that a function $f : \Omega_1 \to \Omega_2$ is *measurable* if for all $A \in \mathcal{A}_2$ we have $f^{-1}(A) \in \mathcal{A}_1$.

The sets $\mathbb{R}$ and $\mathbb{R}^d$ will always be equipped with the Borel σ-field, and the reference measure is the Lebesgue measure. Here are some examples or statements of stability for the measurability of real functions of real variables, which can easily be generalized.

- If $f, g : (\Omega_1, \mathcal{A}_1) \to \mathbb{R}$ are measurable and if $a, b \in \mathbb{R}$, then

  $$af + bg, \quad fg, \quad \max(f,g) \quad \text{and} \quad \min(\mathrm{f,g})$$

  are measurable functions. In particular,

  $$f^+ := \max(f,0), \quad f^- := -\min(f,0) \quad \text{and} \quad |f|$$

  are measurable.
- If $f_1, \ldots, f_d$ are measurable functions defined on an open set U of $\mathbb{R}$, and if $\Phi : \mathbb{R}^d \to \mathbb{R}$ is continuous, then

  $$\Phi(f_1, \ldots, f_d)(x) := \Phi(f_1(x), \ldots, f_d(x))$$

  is measurable.
- If $f(x) \neq 0$ a.e. on an open set U of $\mathbb{R}$, then $x \mapsto 1/f(x)$ is measurable.
- Any continuous function on an open set U of $\mathbb{R}$ is measurable.
- Any piecewise continuous function on an interval $I \subset \mathbb{R}$ is measurable.
- Let $(f_n)_{n \geq 0}$ be a sequence of measurable functions, defined on an open set U of $\mathbb{R}$, and f be such that $f_n \to f$ a.e. on U. Then, f is measurable.

Remark A.1.18 There are functions from $\mathbb{R}$ to $\mathbb{R}$ that are not measurable (their construction is always based on the axiom of choice).

A.2 Probability Measures and Expectation

Probability theory considers analogous concepts but using a specific vocabulary.

The set Ω is called the *fundamental space* or *state space*. An element $\omega \in \Omega$ is called a *possible outcome* of an experiment. The set Ω is equipped with a σ-field $\mathcal{A}$. The elements of the σ-field $\mathcal{A}$ are called *events*.

A.2.1 Probability Measures

We equip the measurable set $(\Omega, \mathcal{A})$ with a measure of total mass 1, as defined below.

Definition A.2.1 Let $(\Omega, \mathcal{A})$ be a measurable space. A *probability measure* on $(\Omega, \mathcal{A})$ is a mapping $\mathbb{P} : \mathcal{A} \to [0,1]$ satisfying the following axioms:

- $\mathbb{P}(\Omega) = 1$;
- If $(A_n)_{n\geq 0}$ is a sequence (finite or countable) of elements of $\mathcal{A}$ that are pairwise disjoint, then

$$\mathbb{P}\Big(\bigcup_{n\geq 0} A_n\Big) = \sum_{n\geq 0} \mathbb{P}(A_n).$$

We call the triplet $(\Omega, \mathcal{A}, \mathbb{P})$ a *probability space*.

A.2.2 Random Variables

Definition A.2.2 A *real random variable* is a measurable mapping from the measurable space $(\Omega, \mathcal{A})$ to $(\mathbb{R}, \mathcal{B}(\mathbb{R}))$.

Random variables are generally denoted by a capital letter (for example, X or Y). The values taken by the random variable, called *realizations* of the random variable, are denoted by lowercase letters. For example, we write $X(\omega) = x$ for $\omega \in \Omega$ and x a realization of X and the subset of Ω defined by $\{\omega, X(\omega) = x\}$ is usually denoted by $\{X = x\}$.

The σ-field generated by a random variable

Let X be a random variable with values in any measurable space $(E, \mathcal{E})$. The *σ-field generated by* X, denoted by $\sigma(X)$, is by definition the smallest σ-field on Ω that makes X measurable:

$$\sigma(X) = \{A = X^{-1}(B) : B \in \mathcal{E}\}.$$

This definition can be generalized to any family $(X_i)_{i\in I}$ of random variables, where X_i has values in $(E_i, \mathcal{E}_i)$. In this case,

$$\sigma(X_i, i \in I) = \{X^{-1}(B_i) : B_i \in \mathcal{E}_i, i \in I\}.$$

Proposition A.2.3 *Let X be a random variable with values in a space $(E, \mathcal{E})$, and let Y be a real random variable. Then the following are equivalent:*

(i) *Y is $\sigma(X)$-measurable;*
(ii) *there exists a measurable function f from $(E, \mathcal{E})$ to $(\mathbb{R}, \mathcal{B}(\mathbb{R}))$ such that $Y = f(X)$.*

A.2.3 Expectation of Random Variables

Simple random variables

In this subsection, we work with a probability space $(\Omega, \mathcal{A}, \mathbb{P})$; in other words, Ω is a set, $\mathcal{A}$ is a σ-field of subsets of Ω, and $\mathbb{P} : \mathcal{A} \to [0,1]$ is a measure of mass 1 on the subsets of Ω belonging to the σ-field $\mathcal{A}$.

Definition A.2.4 A real random variable defined on the measurable space $(\Omega, \mathcal{A})$ is said to be *simple* if it takes only a finite number of values. We denote by $\mathcal{S}^+$ the set of positive simple random variables.

A bounded random variable is written as $\sum_{i\in I} \alpha_i \mathbf{1}_{A_i}$ with I finite and $A_i \in \mathcal{A}$ for all $i \in I$; this decomposition is not unique.

We will use these random variables to define the expectation of random variables. Simple variables are random variables to which an expectation can easily be assigned, and the general case will be obtained by extending the construction by passing to the limit.

Definition A.2.5 Let $X \in \mathcal{S}^+$. The *expectation of X with respect to the probability* $\mathbb{P}$ is the element $\mathbb{E}(X) \in \overline{\mathbb{R}}_+$ defined as the following finite sum:

$$\mathbb{E}(X) = \sum_{\alpha \text{ value of } X} \alpha\, \mathbb{P}(\{X = \alpha\}) \leq +\infty,$$

where $\{X = \alpha\}$ denotes the measurable set $X^{-1}(\{\alpha\})$. We also denote this quantity by $\int X \, \mathrm{d}\mathbb{P}$.

We say that X is *integrable* if $\mathbb{E}(X) < +\infty$.

For an expression of $X \in \mathcal{S}^+$ in the form $X = \sum_{i\in I} \alpha_i \mathbf{1}_{A_i}$ with I finite and the $A_i \in \mathcal{A}$ pairwise disjoint, we have: $\mathbb{E}(X) = \sum_{i\in I} \alpha_i \mathbb{P}(A_i)$.

The expectation of positive simple random variables has the following properties.

1. *Growth*: If $X, Y \in \mathcal{S}^+$ are such that $X \leq Y$, then

$$\mathbb{E}(X) \leq \mathbb{E}(Y).$$

2. *Additivity*: for $X, Y \in \mathcal{S}^+$, we have

$$\mathbb{E}(X+Y) = \mathbb{E}(X) + \mathbb{E}(Y).$$

3. *Homogeneity*: for $X \in \mathcal{S}^+$ and $a \geq 0$, we have

$$\mathbb{E}(aX) = a\,\mathbb{E}(X).$$

Proof Points 1 and 3 are easy. For 2, we write $X = \sum_{i \in I} \alpha_i \mathbf{1}_{A_i}$ with I finite and the $A_i \in \mathcal{A}$ partitioning Ω; similarly for $Y = \sum_{j \in J} \beta_j \mathbf{1}_{B_j}$. This gives

$$X+Y = \sum_{i,j} (\alpha_i + \beta_j) \mathbf{1}_{A_i \cap B_j},$$

which is similar to the previous expressions. Then:

$$\mathbb{E}(X+Y) = \sum_{i,j} (\alpha_i + \beta_j) \mathbb{P}(A_i \cap B_j) = \sum_i \alpha_i \sum_j \mathbb{P}(A_i \cap B_j) + \sum_j \beta_j \sum_i \mathbb{P}(A_i \cap B_j).$$

The first term is equal to $\sum_i \alpha_i \mathbb{P}(A_i) = \mathbb{E}(X)$ since the B_j partition Ω, and similarly, the second term is equal to $\mathbb{E}(Y) = \sum_j \beta_j \mu(B_j)$ since the A_i partition Ω. □

Approximation by simple random variables

Let us now state the approximation result that allows us to extend the construction of the expectation by passing to the limit.

Theorem A.2.6 *Random variables with values in $\mathbb{R}$ (or $\overline{\mathbb{R}_+}$) are random variables that are everywhere limits of sequences of bounded random variables. Furthermore:*

(i) *every bounded random variable is the uniform limit of a sequence of simple random variables;*
(ii) *every positive random variable is the limit of an increasing sequence of positive simple random variables.*

Proof (abridged) All limits in the statement are indeed random variables, as simple limits of sequences of random variables. Let us now consider the converse. The first point is immediate.

(i) Let X be a random variable. If there exists an M such that $0 \leq X(\omega) \leq M$ for all $\omega \in \Omega$, then for each integer $n \geq 0$ and each $k = 0, 1, \ldots, 2^n - 1$, we define $E_k^n = \{2^{-n} kM \leq X < 2^{-n}(k+1)M\}$. The sequence of simple random variables $X_n = \sum_k \frac{kM}{2^n} \mathbf{1}_{E_k^n}$ is increasing and converges uniformly to the random variable X.

(ii) If X is positive with possibly infinite values, for each integer $n \geq 0$ we introduce $Y_n = \inf\{n; X\}$ to obtain a positive bounded random variable. From the above, we know how to find a bounded simple variable Z_n satisfying $Y_n - \frac{1}{n} \leq Z_n \leq Y_n$, which provides a sequence that converges to X. To obtain an increasing sequence, we finally set $X_n = \max\{Z_1, Z_2, \ldots, Z_n\}$. □

We denote by $\mathcal{V}^+$ the set of positive random variables (with values in $\overline{\mathbb{R}}_+$). The random variables in $\mathcal{V}^+$ are the limits of increasing sequences of positive simple random variables.

Definition A.2.7 Let $X \in \mathcal{V}^+$. The *expectation* of X with respect to the probability $\mathbb{P}$, denoted by $\mathbb{E}(X)$ or $\int X\,\mathbb{P}$, is the upper bound, finite or $+\infty$, of the set of integrals of positive simple random variables bounded above by X. We say that X is *integrable* if $\mathbb{E}(X) < +\infty$.

In the case where the random variable is discrete, this coincides with the previous definition. The definition in terms of the upper bound allows us to deduce the growth of this expectation from that of discrete random variables.

First limit theorems

If $X \in \mathcal{S}^+$ and if $A \in \mathcal{A}$, then the product $X\mathbf{1}_A$ is still a simple random variable and we have $\mathbb{E}(X\mathbf{1}_A) = \int_A X\,\mathrm{d}\mathbb{P}$.

If A and B are *disjoint* measurable sets, the additivity of expectation ensures that

$$\mathbb{E}(X\mathbf{1}_{A\cup B}) = \mathbb{E}(X\mathbf{1}_A) + \mathbb{E}(X\mathbf{1}_B).$$

Lemma A.2.8 (Monotone limit of simple random variables) *Let $X = \sum_{i\in I} \alpha_i \mathbf{1}_{A_i}$ be a positive simple random variable and let $\{E_n\}_{n\geq 0}$ be an increasing sequence of measurable sets whose limit (i.e., union) is the entire set Ω. Then we have*

$$\mathbb{E}(X) = \lim_{n\to\infty} \mathbb{E}(X\mathbf{1}_{E_n}).$$

Proof We return to the case where the A_i are pairwise disjoint; then:

$$\mathbb{E}(X\mathbf{1}_{E_n}) = \sum_{i\in I} \alpha_i\,\mathbb{P}(A_i \cap E_n).$$

For each $i \in I$, the increasing sequence of measurable sets $\{E_n \cap A_i\}_{n\geq 0}$ converges to A_i (i.e., its union is equal to A_i), so

$$\mathbb{P}(A_i) = \lim_{n\to\infty} \mathbb{P}(E_n \cap A_i).$$

This ultimately allows us to write:

$$\begin{aligned}\mathbb{E}(X) &= \sum_{i\in I} \alpha_i\,\mathbb{P}(A_i)\\ &= \lim_{n\to\infty} \sum_{i\in I} \alpha_i\,\mathbb{P}(E_n\cap A_i)\\ &= \lim_{n\to\infty} \mathbb{E}(X\,\mathbf{1}_{E_n}),\end{aligned}$$

the interchange between limit and sum being justified since I is finite. □

We deduce Beppo Levi's increasing convergence theorem.

Theorem A.2.9 (Increasing convergence) *Let $(X_n)_{n\geq 0}$ be an increasing sequence of random variables in $\mathcal{V}^+$ converging almost surely to X. Then we have:*

$$\mathbb{E}(X) = \lim_{n\to\infty} \mathbb{E}(X_n) \leq +\infty.$$

Proof The growth of expectation and the inequalities $X_n \leq X$ for all $n \geq 0$ give $\mathbb{E}(X_n) \leq \mathbb{E}(X)$, and therefore $\lim_{n\to\infty} \mathbb{E}(X_n) \leq \mathbb{E}(X)$ by passing to the limit.

For the reverse inequality, by definition of the integral over $\mathcal{V}^+$, it suffices to see that for all $Y \in \mathcal{S}^+$ such that $Y \leq X$, we have:

$$\mathbb{E}(Y) \leq \lim_{n\to\infty} \mathbb{E}(X_n).$$

We then conclude by passing to the upper bound of $\mathbb{E}(Y)$ for $Y \in \mathcal{S}^+$ such that $Y \leq X$.

To prove this last inequality, let $a \in]0,1[$ and for each $n \geq 0$ set $E_n = \{X_n \geq aY\}$. By definition of E_n, we thus have $X_n \geq (aY)\,\mathbf{1}_{E_n}$ for all $n \geq 0$. This implies

$$\mathbb{E}(X_n) \geq \mathbb{E}(\mathbf{1}_{E_n}\,aY) = a\,\mathbb{E}(\mathbf{1}_{E_n}\,Y).$$

Furthermore, the sequence of events $(E_n)_{n\geq 0}$ is increasing by inclusion since the sequence of random variables $(X_n)_{n\geq 0}$ is increasing, and by the hypothesis of pointwise convergence (and since $a < 1$), the E_n tend towards the entire Ω. Thus, the previous lemma implies that

$$\lim_{n\to\infty} \mathbb{E}(\mathbf{1}_{E_n}\,Y) = \mathbb{E}(Y),$$

which allows us to move to the limit in the previous inequality to obtain

$$\lim_{n\to\infty} \mathbb{E}(X_n) \geq a\,\mathbb{E}(Y),$$

and finally the desired inequality, since a is arbitrary in $]0,1[$. □

Properties of the expectation

1. *Growth*: if $X, Y \in \mathcal{V}^+$ are such that $X \leq Y$, then

$$\mathbb{E}(X) \leq \mathbb{E}(Y).$$

2. *Additivity*: for $X, Y \in \mathcal{V}^+$, we have

$$\mathbb{E}(X+Y) = \mathbb{E}(X) + \mathbb{E}(Y).$$

3. *Homogeneity*: for $X \in \mathcal{V}^+$ et $a \geq 0$, we have

$$\mathbb{E}(aX) = a\,\mathbb{E}(X).$$

Proof The first point was seen immediately after the definition of expectation on $\mathcal{V}^+$, and the third is immediate. For the second, we know that there exist increasing sequences of random variables in $\mathcal{S}^+$, say $(X_n)_{n\geq 0}$ and $(Y_n)_{n\geq 0}$, which converge to X and Y respectively (approximation theorem).

Then $(X_n + Y_n)_{n\geq 0}$ is an increasing sequence in $\mathcal{S}^+$ that converges to $X+Y$, and we have

$$\mathbb{E}(X+Y) = \lim_{n\to\infty} \mathbb{E}(X_n + Y_n) = \lim_{n\to\infty} \mathbb{E}(X_n) + \lim_{n\to\infty} \mathbb{E}(Y_n) = \mathbb{E}(X) + \mathbb{E}(Y),$$

Beppo Levi's theorem ensuring the two equalities at the ends of the chain; the middle equality comes from the additivity of the integral over $\mathcal{S}^+$. □

Proposition A.2.10 *Let X be a random variable in $\mathcal{V}^+$. The expectation $\mathbb{E}(X)$ is zero if and only if X is zero almost surely with respect to $\mathbb{P}$, that is, if $\mathbb{P}(X = 0) = 1$.*

Proof This is true for positive bounded random variables. In the general case, the condition is sufficient by definition of the expectation as an upper bound. Conversely, let $X \in \mathcal{M}^+$ with zero expectation. We give ourselves an increasing sequence $(X_n)_{n\geq 0}$ in $\mathcal{S}^+$ that converges to X. Then for all $n \geq 0$, we have $\mathbb{E}(X_n) = 0$ and therefore $X_n = 0$ $\mathbb{P}$-almost surely since X_n is bounded. Since $X(\omega) = \lim_{n\to\infty} X_n(\omega)$, we conclude that X is zero $\mathbb{P}$-almost surely. □

Expectation of real random variables

For $X : \Omega \to \mathbb{R}$ a random variable, X^+ and X^- are in $\mathcal{V}^+$.

Definition A.2.11 Let $X : \Omega \to \mathbb{R}$ be a random variable. We say that X is *integrable* with respect to $\mathbb{P}$ if X^+ and X^- are integrable in the sense given above. We then set

$$\mathbb{E}(X) = \mathbb{E}(X^+) - \mathbb{E}(X^-).$$

We denote by L^1 the set of integrable random variables with real values.

We immediately obtain the following properties.

Proposition A.2.12

(i) *The set* L^1 *is an* $\mathbb{R}$*-vector space.*
(ii) *The mapping* $X \mapsto \mathbb{E}(X)$ *is an* $\mathbb{R}$*-linear form on* L^1.
(iii) *For all* $X \in L^1$, *we have* $|X| \in L^1$ *and*

$$|\mathbb{E}(X)| \leq \mathbb{E}(|X|).$$

Another very useful property is the Markov inequality.

Proposition A.2.13 (Markov inequality) *Let* X *be a positive random variable and* α *be a strictly positive real number. Then*

$$\mathbb{P}(X \geq \alpha) \leq \frac{1}{\alpha}\mathbb{E}(X).$$

Proof Let A be the event $\{X \geq \alpha\} = \{\omega \in \Omega : X(\omega) \geq \alpha\}$. We have $X \geq \alpha \cdot \mathbf{1}_A$ and the result follows by taking the expectation. □

A.2.4 Convergence Theorems

Consider a probability space $(\Omega, \mathcal{A}, \mathbb{P})$.

Beppo Levi's monotone convergence theorem

Proposition A.2.14 *Let* $(X_n)_{n\geq 0}$ *be an increasing sequence of real integrable random variables. Then we have*

$$\mathbb{E}\big(\lim_{n\to\infty} X_n\big) = \lim_{n\to\infty} \mathbb{E}\big(X_n\big) \leq +\infty.$$

Proof Since it is increasing, the sequence $(X_n)_{n\geq 0}$ converges pointwise to a random variable $X : \Omega \to \overline{\mathbb{R}}$ whose negative part X^- is integrable (since $(X_n^-)_n$ is decreasing). Therefore, $\mathbb{E}(X) = \mathbb{E}(X^+) - \mathbb{E}(X^-)$ is well defined. We return to Beppo Levi's theorem by considering the sequence of positive random variables $X_n - X_1$, which are also well-defined because X_1 is integrable and therefore only equals $\pm\infty$ on a negligible set. □

The monotonicity assumption is crucial, as shown in the following example.

Example The probability space is $[0, 1]$ endowed with the Lebesgue measure. Consider the sequence of functions $f_n(x) = 2nx(1 - x^2)^{n-1}$ defined on $[0, 1]$. We verify that $\lim_{n\to+\infty} f_n(x) = 0$ for all $x \in]0, 1[$ and that $\int_0^1 f_n(x)\,\mathrm{d}x = 1$ for all $n \geq 1$. Therefore:

$$0 = \int_0^1 \lim_{n \to +\infty} f_n(x)\,\mathrm{d}x < \lim_{n \to +\infty} \int_0^1 f_n(x)\,\mathrm{d}x = 1.$$

The remaining inequality is explained by the following lemma, which is very useful in practice.

Fatou's Lemma

The following result provides an inequality, not exactly a limit-sum inversion. It can be useful in particular for proving divergences of numerical sequences defined by expectations. In many practical cases, the sequence of random variables converges and the first lower limit is then a true random variable.

Lemma A.2.15 (Fatou's Lemma) *Let $(X_n)_{n \geq 0}$ be a sequence of measurable positive random variables. Then,*

$$\mathbb{E}\big(\liminf_{n \to \infty} X_n\big) \leq \liminf_{n \to +\infty} \mathbb{E}\big(X_n\big).$$

Proof The sequence of measurable random variables $(Y_n = \inf_{k \geq n} X_k)_{n \geq 0}$ is increasing and converges to $\liminf_{n \to +\infty} X_n$. By Beppo Levi's theorem, we therefore have:

$$\mathbb{E}\big(\liminf_{n \to +\infty} X_n\big) = \mathbb{E}\big(\lim_{n \to \infty} Y_n\big) = \lim_{n \to \infty} \mathbb{E}\big(Y_n\big).$$

Finally, the fact that $Y_n \leq X_n$ implies that $\lim_{n \to \infty} \mathbb{E}\big(Y_n\big) \leq \liminf_{n \to +\infty} \mathbb{E}\big(X_n\big)$, hence the result. □

Lebesgue's dominated convergence theorem

The following theorem is fundamental and very helpful.

Theorem A.2.16 (Lebesgue's dominated convergence theorem) *Let $(X_n)_{n \geq 0}$ be a sequence of integrable random variables (belonging to L^1). We assume that:*

- *the sequence $\big(X_n\big)_{n \geq 0}$ converges almost surely to a random variable X;*
- *there exists a $Y \in L^1$ such that almost surely and for all $n \geq 0$, $|X_n| \leq Y$.*

Then, $X \in L^1$ and

$$\lim_{n \to +\infty} \mathbb{E}\big(X_n\big) = \mathbb{E}\big(X\big).$$

This is a very important theorem, applying to a vast class of random variables, and useful in many limit calculations involving expectation. Among its important theoretical applications are general results on continuity and differentiation under expectation.

Appendix B
Poisson Point Measures

In this appendix, we summarize the main definitions and results concerning Poisson point measures. The reader can consult the two main books by Ikeda–Watanabe [42] and by Jacod–Shiryaev [45] for more details.

Definition B.1 Let $(E,\mathcal{E})$ be a measurable space and μ a σ-finite measure on this space. A (homogeneous) *Poisson point measure Q with intensity $\mu(\mathrm{d}h)\mathrm{d}t$ on $\mathbb{R}_+ \times E$* is an $(\mathbb{R}_+ \times E, \mathcal{B}(\mathbb{R}_+)\otimes\mathcal{E})$-random measure defined on a probability space $(\Omega,\mathcal{F},\mathbb{P})$ which satisfies the following properties:

1. Q is a counting measure: $\forall \widehat{A} \in \mathcal{B}(\mathbb{R}_+)\otimes\mathcal{E}$, $\forall\omega\in\Omega$, $Q(\omega,\widehat{A}) \in \mathbb{N}\cup\{+\infty\}$.
2. $\forall\omega\in\Omega$, $Q(\omega,\{0\}\times E) = 0$: no jump at time 0.
3. $\forall \widehat{A} \in \mathcal{B}(\mathbb{R}_+)\otimes\mathcal{E}$, $\mathbb{E}(Q(\widehat{A})) = \nu(\widehat{A})$, where $\nu(\mathrm{d}t,\mathrm{d}h) = \mu(\mathrm{d}h)\mathrm{d}t$.
4. If $\widehat{A}$ and $\widehat{B}$ are disjoint in $\mathcal{B}(\mathbb{R}_+)\otimes\mathcal{E}$ and if $\nu(\widehat{A}) < +\infty, \nu(\hat{B}) < +\infty$, then the random variables $Q(\widehat{A})$ and $Q(\widehat{B})$ are independent.

The existence of such a Poisson point measure with intensity $\mu(\mathrm{d}h)\mathrm{d}t$ is proven in [45], for any σ-finite measure μ on $(E,\mathcal{E})$.

Let us note that for any $A\in\mathcal{E}$ with $\mu(A) < \infty$ the process defined by

$$N_t(A) = Q((0,t]\times A)$$

is a Poisson process with intensity $\mu(A)$.

Definition B.2 The *filtration $(\mathcal{F}_t)_t$ generated by Q* is given by

$$\mathcal{F}_t = \sigma(Q((0,s]\times A), \forall s\le t, \forall A\in\mathcal{E}).$$

If $\widehat{A}\in(s,t]\times\mathcal{E}$ and $\nu(\widehat{A}) < \infty$, then $Q(\widehat{A})$ is independent of $\mathcal{F}_s$.

We first assume that the measure μ is finite on $(E,\mathcal{E})$. Then $(N_t(E), t\ge 0)$ is a Poisson process with intensity $\mu(E)$. The point measure is associated with a compound Poisson process. Indeed, let us write

$$\mu(\mathrm{d}h) = \mu(E)\,\frac{\mu(\mathrm{d}h)}{\mu(E)},$$

S. Méléard, *Random Models in Biology, Ecology and Evolution*, Texts in Applied Mathematics 87, https://doi.org/10.1007/978-3-662-73483-4

the decomposition of the measure μ as the product of the jump rate $\mu(E)$ and the jump amplitude law $\frac{\mu(\mathrm{d}h)}{\mu(E)}$. We fix $T > 0$ and introduce $T_1, \ldots, T_\gamma$, the jump times of the process $(N_t(E), t \geq 0)$ between 0 and T. We know that the jump number γ is a Poisson variable with parameter $T\mu(E)$. Moreover, conditionally on $\gamma, T_1, \ldots, T_\gamma$, the jumps $(U_n)_{n=1,\ldots,\gamma}$ are independent with the same law $\frac{\mu(dh)}{\mu(E)}$. We can write in this case

$$Q(\mathrm{d}t, \mathrm{d}h) = \sum_{n=1}^{\gamma} \delta_{(T_n, U_n)}.$$

Therefore, one can define for any measurable function $G(\omega, s, h)$ defined on $\Omega \times \mathbb{R}_+ \times E$ the random variable

$$\int_0^T \int_E G(\omega, s, h)\, Q(\omega, \mathrm{d}s, \mathrm{d}h) = \sum_{n=1}^{\gamma} G(\omega, T_n, U_n).$$

In the following, we will forget the ω. Let us remark that

$$T \longrightarrow \int_0^T \int_E G(s, h)\, Q(\mathrm{d}s, \mathrm{d}h)$$

is a finite variation process which is increasing if G is positive. A main example is the case where $G(\omega, s, h) = h$. Then

$$\begin{aligned} X_T &= \int_0^T \int_E h\, Q(\mathrm{d}s, \mathrm{d}h) \\ &= \sum_{n=1}^{\gamma} U_n = \sum_{s \leq T} \Delta X_s \end{aligned}$$

is the sum of the jumps between 0 and T.

We now generalize the definition of the integral of G with respect to Q when $\mu(E) = +\infty$. In this case, one can have an accumulation of jumps during the finite time interval $[0, T]$ and the counting measure Q is associated with a countable set of points:

$$Q = \sum_{n \geq 1} \delta_{(T_n, U_n)}.$$

We need additional properties on the process G. Since μ is σ-finite, there exists an increasing sequence $(E_p)_{p \in \mathbb{N}}$ of subsets of E such that $\mu(E_p) < \infty$ for each p and $E = \cup_p E_p$. As seen previously, one easily defines $\int_0^T \int_{E_p} G(s, h)\, Q(\mathrm{d}s, \mathrm{d}h)$ for any p.

We introduce the *predictable σ-field* $\mathcal{P}$ on $\Omega \times \mathbb{R}_+$ (generated by all left-continuous adapted processes) and define a *predictable process* $(G(s, h), s \in \mathbb{R}_+, h \in E)$ as a $\mathcal{P} \otimes \mathcal{E}$ measurable process.

Theorem B.3 *Let $G(s,h)$ be a predictable process and assume that*

$$\mathbb{E}\left(\int_0^T \int_E |G(s,h)|\,\mu(\mathrm{d}h)\mathrm{d}s\right) < +\infty. \tag{B.1}$$

1) *The sequence of random variables* $\left(\int_0^T \int_{E_p} G(s,h)\,Q(\mathrm{d}s,\mathrm{d}h)\right)_p$ *is Cauchy in* $\mathbb{L}^1$ *and converges to an* $\mathbb{L}^1$*-random variable that we denote by* $\int_0^T \int_E G(s,h)\,Q(\mathrm{d}s,\mathrm{d}h)$. *It is an increasing process if G is non-negative. Moreover, we have*

$$\mathbb{E}\left(\int_0^T \int_E G(s,h)\,Q(\mathrm{d}s,\mathrm{d}h)\right) = \mathbb{E}\left(\int_0^T \int_E G(s,h)\,\mu(\mathrm{d}h)\mathrm{d}s\right).$$

2) *The process* $M = (\int_0^t \int_E G(s,h)\,Q(\mathrm{d}s,\mathrm{d}h) - \int_0^t \int_E G(s,h)\,\mu(\mathrm{d}h)\mathrm{d}s, t \le T)$ *is a martingale.*

   The random measure

$$\widetilde{Q}(\mathrm{d}s,\mathrm{d}h) = Q(\mathrm{d}s,\mathrm{d}h) - \mu(\mathrm{d}h)\mathrm{d}s$$

   is called the compensated martingale-measure *of Q.*

3) *If we assume moreover that*

$$\mathbb{E}\left(\int_0^T \int_E G^2(s,h)\,\mu(\mathrm{d}h)\mathrm{d}s\right) < +\infty, \tag{B.2}$$

   then the martingale M is square-integrable with quadratic variation process

$$\langle M\rangle_t = \int_0^t \int_E G^2(s,h)\,\mu(\mathrm{d}h)\mathrm{d}s.$$

Recall that the quadratic variation process of a square-integrable martingale M is the unique predictable process $\langle M\rangle$ such that $M^2 - \langle M\rangle$ is a martingale (see for example [71]).

We can improve the conditions under which the martingale (M_t) can be defined. Indeed, if (B.2) is satisfied but not (B.1), the quantity $\int_0^t \int_E G(s,h)\,Q(\mathrm{d}s,\mathrm{d}h)$ isn't always well defined and we consider directly the compensated martingale-measure.

Theorem B.4 *Let $G(s,h)$ be a predictable process satisfying* (B.2). *Then the process* $M = (\int_0^t \int_E G(s,h)\,\widetilde{Q}(\mathrm{d}s,\mathrm{d}h), t \le T)$ *is a square-integrable martingale with quadratic variation*

$$\langle M\rangle_t = \int_0^t \int_E G^2(s,h)\,\mu(\mathrm{d}h)\mathrm{d}s.$$

The proof of this theorem is tricky and the definition of M comes from an $\mathbb{L}^2$-limiting argument, as for the usual stochastic integrals. This consists in studying the $\mathbb{L}^2$-limit of the sequence of martingales $\int_0^t \int_{E_p} G(s,h)\widetilde{Q}(\mathrm{d}s,\mathrm{d}h)$ as p tends to infinity. Once again, this sequence is Cauchy in $\mathbb{L}^2$ and converges to a limit which is a square-integrable martingale.

Example Let $\alpha \in (0,2)$. A symmetric α-stable process S can be defined as

$$S_t = \int_0^t \int_{\mathbb{R}} h\mathbf{1}_{\{0<|h|<1\}}\, \widetilde{Q}(\mathrm{d}s, \mathrm{d}h) + \int_0^t \int_{\mathbb{R}} h\mathbf{1}_{\{|h|\geq 1\}}\, Q(\mathrm{d}s, \mathrm{d}h), \qquad \text{(B.3)}$$

where $Q(\mathrm{d}s, \mathrm{d}h)$ is a Poisson point measure with intensity $\mu(\mathrm{d}h)\mathrm{d}s = \frac{1}{|h|^{1+\alpha}}\,\mathrm{d}h\mathrm{d}s$. There is an accumulation of small jumps and the first term in the r.h.s. of (B.3) is defined as a compensated martingale. The second term corresponds to the big jumps, which are in finite number on any finite time interval. If $\alpha \in (1,2)$, then $\int h \wedge h^2 \mu(\mathrm{d}h) < \infty$ and the process is integrable. If $\alpha \in (0,1)$, we only have that $\int 1 \wedge h^2 \mu(\mathrm{d}h) < \infty$ and the integrability of the process can fail.

References

1. D. Aldous. Stopping times and tightness. *Ann. Probab.* **6**, 335–340, 1978.
2. L.J.S. Allen. *An Introduction to Stochastic Processes with Applications to Biology*, Second edition. CRC Press, Chapman & Hall/CRC, 2011.
3. E.S. Allman and J.A. Rhodes. *Mathematical Models in Biology: An Introduction*. Cambridge Univ. Press, Cambridge, U.K., 2004.
4. W.J. Anderson. *Continuous-Time Markov Chains*. Springer Series in Statistics: Probability and its Applications. Springer-Verlag, New York, 1991.
5. K.B. Athreya and P.E. Ney. *Branching Processes*. Springer 1972.
6. N. Bacaer. *A Short History of Mathematical Population Dynamics*. Springer, 2011.
7. V. Bansaye and A. Lambert. New approaches to source-sink metapopulations decoupling demography and dispersal. *Theoret. Popul. Biol.* **88**, 31–46, 2013.
8. V. Bansaye and S. Méléard. *Stochastic Models for Structured Populations*. Mathematical Biosciences Institute Lecture Series 1.4. Springer 2015.
9. S. Billiard and C.V. Tran. A general stochastic model for sporophytic self-incompatibility. *J. Math. Biol.* **64**(1), 163–210 (2012).
10. P. Billingsley. *Probability and Measure*. Wiley, New York, 1995.
11. P. Billingsley. *Convergence of Probability Measures*. Third edition, Wiley, New York, 1999.
12. S. Boi, V. Capasso, and D. Morale. Modeling the aggregative behavior of ants of the species Polyergus rufescens. *Nonlinear Analysis* **1**, 163–176, 2000.
13. E. Bouin and V. Calvez. Travelling waves for the cane toads equation with bounded traits. *Nonlinearity* **27**, no. 9, 2233–2253, 2014.
14. J.H. Brown, J.F. Gillooly, A.P. Allen, V.M. Savage, and G.B. West. Toward a metabolic theory of ecology. *Ecology* **85**(7), 1771–1789, 2004.
15. H. Caswell, M. Fujiwara, and S. Brault. Declining survival probability threatens the North Atlantic right whale. *Proc. Natl. Acad. Sci. USA*, Vol. **96**, 3308–3313, 1999.
16. P. Cattiaux, D. Chafaï, and S. Motsch. Asymptotic analysis and diffusion limit of the persistent Turning Walker model. *Asymptot. Anal.* **67**, no. 1-2, 17–31, 2010.
17. P. Cattiaux and S. Méléard. Competitive or weak cooperative stochastic Lotka–Volterra systems conditioned on non-extinction. *J. Math. Biol.* **6**, 797–829, 2010.
18. N. Champagnat. A microscopic interpretation for adaptive dynamics trait substitution sequence models. *Stoch. Proc. Appl.* **116**, 1127–1160 (2006).
19. N. Champagnat, R. Ferrière, and S. Méléard. Unifying evolutionary dynamics: from individual stochastic processes to macroscopic models. *Theoret. Popul. Biol.* **69** 297–321, 2006.
20. N. Champagnat and S. Méléard. Invasion and adaptive evolution for individual-based spatially structured populations. *J. Math. Biol.* **55**, no. 2, 147–188, 2007.
21. J.T. Chang. Recent common ancestors of all present-day individuals. *Adv. Appl. Probab.* **31**, 1002–1026, 1999.

S. Méléard, *Random Models in Biology, Ecology and Evolution*,
Texts in Applied Mathematics 87, https://doi.org/10.1007/978-3-662-73483-4

22. J.R. Chazottes, P. Collet and S. Méléard. Sharp asymptotics for the quasi-stationary distribution of birth-and-death processes. *Probab. Theory Related Fields* **164** (2016), no. 1-2, 285–332.
23. K.L. Chung. *Markov Chains With Stationary Transition Probabilities*. Second Edition, Springer, 1967.
24. P. Collet, S. Martinez, and J. San Martin. *Quasi-Stationary Distributions: Markov Chains, Diffusions and Dynamical Systems*. Probability and Its Applications, Springer, 2013.
25. P. Collet, S. Méléard, and J.A.J. Metz. A rigorous model study of the adaptative dynamics of Mendelian diploids. *J. Math. Biol.* Volume **67**, Issue 3, Page 569–607, 2013.
26. F. Comets and T. Meyre. *Calcul stochastique et modèles de diffusions*. Dunod, 2006.
27. C. Coron. Stochastic modeling of density-dependent diploid populations and extinction vortex. *Adv. Appl. Prob.* **46**, 446–477, 2014.
28. C. Coron. Slow-fast stochastic diffusion dynamics and quasi-stationary distributions for diploid populations with varying size. *J. Math. Biol.* **72**, 171–202, 2015.
29. C. Coron, S. Méléard, E. Porcher, and A. Robert. Quantifying the mutational meltdown in diploid populations. *Am. Nat.* **181**(5): 623–636, 2013.
30. J.N. Darroch and E. Seneta. On quasi-stationary distributions in absorbing discrete-time finite Markov chains. *J. Appl. Prob.* **2**, 88–100, 1965.
31. J.F. Delmas and B. Jourdain. *Modèles aléatoires : applications aux sciences de l'ingénieur et du vivant*. Springer, 2006.
32. R. Douc, E. Moulines, P. Priouret, and P. Soulier. *Markov chains*. Springer, 2018.
33. R. Durrett. *Probability Models for DNA Sequence Evolution*, 2nd edition. Springer, 2007.
34. S.N. Ethier and T.G. Kurtz. *Markov Processes, Characterization and Convergence*. Springer, 1986.
35. S.N. Evans, A. Hening, and S.J. Schreiber. Protected polymorphisms and evolutionary stability of patch-selection strategies in stochastic environments. *J. Math. Biol.* **71**(2), 325–359, 2015.
36. W.J. Ewens. *Mathematical Population Genetics*. Second Edition. Springer, 2004.
37. W. Feller. *An introduction to Probability Theory and its Applications*, 2 Vol. Wiley, 1957.
38. C. Graham. *Chaînes de Markov*. Dunod, 2008.
39. P. Haccou, P. Jagers, and V.A. Vatutin. *Branching Processes: Variation, growth and extinction of populations*. Cambridge University Press, 2005.
40. F. Hoppe. Pólya-like urns and the Ewens' sampling formula. *J. Math. Biol.* **20**, 91–94, 1984.
41. S.P. Hubbell. *The Unified Neutral Theory of Biodiversity and Biogeography*. Princeton University Press, Monographs in population biology, 2001.
42. N. Ikeda and S. Watanabe. *Stochastic Differential Equations and Diffusion Processes*, 2nd Edition. North-Holland Publishing Company, 1989.
43. J. Istas. *Mathematical Modeling for the Life Sciences*. Universitext, Springer, 2006.
44. J. Jacod and P. Protter. *Probability Essentials*. Springer, 2004.
45. J. Jacod and A.N. Shiryaev. *Limit Theorems for Stochastic Processes*. Springer, 1987.
46. T.H. Jukes and C.R. Cantor. Evolution of protein molecules. In *Mammalian Protein Metabolism*. Monro, H.HN (eds), 21–132, Academic Press, New York, 1969.
47. I. Kaj and S. Sagitov. Limit processes for age-dependent branching particle systems. *J. Theoret. Prob.* **11**(1), 225–257, 1998.
48. I. Karatzas and S.E. Shreve. *Brownian Motion and Stochastic Calculus*, Second Edition. Springer, 1998.
49. S. Karlin and H.M. Taylor. *A First Course in Stochastic Processes*. Academic Press [A subsidiary of Harcourt Brace Jovanovich, Publishers], New York-London, second edition, 1975.
50. S. Karlin and H.M. Taylor. *A Second Course in Stochastic Processes*. Academic Press, 1981.
51. M. Kimmel and D.E. Axelrod. *Branching Processes in Biology*. Springer, 2002.
52. J.F.C. Kingman. The coalescent. *Stoch. Process. Appl.* **13**, no. 3, 235–248, 1982.
53. M. Kot. *Elements of Mathematical Biology*. Cambridge, 2001.
54. P. Lafitte-Godillon, K. Raschel and C.V. Tran. Extinction probabilities for distylous plant population modeled by an inhomogeneous random walk on the positive quadrant. *SIAM J. Appl. Math.* **73**, no. 2, 700–722, 2013.
55. T. Lagache and D. Holcman. Effective motion of a virus trafficking inside a biological cell. *SIAM J. Appl. Math.* **68**, no. 4, 1146–1167, 2008.

56. A. Lambert. Population Dynamics and Random Genealogies. *Stoch. Models* **24**, 45–163, 2008.
57. J.-F. Le Gall. *Brownian Motion, Martingales, and Stochastic Calculus*. Springer, Universitext series, 2016.
58. C. Lipow. Limiting diffusions for population size dependent branching processes. *J. Appl. Prob.* **14**, 14–24, 1977.
59. S. Méléard. *Aléatoire: Introduction à la théorie et au calcul des probabilités*. Editions de l'Ecole polytechnique, 2010.
60. S. Méléard. Random modeling of adaptive dynamics and evolutionary branching. In *The Mathematics of Darwin's Legacy*, F. Chalub and J.F. Rodrigues eds., 175–192, Birkhäuser, 2011.
61. S. Méléard and D. Villemonais. Quasi-stationary distributions and population processes. *Probability Surveys*, Vol. **9**, 340–410, 2012.
62. D. Morale, V. Capasso, and K. Oelschläger. An interacting particle system modelling aggregation behavior: from individuals to populations. *J. Math. Biol.* **50**, no. 1, 49–66, 2005.
63. J. Neveu. *Bases mathématiques du calcul des probabilités*, 2nd edition. Masson, 1970.
64. J. Neveu. *Martingales discrètes*. Masson, 1972.
65. J.R. Norris. *Markov Chains*. Cambridge University Press. 1997.
66. D. Nualart and E. Nualart. *Introduction to Malliavin Calculus*. IMS Textbooks, Cambridge, 2018.
67. E. Pardoux, *Markov Processes and Applications*. Dunod, 2007.
68. B.L. Phillips, G.P. Brown, J.K. Webb, and R. Shine. Invasion and evolution of speed in toads. *Nature* **439**, 803, 2006.
69. P.E. Protter. *Stochastic Integration and Differential Equations*, 2nd Edition. Springer, 2004
70. E. Renshaw. *Modelling Biological Populations in Space and Time*. Cambridge University Press, 1991.
71. D. Revuz and M. Yor. *Continuous Martingales and Brownian Motion*. Springer, 1991.
72. W. Rudin. *Real and Complex Analysis*. 1966; 3rd ed., 1987.
73. R. Sainudiin, B.D. Thatte, and A. Veber. Ancestries of a recombining diploid population. *J. Math. Biol.* **72**(1-2), 363–408, 2016.
74. E. Seneta and D. Vere-Jones. On quasi-stationary distributions in discrete-time Markov chains with a denumerable infinity of states. *J. Appl. Prob.* **3**, 403–434, 1966.
75. D. Serre. *Matrices: Theory and Applications*. Second Edition. Springer, 2010.
76. M.L. Silverstein. A new approach to local times. *J. Math. Mech.* **17**, 1023–1054, 1967/1968.
77. T. Tully. A quoi ça sert de vieillir? *La Recherche*, 2007.
78. G.B. West, J.H. Brown, and B.J. Enquist. The fourth dimension of life: fractal geometry and allometric scaling of organisms. *Science* **284** (5420), 1677–1679, 1999.
79. A.M. Yaglom. Certain limit theorems of the theory of branching stochastic processes (in Russian). *Dokl. Akad. Nauk. SSSR* (n.s.) **56**, 795–798, 1947.

Index

S. Méléard, *Random Models in Biology, Ecology and Evolution*, Texts in Applied Mathematics 87, https://doi.org/10.1007/978-3-662-73483-4

MIX
Papier aus verantwortungsvollen Quellen
Paper from responsible sources
FSC® C105338

If you have any concerns about our products,
you can contact us on
ProductSafety@springernature.com

In case Publisher is established outside the EU,
the EU authorized representative is:
Springer Nature Customer Service Center GmbH
Europaplatz 3, 69115 Heidelberg, Germany

Printed by Libri Plureos GmbH
in Hamburg, Germany